Sustainable Utilization of Humic Substances and Organic Waste in Green Agriculture

Sustainable Utilization of Humic Substances and Organic Waste in Green Agriculture

Editor

Maria Roulia

MDPI

Basel • Beijing • Wuhan • Barcelona • Belgrade • Novi Sad • Cluj • Manchester

Editor
Maria Roulia
National and Kapodistrian
University of Athens
Athens
Greece

Editorial Office
MDPI AG
Grosspeteranlage 5
4052 Basel, Switzerland

This is a reprint of articles from the Special Issue published online in the open access journal *Agriculture* (ISSN 2077-0472) (available at: https://www.mdpi.com/journal/agriculture/special_issues/Humic_Substances_Organic_Waste).

For citation purposes, cite each article independently as indicated on the article page online and as indicated below:

Lastname, A.A.; Lastname, B.B. Article Title. *Journal Name* **Year**, *Volume Number*, Page Range.

ISBN 978-3-7258-1989-8 (Hbk)
ISBN 978-3-7258-1990-4 (PDF)
doi.org/10.3390/books978-3-7258-1990-4

Cover image courtesy of Maria Roulia and Alexandros A. Vassiliadis

Contents

Preface

Humans have always relied on available resources for survival. With increased population, urbanization, and anthropogenic activities, billions of tons of waste burden the environment every year. Advanced organic waste management not only reduces the strain on landfills but also offers new perspectives for a greener and more sustainable future. Achieving organic resource reutilization and a circular bioeconomy requires a multidisciplinary effort to shift from waste disposal to waste management awareness.

Maria Roulia
Editor

agriculture

MDPI

Editorial

Sustainable Utilization of Humic Substances and Organic Waste in Green Agriculture

Maria Roulia

Inorganic Chemistry Laboratory, Department of Chemistry, National and Kapodistrian University of Athens, Panepistimiopolis, 157 71 Athens, Greece; roulia@chem.uoa.gr; Tel.: +30-210-7274780

Citation: Roulia, M. Sustainable Utilization of Humic Substances and Organic Waste in Green Agriculture. *Agriculture* **2024**, *14*, 115. https://doi.org/10.3390/agriculture14010115

Received: 22 December 2023
Accepted: 4 January 2024
Published: 11 January 2024

Organic wastes (OW) comprise biodegradable plant, animal, and industrial and municipal waste; billions of tons are generated annually worldwide, and they are continuously produced as a result of prosperity, the increase in population, and the escalation of anthropogenic activities. Conventional indiscriminate dumping and burning dissipate the energy content of OW, creating soil and water pollution and posing threats to the environment and human health. Waste management processes mostly include reduction, reuse, and recycling, as well as physical, biological, and chemical treatments (e.g., composting). During composting, the transformation of organic matter into nutrient-rich humus is mainly achieved by microorganisms; amended OW are rich in humic substances (HS) and extremely beneficial to plant growth and soil fertility. Humic substances—omnipresent in terrestrial and aquatic ecosystems, representing a major source of organic carbon and nitrogen—are redox-active, refractory, dark-colored mixtures of heterogeneous organic compounds produced via physicochemical and microbial processes during the early diagenesis in biomass decay [1].

On this basis, recycling and the subsequent reutilization of OW appear as a benignant and ecologically sound route, contributing to both sustainability and circular economy objectives. Besides nutrient preservation, OW recycling strategies support energy conservation, as electricity and bio-based energy carriers (e.g., biogas and biohydrogen) can be produced.

The sustainable utilization of humic substances and organic waste is particularly advantageous in green agriculture; processed OW may serve as soil conditioners and nutrient pools for plants. Together with HS, they affect plant metabolism; regulate nutrient availability and transport, reducing the need and side effects of chemical fertilizers; support microbial growth; enable pollutants sequestration; and improve the physicochemical properties of degraded soils (e.g., increase pH, improve water retention and cation exchange capacities, and ameliorate bulk density). Thus, they assist in soil restoration and remediation and sustainable plant growth.

The *Agriculture* Journal Special Issue "**Sustainable Utilization of Humic Substances and Organic Waste in Green Agriculture**" focuses on the green production processes, properties, and uses of HS and OW; the interaction/complexation of HS with compounds promoting sustainable agriculture; the impact of HS, HS-containing materials, and organic waste on the environment (soil, plants, and living organisms, domestic animals and cattle included); OW from industrial processes (e.g., molasses, cheese whey, slaughterhouse, leather); physical, chemical, and biological OW treatments and recycling (e.g., retention, adsorption, composting, and decomposer microorganisms) that support green agriculture; and the management of pollutants [2] (e.g., chemicals, pharmaceuticals, drugs, dyes, pesticides, and food additives) that accompany OW.

In the review article "Biostimulant Effects of Waste Derived Biobased Products in the Cultivation of Ornamental and Food Plants", E. Montoneri, A. Baglieri, and G. Fascella describe the application of soluble bio-based substances (SBS), derived from composts and urban waste anaerobic digestate, as sustainable and efficient biostimulants in ornamental and vegetable species. A wide variety of SBS tailored for the cultivation of specific plants may assist plant growth, fruits, and ornamental qualities, and they can be used as alternatives to existing fossil-sourced agrochemicals [3].

The article entitled "Advances in Applications of Cereal Crop Residues in Green Concrete Technology for Environmental Sustainability: A Review" by M. A. Suhail, S. Shrivastava, K. Paritosh, N. Pareek, A. A. Kovalev, D. A. Kovalev, Y. V. Litti, V. Panchenko, V. Bolshev, and V. Vivekanand [4] presents the potential use of waste agricultural residues (with significant mechanical properties) from cereal farming in green concrete manufacturing as partial substitutes for cement, sand, coarse aggregates, and fiber reinforcements. The appropriate methods of treatment, selection, and blending ratios of cereal waste resources, as well as innovations in cereal farming residues that allow their potential use in the green construction industry, which are all compatible with circular bioeconomy strategies, are discussed.

In their article "Taif's Rose (*Rosa damascene* Mill var. *trigentipetala*) Wastes Are a Potential Candidate for Heavy Metals Remediation from Agricultural Soil", T. M. Galal, A. Majrashi, H. M. Al-Yasi, E. A. Farahat, E. M. Eid, and E. F. Ali examine the bioaccumulation of heavy metals in Taif rose shrubs [5]. With the exception of Al, plant stems retained higher quantities of all heavy metals studied than leaves; Co and Ni were mostly contained in the stems of 10- and 12-year-old plants, while Cd, Cr, Cu, Fe, Mn, Pb, and Zn accumulated in older Taif's rose shrubs stems, demonstrating the possibility of using Taif roses as a promising, viable, and safe crop for heavy-metal phytoremediation.

A biochar-based fertilizer (BF) was prepared from distillers grains via oxygen-limited cracking [6]. The application of BF in eggplant cultivations positively affected the yield and quality of fruits, reduced fertilizer utilization, and proved to be economically profitable to farmers as "Biochar-Based Fertilizer Enhances the Production Capacity and Economic Benefit of Open-Field Eggplant in the Karst Region of Southwest China" (M. Zhang, Y. Liu, Q. Wei, L. Liu, X. Gu, and J. Gou). Specifically, these eggplants exhibited higher nutrient uptake and their fruits possessed lower nitrate content and elevated vitamin C and soluble sugars, which are all beneficial for eggplant quality.

The Taif Damask rose's pruning waste from plants of different ages is discussed in "Evaluating the Nutrient Contents and Nutritive Value of Taif's Rose (*Rosa damascene* Mill var. *trigintipetala*) Waste to Be Used as Animal Forage or Soil Organic Fertilizers", authored by T. M. Galal, E. F. Ali, E. M. Eid, H. M. Al-Yasi, A. Magrashi, F. Althobaiti, and E. A. Farahat [7]. The morphological characteristics, the N, P, K, Ca, Mg, Na, and ash contents in the stems and leaves, and parameters associated with the nutritional value of the wastes, i.e., fibers, lipids, carbohydrates, and energy, were measured to demonstrate the inorganic and organic nutrients' abundance, which is useful for the potential application of Taif's rose waste as a fertilizer and/or animal forage.

The topic of the article by M. Lanno, M. Klavins, O. Purmalis, M. Shanskiy, A. Kisand, and M. Kriipsalu is the "Properties of Humic Substances in Composts Comprised of Different Organic Source Material". FTIR and EEM spectroscopic techniques were employed to observe the effect of the compost's origin on humic substance contents; the composts containing animal byproducts proved to be richer in humic substances compared to those from kitchen biowaste. Besides organic matter and nutrient contents, the role of humic substances in the comprehensive evaluation of composts is emphasized, especially when the composts are intended for fertilizing applications [8].

Taif Damask rose organs and floral solid distillation waste (SDW) were examined for their chemical composition and biological functions in "Chemical and Nutritional Characterization of the Different Organs of Taif's Rose (*Rosa damascene* Mill. var. *trigintipetala*) and Possible Recycling of the Solid Distillation Wastes in Taif City, Saudi Arabia" by E.F. Ali, H.M. Al-Yasi, A. Majrashi, E.A. Farahat, E.M. Eid, and T.M. Galal [9]. Soluble carbohydrates, cardiac glycosides, total flavonoid contents, and total phenolic compounds were determined, as well as mineral and organic nutrients, nutritional value, and antimicrobial activity toward bacteria and fungi. Both organs and SDW seem to be suitable for mature dry gestating beef cows and health applications.

Animal byproducts can be used for growing media preparation, as reported by R. Li, H. Wang, E. Duan, J. Fan, and L. Wang in their article "Rabbit Manure Compost for Seedling Nursery Blocks: Suitability and Optimization of the Manufacturing Production Process". Nursery blocks obtained by mixing rabbit manure compost, vermiculite, rice straw, and peat were found to improve the transplanting efficiency and survival rate of seedlings. The effect of cold pressing parameters on block quality was also examined. Thus, eco-friendly resource recycling management with respect to rabbit manure is achieved [10].

"Deteriorating Harmful Effects of Drought in Cucumber by Spraying Glycinebetaine" by E.-S.E. Metwaly, H.M. Al-Yasi, E.F. Ali, H.A. Farouk, and S. Farouk proposes the use of glycinebetaine (GlyBet), a vital osmoprotectant produced in crops, to improve drought tolerance in non-accumulating plants, such as cucumbers, by enhancing their water use efficiency [11]. GlyBet could act as a cost-effective and eco-friendly biostimulant, and exogenous spraying demonstrated a beneficial effect on moderating water deficit damage on plant growth and productivity.

J.L. Villalpando-Aguilar, D.F. Chi-Maas, I. López-Rosas, V.Á. Aquino-Luna, J. Arreola-Enríquez, J.C. Alcudia-Pérez, G. Matos-Pech, R.C. Gómez-García, J.F. Martínez-Puc, and W. Cetzal-Ix suggest the development of sustainable alternatives for edible fruit production in "Urban Agriculture as an Alternative for the Sustainable Production of Maize and Peanut". In this context, maize and peanut plants—cultivated in compost originating from organic residues irrigated with temporary rain—exhibited higher yields compared with plants grown in soil and compost [12].

"Sustainable Utilization Strategy of Organic Waste via Fabrication of Bioelastomer with Antibacterial and Antioxidant Activities Using Mandarin Peel Extracts" by K.H. Lee, Y. Chun, J.H. Lee, J.U. Lee, T. Lee, and H.Y. Yoo [13] aims to incorporate bioactive compounds recovered from mandarin peels into a functional bioelastomer without affecting its physical properties. The bioelastomer displayed radical scavenging activity and antibacterial properties against Gram-positive, Gram-negative, and antibiotic-resistant bacteria, and it is expected to be utilized in the food packaging, pharmaceutical, and medical industries, thus upcycling food waste.

The need for chemical fertilizers substitutes is imminent; in "Response of Maize Yield and Nutrient Uptake to Indigenous Organic Fertilizer from Corn Cobs" by M.T.S. Budiastuti, D. Purnomo, B. Pujiasmanto, and D. Setyaningrum, an organic fertilizer prepared from corn cob waste was applied to suboptimal land to increase the harvested corn area [14]. The prepared corncob fertilizer meets the standards for organic fertilizers, positively affects leaf area and root length, and increases chlorophyll a and b and phosphate uptake, supporting the growth, yield, and nutrient uptake of corn plants.

Another organic fertilizer for corn plants was synthesized from the anaerobic digestion of swine wastewater and tested as a nitrogen substitute in comparison with the conventional chemical treatment. The results of the article "Fertilizer Performance of a Digestate from Swine Wastewater as Synthetic Nitrogen Substitute in Maize Cultivation: Physiological Growth and Yield Responses" by E.L. Buligon, L.A.M. Costa, J. de Lucas, Jr., F.T. Santos, P. Goufo, and M.S.S.M. Costa [15] show that the basal application of this liquid biofertilizer completely substituted the mineral nitrogen in corn plants, allowing the minimization of chemical fertilizers without yield penalties.

"The Effect of Dietary Humic Substances on Cellular Immunity and Blood Characteristics in Piglets" was studied by L. Bujňák, A.H. Šamudovská, D. Mudroňová, P. Nad', S. Marcinčák, I. Maskaľová, M. Harčárová, V. Karaffová, and M. Bartkovský [16]. Natural humic substance supplementation increased the proportion of CD4 + CD8- lymphocytes and serum alkaline phosphatase; the phagocytic activity and engulfing capacity of phagocytes and other lymphocyte subpopulations were slightly increased. These observations demonstrate that humic substances stimulate cellular immunity in piglets without negatively affecting their hematological and biochemical parameters.

I would like to express my deepest appreciation to all authors who selected this Special Issue in which to publish their fine papers.

All these articles summarize scientific progress and imprint recent developments in the field. Therefore, I believe that this Special Issue will meet the needs of researchers who focus on humic substances and organic waste valorization without foreclosing the needs and options of the broad readership devoted to sustainability.

Conflicts of Interest: The author declares no conflicts of interest.

References

1. Roulia, M. Humic Substances: A Novel Eco-Friendly Fertilizer. *Agronomy* **2022**, *12*, 754. [CrossRef]
2. Fragouli, P.G.; Roulia, M.; Vassiliadis, A.A. Macromolecular Size and Architecture of Humic Substances Used in the Dyes' Adsorptive Removal from Water and Soil. *Agronomy* **2023**, *13*, 2926. [CrossRef]
3. Montoneri, E.; Baglieri, A.; Fascella, G. Biostimulant Effects of Waste Derived Biobased Products in the Cultivation of Ornamental and Food Plants. *Agriculture* **2022**, *12*, 994. [CrossRef]
4. Suhail, M.A.; Shrivastava, S.; Paritosh, K.; Pareek, N.; Kovalev, A.A.; Kovalev, D.A.; Litti, Y.V.; Panchenko, V.; Bolshev, V.; Vivekanand, V. Advances in Applications of Cereal Crop Residues in Green Concrete Technology for Environmental Sustainability: A Review. *Agriculture* **2022**, *12*, 1266. [CrossRef]
5. Galal, T.M.; Majrashi, A.; Al-Yasi, H.M.; Farahat, E.A.; Eid, E.M.; Ali, E.F. Taif's Rose (*Rosa damascene* Mill var. *trigentipetala*) Wastes Are a Potential Candidate for Heavy Metals Remediation from Agricultural Soil. *Agriculture* **2022**, *12*, 1319. [CrossRef]
6. Zhang, M.; Liu, Y.; Wei, Q.; Liu, L.; Gu, X.; Gou, J. Biochar-Based Fertilizer Enhances the Production Capacity and Economic Benefit of Open-Field Eggplant in the Karst Region of Southwest China. *Agriculture* **2022**, *12*, 1388. [CrossRef]
7. Galal, T.M.; Ali, E.F.; Eid, E.M.; Al-Yasi, H.M.; Magrashi, A.; Althobaiti, F.; Farahat, E.A. Evaluating the Nutrient Contents and Nutritive Value of Taif's Rose (*Rosa damascene* Mill var. *trigintipetala*) Waste to Be Used as Animal Forage or Soil Organic Fertilizers. *Agriculture* **2022**, *12*, 1481. [CrossRef]
8. Lanno, M.; Klavins, M.; Purmalis, O.; Shanskiy, M.; Kisand, A.; Kriipsalu, M. Properties of Humic Substances in Composts Comprised of Different Organic Source Material. *Agriculture* **2022**, *12*, 1797. [CrossRef]
9. Ali, E.F.; Al-Yasi, H.M.; Majrashi, A.; Farahat, E.A.; Eid, E.M.; Galal, T.M. Chemical and Nutritional Characterization of the Different Organs of Taif's Rose (*Rosa damascene* Mill. var. *trigintipetala*) and Possible Recycling of the Solid Distillation Wastes in Taif City, Saudi Arabia. *Agriculture* **2022**, *12*, 1925. [CrossRef]
10. Li, R.; Wang, H.; Duan, E.; Fan, J.; Wang, L. Rabbit Manure Compost for Seedling Nursery Blocks: Suitability and Optimization of the Manufacturing Production Process. *Agriculture* **2022**, *12*, 2156. [CrossRef]
11. Metwaly, E.-S.E.; Al-Yasi, H.M.; Ali, E.F.; Farouk, H.A.; Farouk, S. Deteriorating Harmful Effects of Drought in Cucumber by Spraying Glycinebetaine. *Agriculture* **2022**, *12*, 2166. [CrossRef]
12. Villalpando-Aguilar, J.L.; Chi-Maas, D.F.; López-Rosas, I.; Aquino-Luna, V.Á.; Arreola-Enríquez, J.; Alcudia-Pérez, J.C.; Matos-Pech, G.; Gómez-García, R.C.; Martínez-Puc, J.F.; Cetzal-Ix, W. Urban Agriculture as an Alternative for the Sustainable Production of Maize and Peanut. *Agriculture* **2023**, *13*, 59. [CrossRef]
13. Lee, K.H.; Chun, Y.; Lee, J.H.; Lee, J.U.; Lee, T.; Yoo, H.Y. Sustainable Utilization Strategy of Organic Waste via Fabrication of Bioelastomer with Antibacterial and Antioxidant Activities Using Mandarin Peel Extracts. *Agriculture* **2023**, *13*, 161. [CrossRef]
14. Budiastuti, M.T.S.; Purnomo, D.; Pujiasmanto, B.; Setyaningrum, D. Response of Maize Yield and Nutrient Uptake to Indigenous Organic Fertilizer from Corn Cobs. *Agriculture* **2023**, *13*, 309. [CrossRef]
15. Buligon, E.L.; Costa, L.A.M.; de Lucas, J., Jr.; Santos, F.T.; Goufo, P.; Costa, M.S.S.M. Fertilizer Performance of a Digestate from Swine Wastewater as Synthetic Nitrogen Substitute in Maize Cultivation: Physiological Growth and Yield Responses. *Agriculture* **2023**, *13*, 565. [CrossRef]
16. Bujňák, L.; Hreško Šamudovská, A.; Mudroňová, D.; Naď, P.; Marcinčák, S.; Maskaľová, I.; Harčárová, M.; Karaffová, V.; Bartkovský, M. The Effect of Dietary Humic Substances on Cellular Immunity and Blood Characteristics in Piglets. *Agriculture* **2023**, *13*, 636. [CrossRef]

agriculture

MDPI

Communication

The Effect of Dietary Humic Substances on Cellular Immunity and Blood Characteristics in Piglets

Lukáš Bujňák [1], Alena Hreško Šamudovská [1], Dagmar Mudroňová [2], Pavel Nad' [1], Slavomír Marcinčák [3], Iveta Maskal'ová [1], Michaela Harčárová [1], Viera Karaffová [4,*] and Martin Bartkovský [3]

1 Department of Animal Nutrition and Husbandry, University of Veterinary Medicine and Pharmacy in Košice, Komenského 73, 041 81 Košice, Slovakia
2 Department of Microbiology and Immunology, University of Veterinary Medicine and Pharmacy in Košice, Komenského 73, 041 81 Košice, Slovakia
3 Department of Food Hygiene, Technology and Safety, University of Veterinary Medicine and Pharmacy in Košice, Komenského 73, 041 81 Košice, Slovakia
4 Department of Morphological Disciplines, University of Veterinary Medicine and Pharmacy in Košice, Komenského 73, 041 81 Košice, Slovakia
* Correspondence: viera.karaffova@uvlf.sk; Tel.: +421-905-871-840

Abstract: This study's objective was to determine the impact of dietary humic substances on immune response and blood profiles in piglets. A total of 24 crossbred piglets (Slovakian White × Landrace; 35 days old; average body weight of 11.67 kg) were allotted to two dietary groups with (experimental; 5 g·kg^{-1}) or without (control; 0 g·kg^{-1}) natural humic substances supplementation. In this study, we observed a significant increase of the proportion of CD4+CD8- lymphocytes ($p < 0.001$) in the experimental group. The results also showed a tendency for an increase of the phagocytic activity and the engulfing capacity of phagocytes and the numbers of the other monitored lymphocyte subpopulations (CD3+, CD21+, CD4-D8+, CD4+CD8+, CD4+CD25+) in piglets in the experimental group compared to the control group. Supplementation of humic substances increased serum alkaline phosphatase compared to the control group ($p < 0.05$). Other monitored blood parameters were not significantly affected by dietary treatment. It concluded that inclusion of humic substances in the diet of piglets could have a stimulating effect on cellular immunity, without a negative effect on haematological and biochemical parameters.

Keywords: humic substances; phagocytosis; lymphocyte; blood biochemistry; haematology; piglets

Citation: Bujňák, L.; Hreško Šamudovská, A.; Mudroňová, D.; Nad', P.; Marcinčák, S.; Maskal'ová, I.; Harčárová, M.; Karaffová, V.; Bartkovský, M. The Effect of Dietary Humic Substances on Cellular Immunity and Blood Characteristics in Piglets. *Agriculture* **2023**, *13*, 636. https://doi.org/10.3390/agriculture13030636

Academic Editor: Maria Grazia Cappai

Received: 14 February 2023
Revised: 28 February 2023
Accepted: 6 March 2023
Published: 8 March 2023

1. Introduction

Humic substances (HS) are a class of non-nutritive natural organic bioactive compounds formed in soils [1]. They are important humus components whose primary function is to transfer nutrients from the soil to living organisms. HS specifically include humic acid, fulvic acid, and humin as their principal constituents [2]. These components are essential for plant growth [3] and can promote the efficient utilization of nutrients by the plant [4].

Due to the prohibition on the use of antibiotics as growth promoters in the European Union, interest in alternative feed additives for animal production has increased [5]. In the last two decades, the interest in the use of HS in animal nutrition has increased [2,6]. Many authors observed an improvement in the production parameters after the addition of HS into feedstuff during their studies. In recent years, it has been shown that HS added to the feed of monogastric animals such as swine, poultry, and rabbits promotes growth [7–10].

Furthermore, HS have been used as immunostimulatory, antidiarrheal, analgesic, and antimicrobial agents in veterinary practices [11–13]. The addition of biologically active supplements of humic nature to the diets of animals stimulated metabolic processes and the digestibility of nutrients, and also activated the absorption of some mineral elements [14]. Other studies found that HS help to reduce ammonia excretion from manure and improves the relative number of blood lymphocytes. The modifications mentioned above help the animals' immunity [2,8,15].

However, due to the different sources and types of HS preparations, as well as the fact that there is no single standard for measuring genuine HS effects, their bio-effect depends on specification. Based on our previous results and experiences from poultry studies (broilers and laying hens) on the effect of humic substances (in concentrations of 0.8% and 0.5%, respectively) on the immune status and immune response, we decided to verify this effect in piglets. We observed a significant increase of phagocyte activity and B cell response, as well as a significant increase of CD4+:CD8+ lymphocyte ratio [16,17]. In general, supplementation as a feed additive in pigs has not been well reported compared to poultry, and scientific studies on the influence of HS supplementation in pigs' diets on immunity and metabolism are still relatively limited. Therefore, this study, as a pilot trial, was carried out to assess the effect of 0.5% HS supplementation on the immune indicators, as well as biochemical and haematological variables in the blood of piglets.

2. Materials and Methods

2.1. Experimental Design

The experiment was carried out in accordance with the "European Directive on the protection of animals used for scientific purposes" [18]. The animals were housed in accredited stables of the Department of Animal Nutrition and Husbandry at the University of Veterinary Medicine in Košice, Slovakia, under required zoohygienic conditions. During the experiment, the average temperature in the stable was 20.2 ± 1.5 °C, and the relative humidity was 68.5 ± 4.8%. The trial was approved by the Ethics Commission of the University of Veterinary Medicine and Pharmacy in Košice (protocol no. EKV/2022-11).

A total of 24 crossbred (Slovakian White x Landrace) 35-days-old piglets were divided into two groups (n = 12; 50% male and 50% female in both groups). Prior to the start of the experiment, an initial average animal body weight (BW) of 11.68 ± 1.35 kg in the control group and 11.65 ± 1.34 kg in the experimental group was recorded. The following experimental groups were included in the study: the control group and the experimental group, where the experimental group diet was supplemented with a 0.5% HS supplement. The experiment lasted for 4 weeks. The same diets for both groups were used in the experiment (Table 1). The introduction of the HS supplement into the diet was realized at the expense of barley in the experimental group. The pigs were fed twice per day with complete feed mixtures. Drinking water was available to the animals ad libitum throughout the experimental period. The complete feed mixtures in the experiment were formulated according to the nutritional requirements by Šimeček et al. [19].

The dietary natural HS supplement (HUMAC® Natur AFM; Humac, Ltd., Košice, Slovakia) was ground and physically purified with Leonardite without chemical treatment.

The complete feed mixtures were analysed for dry matter, crude protein, crude fibre, total ash, starch, and total phosphorus according to EC Commission Regulation 152/2009 [20]. The level of dietary calcium and sodium was analysed using the flame method of an atomic absorption spectrometer (Unicam Solar 939, Camberley, Surrey, UK). The metabolisable energy values of complete feed mixtures were calculated with the formula according to the Šimeček et al. [19].

2.2. Sampling and Measurements

In week 4, in the morning on the last day of the experiment, blood samples were taken by the *orbital sinus* puncture from all pigs individually in both groups for subsequent analysis. Serum was obtained by centrifugation (3000 rpm for 30 min) and stored at −20 °C until analysis. Heparinized blood was used to determine haematological parameters, test phagocyte activity and identify lymphocyte subpopulations.

Table 1. Formula and chemical composition of feed mixtures for pigs (as-fed basis).

		Control Diet	Experimental Diet
Components			
Corn	[%]	25.00	25.00
Wheat	[%]	22.50	22.50
Barley	[%]	28.00	27.50
Soybean meal	[%]	21.00	21.00
Vitamin-mineral premix with 3-phytase [1]	[%]	3.00	3.00
Sodium chloride	[%]	0.12	0.12
L-Lysine	[%]	0.23	0.23
DL-Methionine	[%]	0.06	0.06
L-Threonine	[%]	0.09	0.09
HS supplement [2]	[%]	-	0.50
Composition by analysis			
Dry matter	[$g \cdot kg^{-1}$]	885.2	884.3
Crude protein	[$g \cdot kg^{-1}$]	178.2	177.4
Crude fibre	[$g \cdot kg^{-1}$]	35.4	38.9
Ash	[$g \cdot kg^{-1}$]	54.1	57.6
Starch	[$g \cdot kg^{-1}$]	445.0	438.0
Calcium	[$g \cdot kg^{-1}$]	6.4	7.0
Phosphorus	[$g \cdot kg^{-1}$]	5.6	5.6
Sodium	[$g \cdot kg^{-1}$]	2.7	2.8
Metabolisable energy	[$MJ \cdot kg^{-1}$]	13.07	13.02

[1] Vitamin and mineral premix (per kg): vit. A 330,000 IU; D_3 66,000 IU; E 4000 mg; calcium 210 g; phosphorus 25 g; sodium 45 g; copper 4800 mg; iron 1750 mg; zinc 2700 mg; manganese 880 mg; iodine 55 mg; selenium 12.5 mg. [2] The characteristics of the HS supplement were the following: the size of particles up to 100 μm, pH 5.8, humidity max. 15%, content of humic acids min. 65% in dry matter (DM), fulvic acid 5% (DM).; minerals: calcium 42.28, magnesium 5.11, sodium 7.11, potassium 0.93 $g \cdot kg^{-1}$; and microelements: Fe 19,046; Cu 15; Zn 37; Mn 142; Co 1.24; Se 1.67, as well as Mo 2.7 $mg \cdot kg^{-1}$ DM.

2.2.1. Haematological and Serum Biochemical Parameters

Complete blood count was performed with an automated haematology analyser (scil Vet ABC™ Hematology Analyzer, Germany). The variables evaluated in our study were haematocrit value (HCT), haemoglobin concentration (Hb), red blood cells count (RBC), mean corpuscular volume (MCV), and total white blood cell (WBC) count. Serum biochemical parameters—total protein, albumin, glucose, urea, triglycerides, cholesterol, creatinine, aspartate aminotransferase (AST), alkaline phosphatase (ALP), and phosphorus were measured using a fully automatic random access benchtop analyser (Ellipse, Italy). The concentration of calcium in serum was determined by means of a flame atomic absorption spectrometer (Unicam Solar 939, Camberley, Surrey, UK).

2.2.2. Biomarker of the Lipid Peroxidation

The concentrations of lipid peroxidation products (malondialdehyde levels, MDA) in serum were measured as thiobarbituric acid reactive substances (TBARs) according to the spectrophotometric modification method described by Costa et al. [21]. Briefly, serum samples were mixed with a solution composed of trichloroacetic acid (15%; Merck, Darmstadt, Germany), thiobarbituric acid (0.38%; Sigma-Aldrich, St. Louis, MO, USA), and hydrochloric acid (0.25 N; Mikrochem, Pezinok, Slovakia) and heated for 30 min in a boiling water bath. After cooling in ice water and centrifugation, the absorbance of the supernatant was measured at 535 nm. The concentration of TBARs was determined from the standard curve prepared using 1,1,3,3-tetramethoxypropane (malondialdehyde-bis (dimethyl acetal); Acros Organics, Geel, Belgium). The results were expressed as nmol of MDA/mL of serum.

2.2.3. Phagocyte Activity Testing and Identification of Lymphocyte Subpopulations

A commercial Phagotest® assay (Celonic, Munich, Germany) was used to determine the phagocytic activity and the engulfing capacity of the phagocytes. The manufacturer's instructions were followed when performing the test.

Direct immunostaining assay was applied to identify selected subpopulations of lymphocytes. Two combinations of conjugated mouse anti-porcine monoclonal antibodies: CD3e/CD21 and CD4/CD8a/CD25 were used according to the specifications given in Table 2. Heparinised blood in amount of 50 μL was incubated with monoclonal antibodies for 15 min in the dark at laboratory temperature. After incubation, 1 mL of BD FACS lysis solution was added to the tubes. The tubes were incubated for an additional 20 min in the dark at laboratory temperature and then centrifuged (300× *g* for 5 min). Cell pellets were then washed and centrifuged twice with 1 mL phosphate buffer saline (PBS; MP Biomedicals, Illkirch-Graffenstaden, France) at 300× *g* for 5 min. Finally, cells were resuspended in 200 μL of PBS for subsequent cytometric analysis.

Table 2. Specification and amounts of used mouse anti-porcine monoclonal antibodies.

Type	Fluorochrome	Clone	Amount/50 μL Blood	Producer
anti-CD3e	FITC	BB23-8E6	4 μL	BD Biosciences, Franklin Lakes, NJ, USA
anti-CD4	FITC	MIL 17	4 μL	AbD Serotec, Kidlington, UK
anti-CD8a	R-PE	MIL 12	2 μL	AbD Serotec, Kidlington, UK
anti-CD25	PE-Cy7	PC 61.5	1 μL	eBioscience, San Diego, CA, USA
anti-CD21	R-PE	BB6-11C9.6	2 μL	SuthernBiotech, Homewood, AL, USA

CD—cluster of differentiation; FITC—fluorescein isothiocyanate; R-PE—R-phycoerythrin; PE-Cy7—phycoerythrin-cyanine 7.

Phagocytic activity analysis as well as the identification of lymphocyte subpopulations was performed on a six-colour BD FACSCanto™ flow cytometer (Becton Dickinson Biosciences, San Jose, CA, USA) using BD FACS Diva™ software. The position of the analysed cells was gated in FSC vs. SSC dot plots. Granulocytes, monocytes, and both cell populations together, respectively, were gated for phagocytic activity analysis. Based on the low DNA content in the red fluorescence histogram (FL-2), bacterial aggregates were excluded from further analysis. The percentage of active phagocytes and the mean fluorescence intensity were determined in the green fluorescence histogram (FL-1).

Gated lymphocytes (Figure 1a,b) were used for the identification of lymphocyte subpopulations. CD3+ lymphocytes represent T lymphocytes and CD21+ B lymphocytes (Figure 1c). The CD4+CD8- subpopulation was evaluated as T helper lymphocytes, CD4-CD8+ as cytotoxic lymphocytes, CD4+CD8+ as double positive T lymphocytesy (Figure 1d), and CD4+CD25+ as regulatory T lymphocytes (Figure 1e). Proportions of lymphocytes are expressed in percentage.

2.3. Statistical Analysis

The results were statistically evaluated by an unpaired *t*-test with the statistical software GraphPad Prism 8.0. A value of $p < 0.05$ was considered statistically significant. The results obtained in this experiment were expressed as mean ± standard error of the means (SEM).

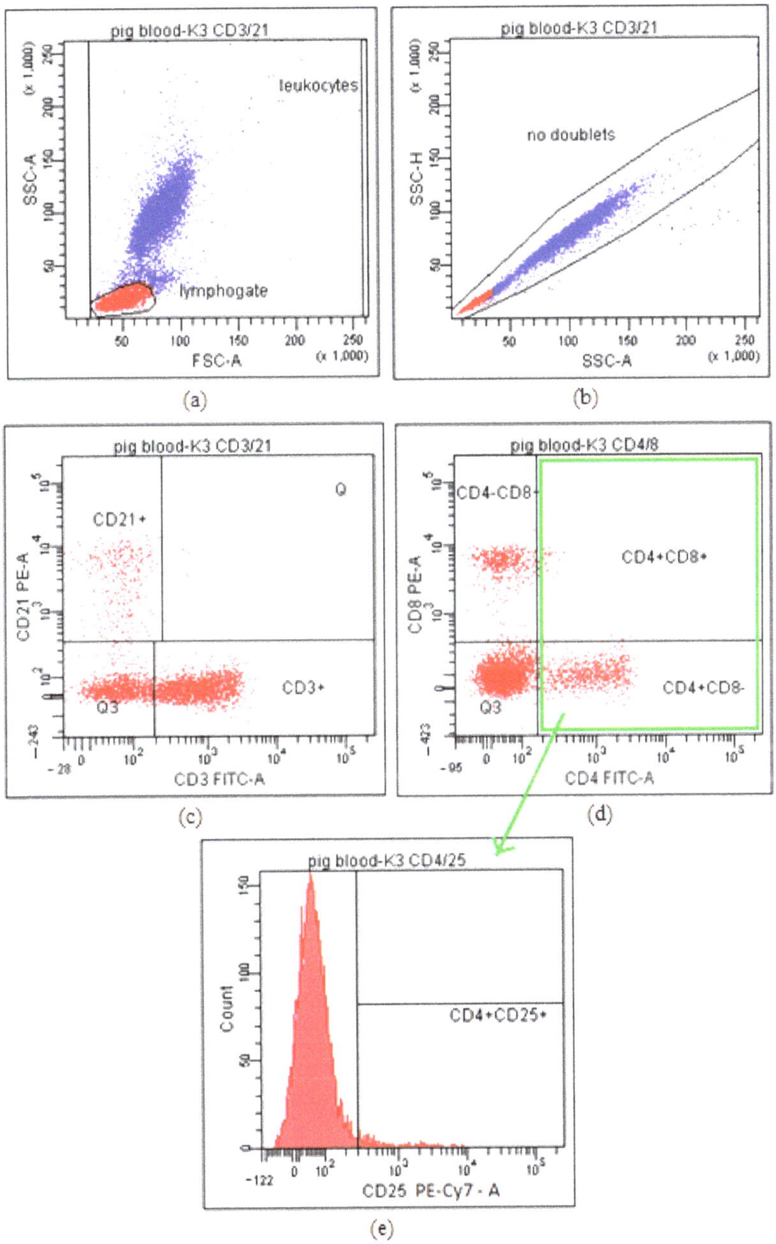

Figure 1. Gating strategy: (**a**) determination of the position of lymphocytes (red) on the basic dot plot (FSC-A versus SSC-A); (**b**) removing doublets and agglomerates from the analysis; (**c**) analysis of the representation of CD3+ and CD21+ lymphocytes; (**d**) analysis of the representation of CD4+ and CD8+ lymphocytes; (**e**) determination of the representation of CD4+CD25+ lymphocytes on histogram from CD4+ positive cells (gated with green).

3. Results

3.1. Haematological and Serum Biochemical Parameters

Effects of HS on haematological and biochemical blood profile are presented in Table 3. The determined blood parameters including content of red blood cells, white blood cells, haemoglobin, haematocrit, mean corpuscular volume, total protein, albumin, glucose, urea, triglycerides, cholesterol, creatinine, AST, Ca, and P were unaffected by the dietary HS treatment. However, dietary supplementation with HS increased activity of ALP ($p < 0.05$).

Table 3. Effect of humic substances on some haematological and serum biochemical parameters in piglets.

		Control	Experimental	*p*-Value
Haematological indices				
Red blood cells	[$T \cdot L^{-1}$]	6.82 ± 0.10	6.81 ± 0.10	0.954
Mean corpuscular volume	[fL]	51.00 ± 0.26	52.00 ± 0.89	0.308
Haematocrit	[$L \cdot L^{-1}$]	0.35 ± 0.01	0.35 ± 0.01	0.739
Haemoglobin	[$g \cdot dL^{-1}$]	11.68 ± 0.25	11.58 ± 0.19	0.755
White blood cells	[$G \cdot L^{-1}$]	10.87 ± 0.20	12.28 ± 1.24	0.287
Serum metabolites				
Total protein	[$g \cdot L^{-1}$]	60.03 ± 0.55	60.41 ± 0.68	0.675
Albumin	[$g \cdot L^{-1}$]	28.78 ± 1.76	32.80 ± 0.87	0.067
Glucose	[$mmol \cdot L^{-1}$]	5.05 ± 0.05	5.19 ± 0.10	0.217
Urea	[$mmol \cdot L^{-1}$]	3.89 ± 0.20	3.73 ± 0.25	0.621
Triglycerides	[$mmol \cdot L^{-1}$]	0.78 ± 0.09	0.89 ± 0.07	0.375
Cholesterol	[$mmol \cdot L^{-1}$]	2.45 ± 0.17	2.47 ± 0.19	0.934
Creatinine	[$\mu mol \cdot L^{-1}$]	79.90 ± 3.55	81.10 ± 3.72	0.820
ALP	[$\mu kat \cdot L^{-1}$]	4.24 ± 0.12	4.60 ± 0.10 *	0.041
AST	[$\mu kat \cdot L^{-1}$]	0.84 ± 0.03	0.88 ± 0.06	0.546
Ca	[$mmol \cdot L^{-1}$]	2.68 ± 0.04	2.82 ± 0.05	0.052
P	[$mmol \cdot L^{-1}$]	2.09 ± 0.02	2.11 ± 0.04	0.692
TBARs	[$nmol\ MDA \cdot mL^{-1}$]	0.72 ± 0.03	0.69 ± 0.03	0.562

ALP—alkaline phosphatase, AST—aspartate aminotransferase, TBARs—thiobarbituric acid reactive substances. * Column labelled with asterisk is significantly different from the control ($p < 0.05$).

Serum TBARs level, secondary products of lipid peroxidation (which are an important parameter used in the determination of lipid peroxidation), was not affected by the addition of HS.

3.2. Cellular Immune Response

The addition of HS to piglets' feed did not have a significant effect on the percentage of active phagocytes (Table 4) and the engulfing capacity of the phagocytes (Table 5). The higher values of these innate immune response parameters were found in the experimental group compared to the control group.

Table 4. Effect of humic substances on the phagocyte activity in piglets' blood evaluated as percentage of active phagocytes—phagocytic activity.

	PA_{total} [%]	PA_{Neu} [%]	PA_{Mo} [%]
Control	83.08 ± 0.62	85.68 ± 0.60	77.18 ± 1.76
Experimental	84.90 ± 1.33	88.04 ± 0.94	78.08 ± 2.10
p-Value	0.251	0.067	0.751

PA_{total}—total phagocytic activity, PA_{Neu}—phagocytic activity of neutrophils, PA_{Mo}—phagocytic activity of monocytes.

Table 5. Effect of humic substances on the engulfing capacity of the phagocytes expressed as mean fluorescence intensity (MFI).

	MFI_{total}	MFI_{Neu}	MFI_{Mo}
Control	31,775 ± 3010	34,783 ± 3223	18,744 ± 1858
Experimental	36,057 ± 1097	39,560 ± 1220	19,041 ± 929
p-Value	0.218	0.203	0.889

MFI_{total}—total mean fluorescence intensity, MFI_{Neu}—mean fluorescence intensity of neutrophils, MFI_{Mo}—mean fluorescence intensity of monocytes.

The percentages of selected lymphocyte subpopulations are presented in Figure 2a–f. It was found that the addition of HS to the diets of piglets significantly increased the proportion of T helper cells (CD4+CD8-) ($p < 0.001$). The results also showed a tendency towards an increase in the number of other monitored lymphocyte subpopulations (CD3+, CD21+, CD4-CD8+, CD4+CD8+, CD4+CD25+) in the piglets in the experimental group compared to the control group, but without a significant difference. Additionally, the ratio of CD4+ and CD8+ lymphocytes (Figure 2g), as a marker of immune stimulation, showed a non-significant increase.

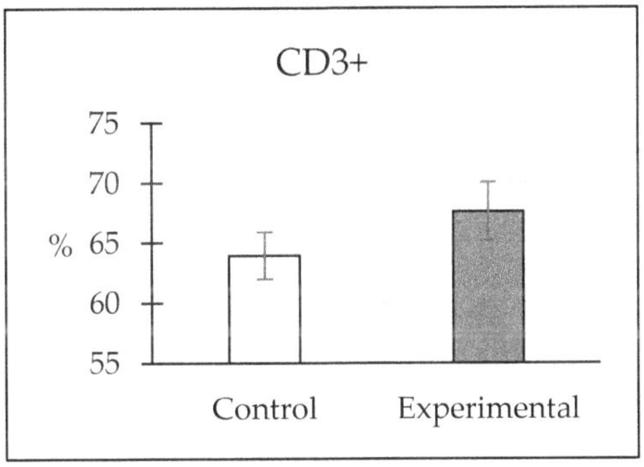

(**a**)

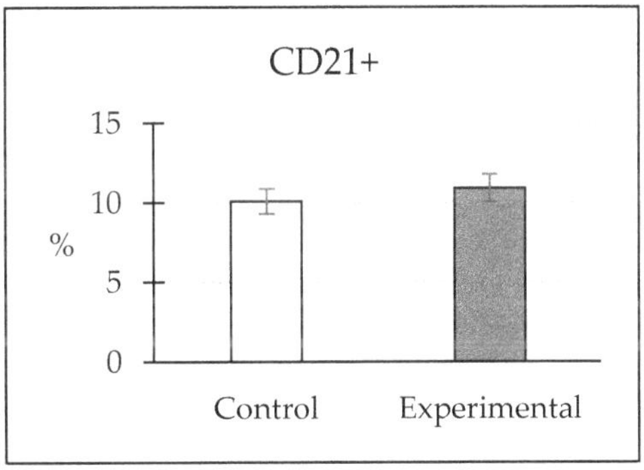

(**b**)

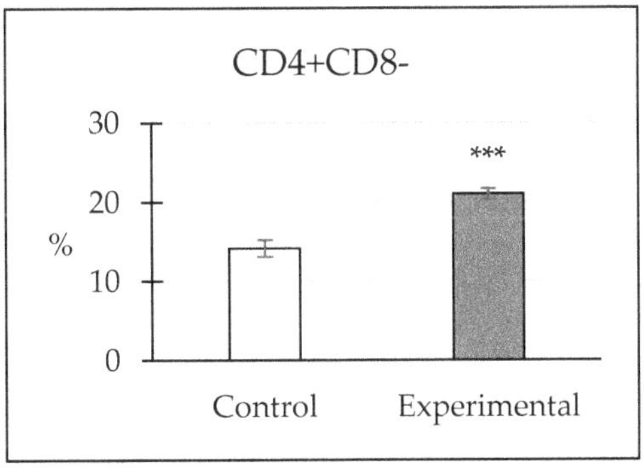

(**c**)

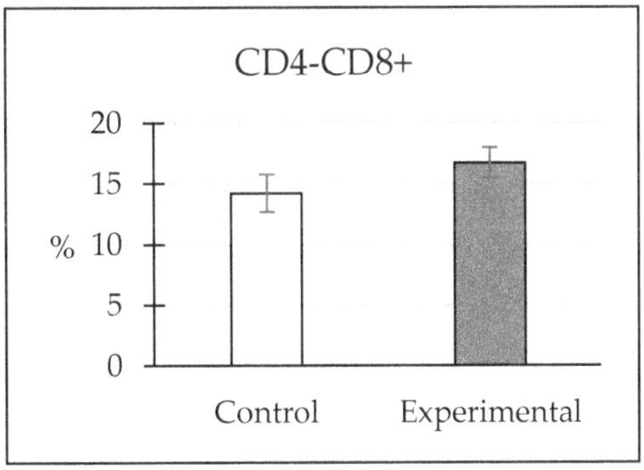

(**d**)

Figure 2. *Cont.*

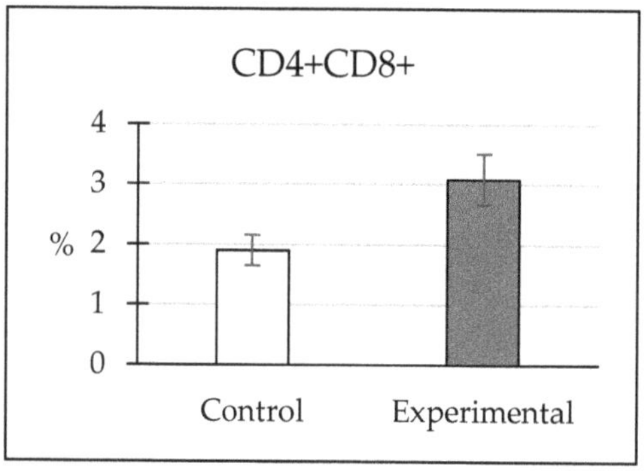

(e)

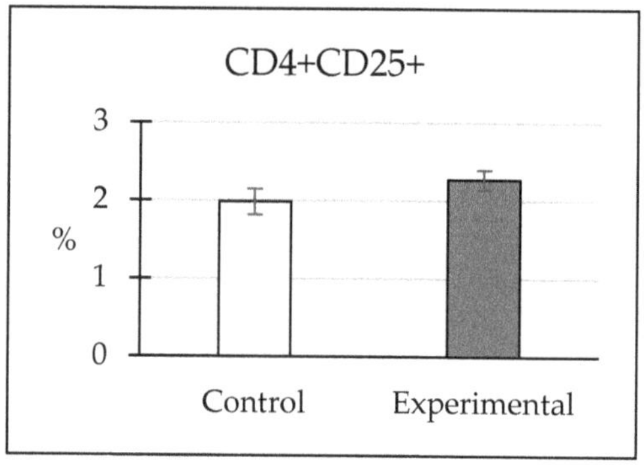

(f)

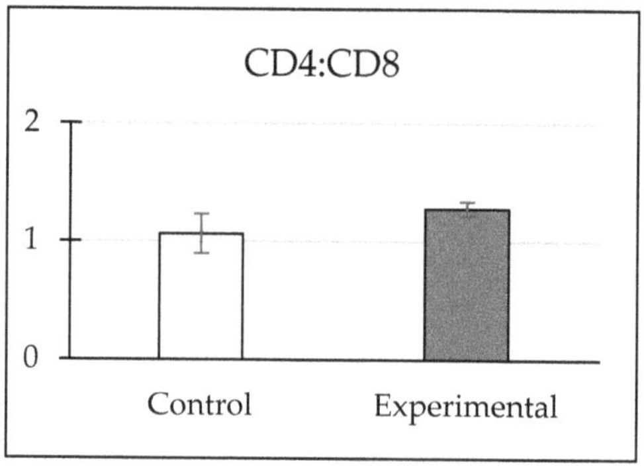

(g)

Figure 2. Effect of humic substances on the percentage of lymphocytes: (**a**) CD3+; (**b**) CD21+; (**c**) CD4+CD8-; (**d**) CD4-CD8+; (**e**) CD4+CD8+; (**f**) CD4+CD25+; and (**g**) ratio of CD4:CD8 lymphocytes in the blood of the piglets. Column labelled with asterisk is significantly different from the control (*** $p < 0.001$).

4. Discussion

4.1. Cellular Immune Response

Our study's major objective was to track how humic substances (HS) affected a few different cellular immunity indicators. The percentage of active phagocytes and their engulfing capacity were selected as indicators for monitoring the effect on innate cellular immunity. In our previous study, it was found that the addition of HS to the diets (0.8%) of broilers significantly increased phagocytic activity as well as mean fluorescence intensity [16]. Additionally, ELnaggar and El-Kelawy [22] observed an increase in phagocytic activity and phagocytic index after humic acid supplementation to the diet (0.1, 0.2, and 0.4%) of Sacco chickens. Similar findings were obtained with laying hens [17]. In the current study, the supplementation of HS to the diet of piglets had no significant effect on phagocytosis. However, the percentage of active phagocytes as well as their engulfing capacity were numerically higher in piglets supplemented with HS than in piglets in the control group. According to Sanmiguel and Rondón [23], the effect of HS on phagocytes is time-dependent. They found that the supplementation of HS to the diets (0.1 and 0.2%) of laying hens increased the phagotic index on day 8 and 30 of application, but on day 60, it was lower than the control group.

The representation of the selected lymphocyte subpopulations in the blood was selected as a parameter for monitoring the effect of HS on the acquired cellular immunity. Results of the present study showed a significant increase in T helper lymphocytes (CD4+CD8-) when 0.5% HS were added to the diet of piglets. The outcomes can be partially compared to those of Wang et al. [2], who found that pigs whose feed had 10% HS added to it, had higher relative lymphocyte counts. Similar results were obtained for poultry, which were fed a diet supplemented with 0.15% humate [24], or were supplemented with 20 mg humic acid/kg of body weight in drinking water [25].

According to Cetin et al. [24], the supplementation of laying hens' diets with humic compounds results in a significant increase in the lymphocyte counts, which can be attributed to increased production of IL-2, as well as expression of the IL-2 receptor on lymphocytes. Humic substances seem to enhance the activity of IL-2 producing cells.

However, other authors did not observe any significant effect on lymphocyte count in pigs that were fed with humic acid supplemented feed [26] as well as in chickens that were given humic acids in their drinking water [27] or their diets [22].

Our results are similar with the observations of Mudroňová et al. [16], who found that the percentage of CD4 lymphocytes was significantly increased for broilers fed HS supplementation (0.8%) compared with the control group. They also noted a significant decrease in CD8+ lymphocytes (T cytotoxic lymphocytes), which resulted in a statistically higher CD4:CD8 ratio, which is used as a marker of immune stimulation. On the other hand, feeding laying hens with a diet of 0.5% HS significantly increased the proportion of IgM+ lymphocytes that represent a subpopulation of B lymphocytes and significantly reduced the proportion of CD3+ lymphocytes that represent total T lymphocytes. The proportion of T helper and T cytotoxic lymphocytes was not affected [17].

Based on our results, as well as various studies, it follows that HS can have an immunostimulatory effect which can be influenced by the composition and quantity of humates used, the method of their administration, the animal species, and, according to ELnaggar and El-Kelawy [22], by the rearing of animals in various regions of the world, differing in climate. The immunomodulation of HS may theoretically consist of the formation of complexes of humates with saccharides. These complexes bind to the surface of T lymphocytes and NK cells and affect their function, including the production of cytokines, which further influence other cells of the immune system [28].

4.2. Haematological and Serum Biochemical Parameters

Another objective of this study was to evaluate selected haematological and biochemical blood parameters of piglets fed diets supplemented with HS. No significant difference in selected variables of protein, energy, and mineral metabolism, as well as in haematological indices, was found between the experimental group of animals with dietary HS supplement and the control group, except for ALP activity in the current study. Biochemical indicators in our study were within the reference range in all piglets, according to Doubek et al. [29] and Kraft and Dürr [30].

Our findings are in line with prior work by Wang et al. [2], who found no significant alterations in the red and white blood cells during the course of the trial in their study to find out the effects of HS on blood variables in finishing pigs.

After treatment with an HS-containing diet, Herzig et al. [31], and also Šamudovská and Demeterová [32], observed non-significantly increased ALP activity in chickens. In our investigation, we found that the experimental group had significantly higher ALP activity than the control group. Their values, meanwhile, remained within the reference range for pig ALP levels (2–17 $\text{kat}\cdot\text{L}^{-1}$) [29,30].

On the other hand, Jad'uttová et al. [33] and Rath et al. [34] noticed a significant reduction in ALP activity in chickens following feeding with HS. The blood levels of ALP in broilers given a 1.00% HS feed supplement decreased, according to these authors. Although the decreased values of ALP activity in the Rath et al. [34] study were statistically different than the controls, they did not reflect any toxic effect of HS on selective organs (muscle, kidney, heart, or liver).

The amount of HS provided by the diet can affect the concentration of minerals in the blood. The levels of calcium and phosphorus in the piglets' blood serum were unaffected by the administration of HS at a concentration of 0.5%. The ability of HS to chelate metals, which is influenced by a significant amount of carboxylic acid side-chains, could be the reason for the decrease in mineral serum concentrations [34]. In our study, the concentrations of calcium and phosphorus in piglets' blood were not decreased below reference values by the addition of HS to the feed mixture of the experimental group.

MDA is the product of lipid peroxidation caused by oxygen free radicals, which can be used as a marker to assess antioxidant status and lipid peroxidation [35]. MDA was determined as a biomarker of oxidative stress by measuring the levels of thiobarbituric acid reactive substances (TBARs) in the blood. A decrease in oxidative stress is characterized by a reduction in MDA levels.

Previous research has shown that HS may have antioxidant properties, protecting against a variety of disorders connected to the oxidative stress that free radicals typically cause. For instance, Wang et al. [15] found that dietary supplementation of sodium humate (2000 $mg \cdot kg^{-1}$) could improve the antioxidant status of weaned piglets. They observed the significant reducing content of MDA in the serum, as well as a significant increase in total antioxidant capacity. However, HS are natural materials with a variable composition, which can cause different effectiveness, which can also depend on their administered amount. In the current experiment, no variation was observed in serum MDA (TBARs) levels of piglets fed HS in comparison to the control group. Our results are consistent with Zhang et al. [36], who found that the addition of sodium humate to the diets (0.1, 0.3, and 0.5%) of laying hens had no effect on serum total antioxidant capacity and MDA values.

5. Conclusions

Based on the results obtained in this study, it may be concluded that a diet supplementation with 0.5% humic substances could have a stimulatory effect on some immune cells in piglets. There was a significant increase in the proportion of CD4+CD8- lymphocytes in the blood. Supplementation with humic substances increased serum ALP. However, these values were still within the reference range. The results showed that even such a low concentration of HS can positively affect cellular immunity in piglets. Due to the fact, that the concentration of humic substances we chose had a significant effect only on some indicators of cellular immunity, further studies will be necessary to choose an appropriate concentration and confirm the stimulating effect on immunity in this category of animals.

Author Contributions: Conceptualization, L.B., A.H.Š., D.M., V.K. and S.M.; methodology, L.B., A.H.Š., I.M. and D.M.; formal analysis, L.B. and A.H.Š.; investigation, L.B., M.H. and M.B.; data curation, A.H.Š., L.B. and D.M.; writing—original draft preparation L.B. and A.H.Š.; writing—review and editing, L.B., A.H.Š., D.M. and S.M.; visualization, L.B. and A.H.Š.; supervision, P.N.; project administration, L.B. and A.H.Š.; funding acquisition, P.N. and S.M. All authors have read and agreed to the published version of the manuscript.

Funding: This publication was supported by the Operational Program Integrated Infrastructure within the project: Demand-Driven Research for Sustainable and Innovative Food, Drive4SIFood 313011V336, co-financed by the European Regional Development Fund (50%) and by the Ministry of Education, Science, Research and Sport of the Slovak Republic (Project VEGA No. 1/0402/20 "Effect of additives in nutrition of monogastric animals on production health, production parameters, product quality and the environment") (50%).

Institutional Review Board Statement: All procedures in the present study were performed in accordance with the principles of the European Union and Slovak Law on Animal Protection.

Data Availability Statement: The data presented in this study are available upon request from the corresponding author.

Acknowledgments: The authors gratefully acknowledge Humac, Ltd., Košice, Slovakia, for providing humic substances for the experiment.

Conflicts of Interest: The authors declare no conflict of interest.

References

1. Peña-Méndez, E.M.; Havel, J.; Patočka, J. Humic substances-compounds of still unknown structure: Applications in agriculture, industry, environment, and biomedicine. *J. Appl. Biomed.* **2005**, *3*, 13–24. [CrossRef]
2. Wang, Q.; Chen, Y.J.; Yoo, J.S.; Kim, H.J.; Cho, J.H.; Kim, I.H. Effects of supplemental humic substances on growth performance, blood characteristics and meat quality in finishing pigs. *Livest. Sci.* **2008**, *117*, 270–274. [CrossRef]
3. Visser, S.A. Effect of humic substances on higher animals and man, the possible use of humic compounds in medical treatments. In Proceedings of the International Humic Acid Society Meeting, Sevilla, Spain, 1988; p. 27.
4. Stevenson, F.J. *Humus Chemistry: Genesis, Composition, Reactions*; Wiley-Inter-Science: New York, NY, USA, 1994; pp. 34–41.
5. Hassan, H.M.A.; Mohamed, M.A.; Youssef, A.W.; Hassan, E.R. Effect of using organic acids to substitute antibiotic growth promoters on performance and intestinal microflora of broilers. *Asian-Aust. J. Anim. Sci.* **2010**, *23*, 1348–1353. [CrossRef]
6. Gálik, B.; Hrnčár, C.; Gašparovič, M.; Rolinec, M.; Hanušovský, O.; Juráček, M.; Šimko, M.; Zábranský, L.; Kovacik, A. The Effect of Humic Substances on the Meat Quality in the Fattening of Farm Pheasants (*Phasianus colchicus*). *Agriculture* **2023**, *13*, 295. [CrossRef]
7. Karaoglu, M.; Macit, M.; Esenbuga, N.; Durdag, H.; Turgut, L.; Bilgin, O.C. Effect of supplemental humate at different levels on the growth performance, slaughter and carcass traits of broilers. *Int. J. Poult. Sci.* **2004**, *3*, 406–410.
8. Ji, F.; McGlone, J.J.; Kim, S.W. Effects of dietary humic substances on pig growth performance, carcass characteristics, and ammonia emission. *J. Anim. Sci.* **2006**, *84*, 2482–2490. [CrossRef]
9. Lacková, Z.; Zigo, F.; Farkašová, Z.; Ondrašovičová, S. The Effect of Humic Substances as an Organic Supplement on the Fattening Performance, Quality of Meat, and Selected Biochemical Parameters of Rabbits. *Life* **2022**, *12*, 1016. [CrossRef]
10. Kaevska, M.; Lorencova, A.; Videnska, P.; Sedlar, K.; Provaznik, I.; Trckova, M. Effect of sodium humate and zinc oxide used in prophylaxis of post-weaning diarrhoea on faecal microbiota composition in weaned piglets. *Vet. Med.* **2016**, *61*, 328–336. [CrossRef]
11. Trckova, M.; Matlova, L.; Hudcova, H.; Martin, F.; Zraly, Z.; Dvorska, L.; Beran, V.; Pavlik, I. Peat as a feed supplement for animals: A review. *Vet. Med. -Czech.* **2005**, *50*, 361–377. [CrossRef]
12. Weber, T.E.; van Sambeek, D.M.; Gabler, N.K.; Kerr, B.J.; Moreland, S.; Johal, S.; Edmonds, M.S. Effects of dietary humic and butyric acid on growth performance and response to lipopolysaccharide in young pigs. *J. Anim. Sci.* **2014**, *92*, 4172–4179. [CrossRef]
13. Mao, Y. Modulation of the growth performance, meat composition, oxidative status, and immunity of broilers by dietary fulvic acids. *Poult. Sci.* **2019**, *98*, 4509–4513. [CrossRef] [PubMed]
14. Stepchenko, L.M. The role of humic preparations in the management of metabolic processes in the formation of biological products of agricultural animals. In *Achievements and Prospects for the Use of Humic Substances in Agriculture*; Publishing House of the Agricultural Institute: Dnepropetrovsk, Ukraine, 2008; pp. 70–74. (In Russian)
15. Wang, Q.; Ying, J.; Zou, P.; Zhou, Y.; Wang, B.; Yu, D.; Li, W.; Zhan, X. Effects of dietary supplementation of humic acid sodium and zinc oxide on growth performance, immune status and antioxidant capacity of weaned piglets. *Animals* **2020**, *10*, 2104. [CrossRef] [PubMed]
16. Mudroňová, D.; Karaffová, V.; Pešulová, T.; Koščová, J.; Maruščáková, I.C.; Bartkovský, M.; Marcinčáková, D.; Ševčíková, Z.; Marcinčák, S. The effect of humic substances on gut microbiota and immune response of broilers. *Food Agric. Immunol.* **2020**, *31*, 137–149. [CrossRef]
17. Mudroňová, D.; Karaffová, V.; Semjon, B.; Nad', P.; Koščová, J.; Bartkovský, M.; Makiš, A.; Bujňák, L.; Nagy, J.; Mojžišová, J.; et al. Effects of dietary supplementation of humic substances on production parameters, immune status and gut microbiota of laying hens. *Agriculture* **2021**, *11*, 744. [CrossRef]
18. European Commission. Directive 2010/63/EU of the European Parliament and of the Council of 22 September 2010 on the protection of animals used for scientific purposes. *Off. J. Eur. Union* **2010**, *276*, 33–79.
19. Šimeček, K.; Zeman, L.; Heger, J. *Nutrient requirement and tables of nutritional values of pig feeds*, 3rd ed.; Czech Academy of Agricultural Sciences: Brno, Czech Republic, 2000; 125p. (In Czech)
20. European Commission. Commission Regulation (EC) No 152/2009 of 27 January 2009 laying down the methods of sampling and analysis for the official control of feed. *Off. J. Eur. Union.* **2009**, *54*, 2–54.
21. Costa, C.M.; Santos, R.C.C.; Lima, E.S. A simple automated procedure for thiol measurement in human serum samples. *J. Bras. Patol. Med. Lab.* **2006**, *42*, 345–350. [CrossRef]
22. ELnaggar, A.S.; El-Kelawy, M.I. Effect of humic acid supplementation on productive performance, blood constituents, immune response and carcass characteristics of sasso chicken. *Egypt. J. Anim. Prod.* **2018**, *55*, 75–84.
23. Sanmiguel, P.R.; Rondón, B.I. Supplementation with humic substances affects the innate immunity in layer hens in posfasting phase. *Rev. MVZ Córdoba* **2016**, *21*, 5198–5210. [CrossRef]
24. Cetin, E.; Guclu, B.K.; Cetin, N. Effect of dietary humate and organic acid supplementation on social stress induced by high stocking density in laying hens. *J. Anim. Vet. Adv.* **2011**, *10*, 2402–2407.
25. Salah, H.; Mansour, E.; Reham, R.R.; Abd El Hamid, E.S. Study on the effect of humic acid on growth performance, immunological, some blood parameters and control intestinal Closterdium in broiler chickens. *Zag. Vet. J.* **2015**, *43*, 102–109. [CrossRef]
26. Kim, K.; Bae, I.; Cho, J.; Choi, Y.; Ha, J.; Choi, J. Effects of humic acid and blueberry leaf powder supplementation in feeds on the productivity, blood and meat quality of finishing pigs. *Food Sci. Anim. Resour.* **2019**, *39*, 276–285. [CrossRef] [PubMed]

27. Lala, A.O.; Okwelum, N.; Irekhore, O.T.; Ogunlade, B.A.; Adigun, A.A.; Elegbede, L.A.; Oyedeji, M.M. Haematological and biochemical parameters of broiler chicken supplemented with humic acid in the drinking water. *Appl. Trop. Agric.* **2016**, *21*, 74–78.
28. Terratol, L.L.C. Effects of Humic Acid on Animals and Humans. Available online: https://fulvic.info/wp-content/uploads/2019/01/M-Terratrol_Article-1.pdf (accessed on 10 August 2022).
29. Doubek, J.; Šlosárková, S.; Řeháková, K.; Bouda, J.; Scheer, P.; Piperisová, I.; Tomenendálová, J.; Matalová, E. *Interpretation of Basic Biochemical and Hematological Findings in Animals*, 2nd ed.; Noviko: Brno, Czech Republic, 2010; p. 102. (In Czech)
30. Kraft, W.; Dürr, U.M. *Klinická Laboratórna Diagnostika vo Veterinárnej Medicíne*, 1st ed.; Hajko & Hajková: Bratislava, Slovakia, 2001; pp. 1–380. (In Slovak)
31. Herzig, I.; Navrátilová, M.; Totušek, J.; Suchý, P.; Večerek, V.; Blahová, J.; Zralý, Z. The effect of humic acid on zinc accumulation in chicken broiler tissues. *Czech J. Anim. Sci.* **2009**, *54*, 121–127. [CrossRef]
32. Šamudovská, A.; Demeterová, M. Effect of diet supplemented with natural humic compounds and sodium humate on performance and selected metabolic variables in broiler chickens. *Acta Vet. Brno* **2010**, *79*, 385–393. [CrossRef]
33. Jaďuttová, I.; Marcinčáková, D.; Bartkovský, M.; Semjon, B.; Harčárová, M.; Nagyová, A.; Váczi, P.; Marcinčák, S. The effect of dietary humic substances on the fattening performance, carcass yield, blood biochemistry parameters and bone mineral profile of broiler chickens. *Acta Vet. Brno* **2019**, *88*, 307–313. [CrossRef]
34. Rath, N.C.; Huff, W.E.; Huff, G.R. Effects of humic acid on broiler chickens. *Poult. Sci.* **2006**, *85*, 410–414. [CrossRef]
35. Sumida, S.; Tanaka, K.; Kitao, H.; Nakadomo, F. Exercise-induced lipid peroxidation and leakage of enzymes before and after vitamin E supplementation. *Int. J. Biochem.* **1989**, *21*, 835–838. [CrossRef]
36. Zhang, A.R.; Pirzado, S.A.; Liu, G.H.; Chen, Z.M.; Chang, W.H.; Cai, H.Y.; Bryden, W.L.; Zheng, A.J. Dietary supplementation with sodium humate improves egg quality and immune function of laying hens. *J. Appl. Anim. Nutr.* **2020**, *8*, 93–99. [CrossRef]

agriculture

MDPI

Article

Fertilizer Performance of a Digestate from Swine Wastewater as Synthetic Nitrogen Substitute in Maize Cultivation: Physiological Growth and Yield Responses

Eduardo L. Buligon [1], Luiz A. M. Costa [1], Jorge de Lucas, Jr. [2], Francielly T. Santos [1], Piebiep Goufo [3,*] and Monica S. S. M. Costa [1,*]

1 Agricultural Engineering Graduate Program, Universidade Estadual do Oeste do Paraná, Street University 2069, Cep, Cascavel 85819-110, Brazil

2 Departamento de Engenharia Rural, Universidade Estadual Paulista "Júlio de Mesquita Filho", Via de Acesso Prof. Paulo Donato Castellane s/n—Jaboticabal/SP—Cep, São Paulo 14884-900, Brazil

3 CITAB—Centro de Investigação e Tecnologias Agroambientais e Biológicas, Universidade de Trás-os-Montes e Alto Douro, Quinta de Prados, 5000–801 Vila Real, Portugal

* Correspondence: pgoufo@utad.pt (P.G.); monicasarollisilva@gmail.com (M.S.S.M.C.)

Abstract: Nitrogen (N) is the primary nutrient required for plant growth. During the last few decades, there has been extensive use of synthetic N-containing fertilizers in agriculture, resulting in increased environmental pollution. In this study, the feasibility of replacing synthetic N with biofertilizer in maize cultivation was investigated. A liquid biofertilizer (digestate obtained from the anaerobic digestion of swine wastewater) was obtained and applied to large plots as a total (100%) or partial (50%) substitute for synthetic N fertilizer. Moreover, the most efficient fertilization mode, i.e., basal versus foliar application, was studied. Physiological growth indices, leaf nutritional status, and grain yield were assessed for each biofertilization treatment and compared with the conventional treatment with synthetic minerals. Compared with the conventional treatment, the total substitution of synthetic N by the biofertilizer (basal application) did not affect the growth parameters and grain yield of maize; the other treatments usually resulted in lower growth rates and yields, although not statistically significant ($p \geq 0.05$). No difference was observed among the treatments for the contents of N, P, K, or Mg in the leaves. Generally, the highest means for Fe, Ca, Cu, Zn, and Mn contents in leaves were observed after in-row broadcast of synthetic fertilizers or basal application of the digestate as a total substitute for synthetic N, with a significant effect for Fe ($p < 0.05$). The mode of the biofertilizer application did not have any significant effect on either growth parameters or leaf nutrients. The data show that under the specific conditions of the study, the total substitution of mineral N with basal application of biofertilizer is the best strategy for minimizing the use of synthetic chemicals in maize cultivation without yield penalties.

Keywords: waste valorization; methane-rich biogas; bio-based fertilizers; soil amendments; organic fertilizers; foliar application; anaerobic digestion; corn growth dynamic; nutrient availability; fertilizer application rate/dose; drought stress mitigation

Citation: Buligon, E.L.; Costa, L.A.M.; de Lucas, J., Jr.; Santos, F.T.; Goufo, P.; Costa, M.S.S.M. Fertilizer Performance of a Digestate from Swine Wastewater as Synthetic Nitrogen Substitute in Maize Cultivation: Physiological Growth and Yield Responses. *Agriculture* **2023**, *13*, 565. https://doi.org/10.3390/agriculture13030565

Academic Editor: Maria Roulia

Received: 8 December 2022
Revised: 20 February 2023
Accepted: 22 February 2023
Published: 26 February 2023

1. Introduction

Brazil is nowadays a major player in international agricultural production and export [1,2]. However, Brazil is heavily dependent on the import of NPK (nitrogen, phosphorus, and potassium) fertilizers. It is estimated that 85% of the fertilizers used in Brazil are imported from the global market [1]. These data have prompted research on alternative local nutrient sources in agriculture, such as stabilized organic wastes, e.g., compost, vermicomposto, and biofertilizer. In 2020, manure production in Brazil was estimated at 61.8 million tons, of which 14.9 million tons was from poultry litter and 46.9 million tons from swine and cattle manure [3].

Before application to agricultural soils as a source of nutrients, organic waste needs to be stabilized, i.e., active biological decomposition processes need to be stopped to minimize disruption of the soil-plant-root interface as much as possible [4–7]. Stabilization can be achieved through several processes such as composting, vermicomposting, mechanical-biological treatments, or anaerobic digestion to produce bio-based products called "biofertilizers" [6,8–11]. Anaerobic digestion involves the degradation of organic matter in an oxygen-free environment to release a gas known as biogas and an organic effluent + residue called digestate [12,13].

According to Koszel and Lorencowicz [14], the main factors determining the use of digestates as biofertilizers are the physicochemical characteristics of the effluent and residue, the edapho-climatic conditions, and local regulations. Locoli et al. [15] found that digestates obtained from the anaerobic digestion of cattle manure, chicken litter, swine manure, and onion wastes had chemical and spectroscopic characteristics (e.g., C/N ratio, ammonium (NH_4^+) to nitrogen ratio (NH_4^+/N), and proportion of short-chain organic acids) similar to those of untreated wastes. Soils amended with the digestates in the study by Locoli et al. [15] emitted less CO_2 than soils amended with manure, and the fertilizing effect of the digestates on lettuce growth was related to the content of NH_4^+ [15]. However, recent review papers have found that feedstock, processing technology, and process operating conditions strongly influence the characteristics of digestates, and that without comprehensive management strategies, digestates can contribute to nutrient pollution [6,13]. Thus, each locally produced digestate should be assessed on its own merit based on optimal parameters suitable for adequate microbial activity such as the C/N ratio (15–30), the psychrophilic temperature (<20 °C), the mesophilic temperature (35–37 °C), and the thermophilic temperature (55 °C) [13].

Using indicators such as N uptake, P recovery rate, soil microbial stimulation, or yield, the biofertilizing effect of various digestates on different plants in different climates and soils has been confirmed. Results from a trial in Belgium indicated that the liquid fraction of a digestate obtained from swine manure could substitute synthetic N fertilizers without maize yield losses [16]. Tsachidou et al. [17] reported that the partial substitution of chemical fertilizers with a raw digestate from bovine manure as the sole source of N reduced the concentration of NO^{3-} in the soil without impacting biomass yield and N content in a pasture system. Zilio et al. [18] concluded that the maize grain yield obtained from plants grown in a soil that received a sewage sludge-based digestate is equivalent to the yield obtained from plants grown using urea. The abovementioned three examples show that a stable digestate can be used as a bio-based fertilizer to replace mineral N fertilizers without yield loss or without increasing the risk of environmental pollution. However, only a few large-scale studies have been conducted on the effects of digestates on crop growth. This highlights the need for more studies that contribute different variables to decision making on the use of biofertilizers in agriculture. Further studies in different locations are essential to gather the necessary information to formulate standardized international protocols for researchers and operators, to develop best management practices for farmers, and to promote digestate product commercialization as part of the organic waste circular economy paradigm.

The study reported in this paper aimed to contribute to the topic by testing different biofertilizer doses and application techniques. Specifically, the effects of a digestate from swine wastewater were evaluated on maize physiology, considering four application scenarios: (i) soil application with a total (100%) replacement of the amount of recommended mineral N; (ii) foliar application with a total (100%) replacement of the amount of recommended mineral N; (iii) soil application with a partial (50%) replacement of the amount of recommended mineral N; and (iv) foliar application with a partial (50%) replacement of the amount of recommended mineral N. The impact of the digestate on maize production was compared to that of the conventional fertilization practice using synthetic fertilizers. The applied treatments were evaluated over six months, with a focus on leaf morphology, physiological indices, dry matter (DM) content in maize, and grain yield.

2. Materials and Methods

2.1. Study Area

The field experiment was conducted at the Experimental Station of the Agricultural Engineering Department of the Western Paraná State University (UNIOESTE) in Cascavel municipality, PR, Brazil. Cascavel is located geographically between 24°57′21″ S and 53°27′19″ W. According to the Köppen–Geiger classification system, the predominant climate in Cascavel is Cfa. The Cfa climate is characterized by infrequent frosts, hot summers, and a trend of rainfall concentration in the summer [19]. The average annual temperature, atmospheric pressure, and rainfall at the experimental site were 20 °C, 936.34 hPa, and 1841 mm, respectively. Soil samples at the experimental site were collected from 0–20 cm depths. The samples were bulked, air dried, gently crushed, and sieved through a 2 mm sieve before analyses by Solanalise Central De Analises Ltd., an accredited laboratory in Cascavel. The clay-textured soil at the site was classified as Dystroferric Red Latosol (Oxisol), and its chemical characteristics are shown in Table 1.

Table 1. Main chemical characteristics of the soil in the experimental area before fertilization. Contents were classified as low, medium, or high based on the ranges specified in the Paraná State Handbook for Fertilization and Liming [20].

Nutrient	Unit	Content	Classification
Ca^{2+}	$cmol_c\ dm^{-3}$	4.59	High
Mg^{2+}	$cmol_c\ dm^{-3}$	1.61	High
PO_4^{3-}	$mg\ dm^{-3}$	6.38	Medium
K^+	$cmol_c\ dm^{-3}$	0.36	High
Al^{+3}	$cmol_c\ dm^{-3}$	0.31	Low
H + Al	$cmol_c\ dm^{-3}$	9.01	High
Sum of bases	$cmol_c\ dm^{-3}$	6.56	High
Cation exchange capacity at pH 7.0	$cmol_c\ dm^{-3}$	15.57	High
Cation exchange capacity efective	$cmol_c\ dm^{-3}$	6.87	High
Carbon	$g\ dm^{-3}$	25.75	High
Organic matter	$g\ dm^{-3}$	44.29	High
Aluminum saturation	%	4.51	Low
Base saturation	%	42.13	Low
B	$mg\ dm^{-3}$	0.24	Low
S	$mg\ dm^{-3}$	4.59	Low
Fe^{+2}	$mg\ dm^{-3}$	27.40	Medium
Mn^{+2}	$mg\ dm^{-3}$	43.40	High
Cu^{+2}	$mg\ dm^{-3}$	5.20	High
Zn^{+2}	$mg\ dm^{-3}$	1.80	Medium
pH ($CaCl_2$)	NA	4.60	NA

NA = not available or not applicable.

2.2. Materials

2.2.1. Test Crop

The maize cultivar used in this study was the hybrid P3380HR. The preceding crop was soybean. Following soybean harvesting, atrazine 500 SC (Nortox, Arapongas, Minas Gerais, Brazil) was applied at 1.0 kg ha^{-1} to control the remaining crops and weeds before maize planting. Maize was mechanically broadcast-seeded in March 2021 at a density of 2.8 seeds per m^2. After seedling emergence 5 d after sowing (DAS), plants were thinned to one plant per hole to homogenize the planting and avoid any influence of spacing and shading on nutrient absorption by individual plants. Harvesting occurred manually in August 2021. Insect control focused on the maize leafhopper, *Dalbulus maidis*, with three applications of the insecticide acephate 750 (Ameribrás, Cotia, São Paulo, Brazil) at 1.0 kg ha^{-1} and Galil 300 SC (Adama, Londrina, Paraná, Brazil) at 250 mL ha^{-1}. No other standard agricultural practices, e.g., irrigation, were implemented.

2.2.2. Mineral Fertilizers

The mineral fertilizers used in this study were urea (46% N), potassium chloride (60% K_2O), and single superphosphate (18% P_2O_5, 17.3% Ca, 3.3% Mg, and 5% S). Mineral fertilizers were obtained from a local vendor. The nutritional recommendations for maize were calculated based on the soil characteristics in Table 1, following the methodology proposed by the Paraná State Handbook for Fertilization and Liming [20]: 120 kg ha^{-1} of N, 110 kg ha^{-1} of P_2O_5, and 70 kg ha^{-1} of K_2O.

2.2.3. Biofertilizer

The biofertilizer was produced by the Laboratory of Agroindustrial Waste Analysis of UNIOESTE using swine wastewater (1.78% of total solids). Anaerobic digestion of the swine wastewater was performed in a horizontal tubular benchtop reactor tank operated in a semi-continuous flow with a volume of 60 L, a hydraulic retention time of 30 d, and a mesophilic-controlled temperature of 35 ± 1 °C. The obtained liquid digestate was used as biofertilizer without any post-processing treatment. The biofertilizer was stored in closed plastic barrels at room temperature until further use.

The biofertilizer was sampled and analyzed to determine its physicochemical characteristics and calculate the amount to be applied to the field. All chemicals used for the analyses were obtained from Química Moderna (Barueri, São Paulo, Brazil). Total Kjeldahl N was determined by digesting the samples with sulfuric acid, followed by distillation and titration using 0.0025 mol of H_2SO_4 [21]. The concentrations of P and K were determined by digesting the samples in a nitric-perchloric acid solution (3:1) with an external heat source, followed by dilution and filtration. Phosphorus was detected by measuring the absorbance at a wavelength of 725 nm in a 700 Plus UV/Vis spectrophotometer (Femto Indústria e Comércio de Instrumentos, São Paulo, São Paulo, Brazil) using the ascorbic acid method. Potassium was quantified using a DM-62 flame photometer (Digimed, Campo Grande, Mato Grosso do Sul, Brazil) as described by Malavolta et al. [21]. The levels of micronutrients (Fe, Zn, Cu, and Mn) and secondary macronutrients (Mg and Ca) were determined by atomic absorption spectroscopy (Shimadzu AA6300, Tokyo, Japan) prior to the digestion of samples with a nitric-perchloric solution (3:1) [22]. The results of the analysis are presented in Table 2.

Table 2. Main chemical characteristics of the liquid biofertilizer (digestate of swine wastewater) tested in the study.

Nutrient	**N**	**P_2O_5**	**K_2O**	**Ca**	**Mg**	**Cu**	**Fe**	**Mn**	**Zn**
Unit		**g L^{-1}**					**mg L^{-1}**		
Content	2.50 ± 0.11	0.45 ± 0.15	0.15 ± 0.0	77.42 ± 5.8	0.08 ± 0.08	3.92 ± 0.02	7.45 ± 0.11	1.36 ± 0.08	2.08 ± 0.01

2.3. *Experimental Design and Treatments*

The experiment was conducted in a completely randomized block design with five treatments and four replications, totaling 20 plots spread across the field. Each of these plots was 3.2 m wide and 10 m long. The total area occupied by the experiment was 520 m^2, 13 m wide and 40 m long. The treatments consisted of two doses of biofertilizer (100 and 50%) established based on the recommended rate of N for maize cultivation in Paraná State [20], two forms of application (basal application on the soil surface and foliar application on the whole plant), and one control treatment (synthetic urea, potassium chloride, and single superphosphate as mineral fertilizers). A detailed description of these treatments is provided in Table 3.

Based on the chemical characterization of the digestate and the composition of synthetic chemicals, a combination of biofertilizers and mineral fertilizers was prepared. All combinations were made ensuring no limiting nutrients and the same N content for all treatments but with 100 and 50% reduced synthetic N amounts for T_1 and T_2 as well as T_3 and T_4, respectively, as shown in Table 4.

Table 3. Description of the five fertilization treatments used in the experiment.

Treatments	Application Mode	Description
T_1	Basal	Dose of biofertilizer corresponding to 100% of the amount of recommended mineral N + P and K supplementation with synthetic fertilizers
T_2	Foliar	Dose of biofertilizer corresponding to 100% of the amount of recommended mineral N + P and K supplementation with synthetic fertilizers
T_3	Basal	Dose of biofertilizer corresponding to 50% of the amount of recommended mineral N + N, P, and K supplementation with synthetic fertilizers
T_4	Foliar	Dosage of biofertilizer corresponding to 50% of the amount of recommended mineral N + N, P, and K supplementation with synthetic fertilizers
T_5	In-row broadcasting	Mineral fertilization as recommended for the maize crop in the Paraná State of Brazil

Table 4. Nutrient composition and application rates (L $Plot^{-1}$) for the different fertilization treatments.

Treatments	Biofertilizer (L)	Nutrient Composition of the Biofertilizer (g)			Synthetic Nutrients Applied Directly to Soil (g)		
		N	P_2O_5	K_2O	N	P_2O_5	K_2O
T_1	122.88	307.20	56.30	19.00	0	225.30	160.20
T_2	122.88	307.20	56.30	19.00	0	225.30	160.20
T_3	61.44	153.60	28.10	9.50	153.60	253.50	169.70
T_4	61.44	153.60	28.10	9.50	153.60	253.50	169.70
T_5	0	0	0	0	307.20	281.60	179.20

The precise volume of biofertilizer for treatments T_1–T_4 was dispersed to the soil around the plants or on the leaves and all over the plants using watering cans. To satisfy crop nutrient requirements for treatments T_1–T_4, an adequate amount of mineral N-urea, P_2O_5, and K_2O powder was mixed and manually broadcast evenly over the soil surface. In the case of treatments T_3–T_4, mineral fertilizers were watered into the soil with 61.44 L $Plot^{-1}$ of tap water. As a reference treatment (T_5), mineral fertilizers were mixed, manually broadcast evenly over the soil surface, and watered into the soil with 122.88 L $Plot^{-1}$ of tap water, a volume equivalent to the volume of the digestate in treatments T_1 and T_2. The treatments were applied at two distinct phases of growth: at the beginning of the growing period to boost the development of plants, and during the period of high demand for nutrients and water for plants to enter reproductive growth. Specifically, 30% of fertilizer was applied at vegetative stage V1 (7 DAS), when plants had one visible leaf collar, and 70% at vegetative stage V10 (45 DAS), when plants began steady and rapid periods of growth and DM accumulation.

2.4. *Parameter Measurements*

2.4.1. Morphological and Physiological Parameters

Data on growth traits were collected at 20, 40, 60, 80, and 100 days after seedling emergence (DAE) at 20-day intervals to cover the main vegetative and reproductive stages of maize. During each sampling period, two plants were selected randomly, cut off at the base with a knife, and separated into leaves, stems, and ears, when present. Then, the number of leaves was counted. The width and length of all leaves were measured in centimeters using a ruler, averaged, and the leaf area was calculated and expressed using the formula proposed by Guimarães et al. [23] (Table 5). The leaves, stems, and ears were oven-dried separately at 105 °C until they reached a constant weight, recorded as DM. Total

DM and the ear DM: Total DM ratios were calculated. Based on leaf area and DM data, physiological indices were estimated using the mathematical equations (Table 5) proposed by Benincasa [24].

Table 5. Morphological and physiological parameters assessed in this study pertaining to maize growth and the respective equations used for their determination as described in Guimarães et al. [23] and in Benicasa [24].

Physiological Indices	Equation	Description of Abbreviations
Total dry matter (W)	$W = W_L + W_S + W_E$	W_L = leaf dry matter (g) W_S = stem dry matter (g) W_E = ear dry matter (g)
Total leaf area (TLA)	$TLA = 0.7458 \times L_W \times L_L$	L_W = leaf width (cm) L_L = leaf length (cm)
Leaf area index (LAI)	$LAI = TLA/S$	TLA = total leaf area (cm^2) S = soil surface area (cm^2)
Absolute growth rate (AGR)	$AGR = (W_2 - W_1)/(t_2 - t_1)$	W = total dry matter (g) t = time (d) 1,2 = two successive sampling periods
Relative growth rate (RGR)	$RGR = (\ln W_2 - \ln W_1)/(t_2 - t_1)$	Ln = Naperian logarithm W = total dry matter (g) t = time (d) 1,2 = two successive sampling periods
Leaf area relative growth rate (RGR_{LA})	$RGR_{LA} = (\ln TLA_2 - \ln TLA_1)/(t_2 - t_1)$	Ln = Naperian logarithm TLA = total leaf area W = total dry matter (g) t = time (d) 1,2 = two successive sampling periods
Net assimilation rate (NAR)	$NAR = [(W_2 - W_1)/(t_2 - t_1)] \times [(\ln TLA_2 - \ln TLA_1)/(TLA_2 - TLA_1)]$	TLA = total leaf area W = total dry matter (g) t = time (d) 1,2 = two successive sampling periods
Leaf area ratio (LAR)	$LAR = (TLA/W)$	TLA = total leaf area W = total dry matter (g)
Specific leaf area (SLA)	$SLA = (TLA/W_E)$	TLA = total leaf area W = total dry matter (g)

2.4.2. Leaf Nutritional Status

Maize leaves were collected at silking stage R1 when female inflorescences were visible on 50–75% of the plants (80 DAE). The R1 stage is characterized by complete K uptake and rapid N and P uptake, making R1 the most critical stage in determining the yield potential [25,26]. The third leaf, counted from the base and below the first (upper) ear, was cut with a knife, and the midrib was removed and discarded. The remaining leaf portion was washed with distilled water and oven-dried at 50 °C until a constant weight was achieved. The nutrient composition of leaves was determined using the method proposed by Martinez et al. [27]. Briefly, leaf samples were powdered in a home mixer and sieved. N in 1 g of powdered sample was estimated using the micro-Kjeldahl method, whereas P and K were analyzed after digestion of the samples in a nitric-perchloric acid solution using a UV/Vis spectrophotometer and a flame photometer, respectively (as described in Section 2.2.3). Ca, Mg, Cu, Fe, Mn, and Zn were analyzed in the nitric-perchloric extract using an atomic absorption spectrophotometer. The nutrient content of the leaves was expressed on a dry-weight basis. The nutritional values obtained were compared with the reference values in mg kg^{-1} considered suitable for maize in the State of Paraná in Brazil [20], that is, N (27,000–35,000), P (1900–4000), K (17,000–35,000), Ca (2300–8000), Mg (1500–5000), Cu (6–20), Fe (30–250), Mn (20–200), and Zn (15–100).

2.4.3. Maize Grain Yield

All ears were harvested at the full-grain maturity stage (120 DAS), at approximately 25% moisture content; the ears were collected manually from the plants in each plot, excluding border rows. The ears were mechanically dehusked and threshed to collect grains using a maize sheller coupled to a tractor. The grains were oven-dried at 105 °C until a moisture content of 13% was attained, which is a safe level for storage and commercialization. The mean weight of the samples was recorded using a scale and was expressed in kg ha^{-1}.

2.5. Statistical Analyses

All data were subjected to an analysis of variance using the statistical program Sisvar 5.6-Build 86 [28]. All variables satisfied the requirements of normal distribution and homoscedastic assumption of variance after examination using the Shapiro–Wilk test and Levene's test, respectively. Therefore, the mean value (n = 4) was calculated without data transformation. Differences between means were evaluated using Tukey's honest significant difference (LSD) test at the 5% probability level. The coefficient of variation (CV) was calculated as the ratio of the standard deviation to the mean to show the extent of variability concerning the mean for all treatments.

3. Results and Discussion

3.1. Effects of the Biofertilizer on Maize Grain Yield

In comparison with T_5 (mineral fertilizers), maize grain yield was not affected ($p < 0.05$) by the use of biofertilizers, although T_3 and T_4 (partial substitution of chemical fertilizers by the digestate) tended to lead to lower yields (Figure 1). The average grain yield of all treatments was 1274.52 kg ha^{-1}, which was well below the 5370 kg ha^{-1} average in Brazil in the experimental year [29]. The low maize yield in the testing field could be explained by the drought event in 2021, specifically acute in June, which coincided with the flowering and pollination stages of maize [25].

The data in Figure 1 expressly indicated the beneficial fertilizing properties of the biofertilizer. A major downside limiting the widespread adoption of biofertilizers by farmers is the long time it takes for organic matter to be oxidized into easily available nutrients [6,13]. However, bio-based fertilizers are known for their high variability in nutrients, and the starting material and processing conditions strongly influence the characteristics of the final product [4,12]. The yield data obtained in the present study could be explained by (i) a high level of NH_4^+ in the swine wastewater-based digestate and/or (ii) a fast and efficient conversion of organic N in the digestate into NH_4^+, an inorganic form of N easily absorbed by the plant root system. Although NH_4^+ levels were not determined in the present study, a reasonable correlation was established between vegetal growth and the level of NH_4^+ in a digestate obtained from cattle manure, poultry litter, and pig slurry [15]. Moreover, Costa et al. [4] observed that more than 60% of the total N in biofertilizers produced with beef cattle manure was NH_4^+. Results from trials in Argentina, Italy, and Belgium have also indicated that digestates applied at adequate dosages to the soil may substitute synthetic N fertilizers without crop yield losses [15–18]. In the former example, digestate application to soil produced a fast and short microbial stimulation [15]. In the latter examples, the digestates contributed to a short-term renewal of soil organic matter [16–18]. The results of the study presented in this paper further indicate that the N mineralization rate of some biofertilizers can be very fast, resulting in noticeable effects within months.

3.2. Effects of the Biofertilizer on the nutritional Value of Maize Leaves

There was a significant difference ($p < 0.05$) in the Fe content of maize leaves grown with biofertilizers and mineral fertilizers (Table 6). The lowest Fe content was observed in leaves from T_3 at 217.66 mg kg^{-1}, statistically different from the content in leaves from T_1. Overall, treatments T_1 and T_5 resulted in the highest Fe contents of 540.68 and 326.55 mg kg^{-1}, respectively. Although not significant ($p > 0.05$), higher Ca, Cu, Zn, and Mn contents were observed in leaves from T_1 and T_5 than in leaves from the other treatments. A lack of treatment effect on the contents of N, P, and K in maize leaves (Table 6) reinforces the grain yield results (Figure 1) and underlines the suitability of biofertilizers as partial or total substitutes for chemical fertilizers in the cultivation of maize. However, data from Figure 1 and Table 6 indicate better results with the total replacement of mineral fertilizers than with partial replacement. In a previous study [30], foliar application of a biofertilizer obtained from sewage sludge increased the contents of macro- and micronutrients in maize leaves; when the biofertilizer was applied directly to the soil, however, the contents of nutrients analyzed in both soil and leaves were not affected.

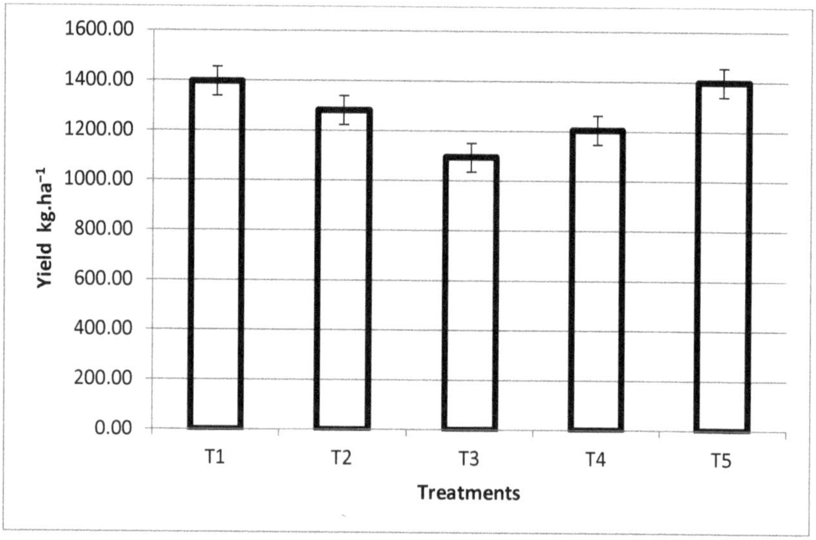

Figure 1. Effect of fertilization treatments on maize grain yield. T_1 = basal application of the digestate from swine wastewater as total substitute (100%) for synthetic N; T_2 = foliar application of the digestate as total substitute (100%) for synthetic N; T_3 = basal application of the digestate as partial substitute (50%) for synthetic N; T_4 = foliar application of the digestate as partial substitute (50%) for synthetic N; T_5 = in-row broadcast of mineral fertilizers. The treatments are fully described in Table 3. Mean bars with no letter or same letters are not statistically different (Tukey's HSD, $p < 0.05$, $n = 4$). Error bars indicate standard deviations.

Leaf nutrient analysis is an effective tool for diagnosing nutrient deficiency or excess in plants and for visualizing the capacity of plants to absorb nutrients from different fertilizers [21,27]. The Ca, Zn, and Mn contents in the leaves were within the reference range for maize in Paraná State; the N, Cu, and Fe contents in the leaves were above the reference values, whereas the P, K, and Mg contents were much lower than the minimum values reported in the literature [20] (Table 6). Cu [31], Fe ([32], Mn [33], and Zn [34] are among the micronutrients that directly affect photosynthesis. The biofertilizer contained a significant amount of Cu and Fe (Table 2), whereas mineral fertilizers were devoid of these two nutrients. However, the soil Cu content was considered high (Table 1), and the Dystroferric Red Latosol used in the experiment was characterized by the presence of high levels of Fe and aluminum oxide [35]; these two observations probably explain, respectively, the high contents of Cu and Fe in leaves following plant uptake. The P content was approximately 20 times less than the reference value, indicating a deficiency [26]. The low P level was however inconsistent with that in the soil (Table 1) or the digestate (Table 2), and the possible reason for this is that more Cu and Fe ions in the leaves affected P absorption and utilization [36,37]. Moreover, the low soil moisture resulting from the drought that occurred during maize growth could have damaged the root structure and reduced water and nutrient absorption, including P [37]. The same explanation holds for the K and Mg contents.

Table 6. Effect of fertilization treatments on the contents of nutrients in maize leaves at the R1 silking stage. T_1 = basal application of the digestate from swine wastewater as total substitute (100%) for synthetic N; T_2 = foliar application of the digestate as total substitute (100%) for synthetic N; T_3 = basal application of the digestate as partial substitute (50%) for synthetic N; T_4 = foliar application of the digestate as partial substitute (50%) for synthetic N; T_5 = in-row broadcast of mineral fertilizer. The treatments are fully described in Table 3.

Treatments	T_1	T_2	T_3	T_4	T_5	CV (%)	Reference
N (g kg^{-1})	40.60	38.60	42.00	40.50	37.80	12.40	27–35
P (g kg^{-1})	0.11	0.08	0.09	0.08	0.09	32.83	1.9–4.0
K (g kg^{-1})	11.40	12.10	8.40	10.30	9.80	18.88	17–35
Ca (g kg^{-1})	6.80	5.80	8.50	6.30	8.50	50.53	2.3–8.0
Mg (g kg^{-1})	0.035	0.029	0.041	0.034	0.037	30.11	1.5–5.0
Cu (mg kg^{-1})	140.90	68.61	69.34	85.27	172.00	66.19	6–20
Zn (mg kg^{-1})	67.54	54.43	54.73	50.25	71.24	30.99	15–100
Fe (mg kg^{-1})	540.68 ^{A}	313.03 ^{AB}	217.66 ^{B}	300.97 ^{AB}	326.55 ^{AB}	41.23	30–250
Mn (mg kg^{-1})	36.94	31.38	26.44	29.31	37.61	56.07	20–200

Mean within a row followed with no letter or same letters are not statistically different (Tukey's HSD, $p < 0.05$, $n = 4$). CV = coefficient of variation.

3.3. Effects of the Biofertilizer on the Growth Parameters of Maize

Data from samples collected at 100 DAE showed that the fertilization treatments influenced ($p < 0.05$) both leaf and stem productivity, as shown in Table 7. Overall, the highest values of total leaf area (TLA), leaf DM, stem DM, and total DM were observed in T_1 and T_5. The treatments had no effect on the number of leaves (average of 14 leaves), ear DM (average of 29.24 g), or ear DM: total DM ratio (average of 0.20). The key benefits of biofertilizers are their ability to increase the soil concentration of mineralized or partially available macronutrients and micronutrients, make the soil biologically alive with the presence of a wide diversity of beneficial microorganisms, build soil organic matter, and boost the amounts of humic substances [12,18,38,39]. These attributes allow biofertilizers to improve soil health and restore normal fertility through positive effects on the physical, chemical, and biological qualities of the soil system, thereby stimulating plant growth [14–18,30]. The data in Figure 1 and Tables 6 and 7 show that applying synthetic N to soil together with biofertilizers adversely affected the attributes of biofertilizers, as seen by the low performance of treatments T_3 and T_4, in which synthetic N was only partially replaced.

Table 7. Effect of fertilization treatments on the leaf morphology and dry matter of maize at the R5 dent stage. T_1 = basal application of the digestate from swine wastewater as total substitute (100%) for synthetic N; T_2 = foliar application of the digestate as total substitute (100%) for synthetic N; T_3 = basal application of the digestate as partial substitute (50%) for synthetic N; T_4 = foliar application of the digestate as partial substitute (50%) for synthetic N; T_5 = in-row broadcast of mineral fertilizer. The treatments are fully described in Table 3.

Treatments	Total Leaf Area (cm^2)	N° of Leaves	Leaf DM (g)	Stem DM (g)	Ear DM (g)	Total DM (g)	Ear DM: Total DM
T_1	5874 ^{AB}	14	42.54 ^{AB}	97.27 ^{A}	35.47	175.28 ^{A}	0.21
T_2	5709 ^{AB}	14	39.11 ^{AB}	76.22 ^{AB}	29.08	144.41 ^{AB}	0.20
T_3	5006 ^{B}	14	34.69 ^{AB}	72.71 ^{AB}	23.00	130.40 ^{AB}	0.18
T_4	5150 ^{B}	14	34.09 ^{B}	65.94 ^{B}	22.39	122.42 ^{B}	0.18
T_5	6424 ^{A}	14	44.15 ^{A}	86.63 ^{AB}	35.25	166.03 ^{AB}	0.22
CV (%)	13.43	7.22	16.05	22.7	36.68	21.61	NA

Mean within a column followed with no letter or same letters are not statistically different (Tukey's HSD, $p < 0.05$, $n = 4$). CV = coefficient of variation. DM = dry matter. NA = not applicable.

The fertilization treatments did not show statistical differences ($p > 0.05$) concerning phytometric parameters, except for a considerable decrease in relative growth rate (RGR) and absolute growth rate (AGR) 100 DAE in plants exposed to treatments T_2, T_3, and T_4 compared to T_5 (Table 8). The RGR represents the increase in DM of a plant or its organs relative to the existing DM when the observation period begins [24]. RGR is dependent on the leaf area ratio (LAR) and the net assimilation rate (NAR, the gross photosynthetic rate discounting respiration), and can also be expressed by the equation RGR = NAR × LAR [40,41]. RGR values decreased with sampling time (Table 8) because of an increase in the plant DM. The AGR represents the variation in DM with time, that is, the average growth rate over the observation period [24]. Maize growth behavior as a function of biomass accumulation was similar among the treatments up to 60 DAE, as shown by the variation in total DM (Figure 2a). At 100 DAE when the crop was close to physiological maturity, a treatment effect was observed, and the highest AGR values ($p < 0.05$) were calculated for T_5 and T_1, 4.53 and 4.05 g day^{-1}, respectively (Table 8).

Table 8. Effect of fertilization treatments on the phytometric parameters of maize. T_1 = basal application of the digestate from swine wastewater as total substitute (100%) for synthetic N; T_2 = foliar application of the digestate as total substitute (100%) for synthetic N; T_3 = basal application of the digestate as partial substitute (50%) for synthetic N; T_4 = foliar application of the digestate as partial substitute (50%) for synthetic N; T_5 = in-row broadcast of mineral fertilizer. The treatments are fully described in Table 3. CV = coefficient of variation. DAE = days after emergency; AGR = absolute growth rate; RGR = relative growth rate; RGRLA = leaf area relative growth rate; NAR = net assimilation rate; LAR = leaf area ratio; SLA = specific leaf area.

Treatments	DAE [1]	AGR	RGR	RGR_{LA}	NAR	LAR	SLA
Unit	(d)	(g d^{-1})	(g g^{-1}d^{-1})	(dm^{2}dm^{-2} d^{-1})	(g m^{-2} d^{-1})	(m^2 g^{-1})	(m^2 g^{-1})
T_1	40	1.25	0.10	0.08	8.75	1.60	0.02
	60	1.40	0.03	0.02	4.07	1.01	0.01
	80	1.88	0.03	0.00	3.87	0.86	0.01
	100	4.05 ^{AB}	0.03 ^{AB}	0.01	7.37	0.56	0.01
T_2	40	1.05	0.09	0.07	7.37	1.68	0.02
	60	2.11	0.05	0.04	5.48	1.05	0.01
	80	1.84	0.02	0.00	3.27	0.83	0.01
	100	2.02 ^{B}	0.02 ^{B}	0.00	3.64	0.53	0.01
T_3	40	0.90	0.09	0.07	7.59	1.58	0.02
	60	1.92	0.05	0.04	5.61	1.07	0.01
	80	1.03	0.01	0.00	2.20	0.82	0.01
	100	2.50 ^{B}	0.02 ^{B}	0.01	5.32	0.56	0.01
T_4	40	1.30	0.10	0.07	8.05	1.63	0.02
	60	1.88	0.04	0.03	4.70	1.01	0.01
	80	0.72	0.01	0.00	1.30	0.81	0.01
	100	2.00 ^{B}	0.02 ^{B}	0.00	4.16	0.63	0.01
T_5	40	1.26	0.10	0.08	8.54	1.64	0.02
	60	2.07	0.04	0.03	4.92	1.02	0.01
	80	0.26	0.00	0.00	0.59	0.80	0.01
	100	4.53 ^{A}	0.04 ^{A}	0.01	8.05	0.67	0.01

[1] Differences between data values of two successive sampling periods were used for the calculations. Since no data was collected at 0 DAE, data for 20 DAE are not applicable to the study. Mean within a column followed with no letter or same letters are not statistically different (Tukey's HSD, $p < 0.05$, $n = 4$).

Variations in leaf area index (LAI) according to the fertilization treatments (Figure 2b) were very similar to variations in AGR and RGR, with a statistical effect ($p < 0.05$) observed at 80 DAE and the highest values calculated for treatments T_5 and T_1 at 100 DAE. The LAI was calculated as the ratio of leaf area per plant to the soil area occupied by the plant. Remarkably reasonable correlations were observed between LAI and total DM at all sampling stages with the following equations: $y = -0.0030x^2 + 0.8231x + 5.3175$ ($R^2 = 0.9704$ for T_1), $y = -0.0044x^2 + 0.9918x + 3.8876$ ($R^2 = 0.9715$ for T_2), $y = -0.0047x^2 + 0.9609x + 3.4542$ ($R^2 = 0.9644$ for T_3), $y = -0.0056x^2 + 1.0759x + 2.8235$ ($R^2 = 0.9808$ for T_4), and $y = -0.0035x^2 + 0.943x + 3.6523$ ($R^2 = 0.9849$ for T_5).

RGR, AGR (Table 8), total DM (Figure 2a), and LAI (Figure 2b) results demonstrate that treatments T_1 and T_5 are highly comparable. The expectation was that, because of their high solubility, synthetic fertilizers in T_5 will promote higher growth rates and yields than biofertilizers in T_1, T_2, T_3, and T_4, especially under water deficit conditions, i.e., at 100 DAE. Rainfall levels recorded during the sampling period were 45, 67, 95, 270, and 0 mm at 20, 40, 60, 80, and 100 DAE, respectively, for a total of 477 mm of accumulated precipitation (data not shown). Soil application of biofertilizers with a total replacement of the amount of recommended synthetic N (treatment T_1) might have been beneficial to plants in coping with drought stress between 80 and 100 DAE when there was no precipitation. The beneficial microorganisms in the biofertilizer competitively colonizing the roots might have produced a more robust root system, which allowed the plant to seek water and nutrients in deeper soil layers [42].

(a)

(b)

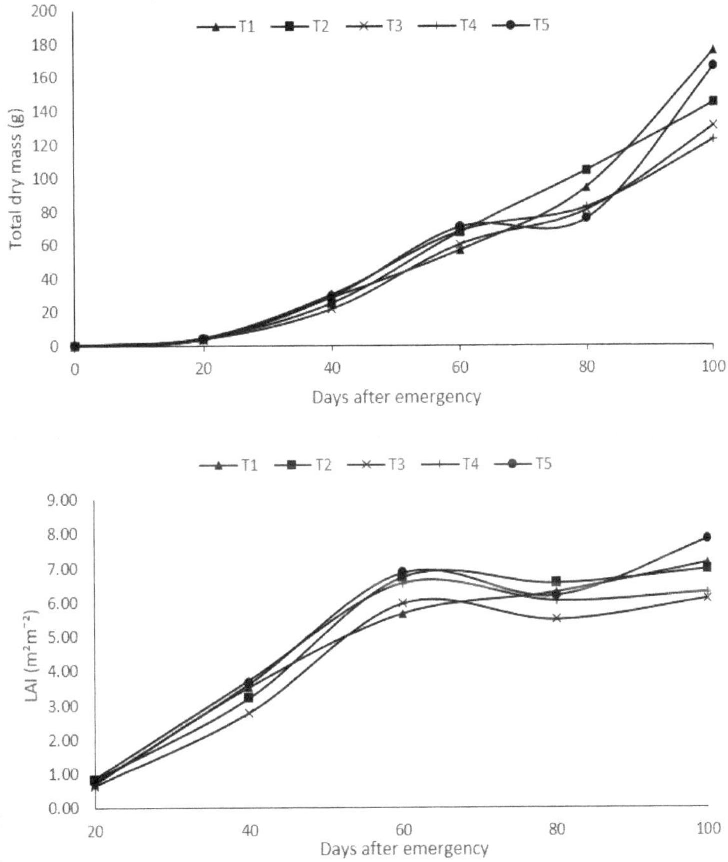

Figure 2. Variations in total dry matter DM (**a**) and leaf area index LAI (**b**) with maize growth development under different fertilization treatments. T_1 = basal application of the digestate from swine wastewater as total substitute (100%) for synthetic N; T_2 = foliar application of the digestate as total substitute (100%) for synthetic N; T_3 = basal application of the digestate as partial substitute (50%) for synthetic N; T_4 = foliar application of the digestate as partial substitute (50%) for synthetic N; T_5 = in-row broadcast of mineral fertilizer. The treatments are fully described in Table 3. Statistically, differences are not shown.

4. Conclusions

In this study, physiological growth and leaf nutrient parameters were measured at five different periods during the vegetative and reproductive stages of maize plants fertilized with biofertilizers (digestate from swine wastewater) or synthetic fertilizers. The experiment was conducted during the off-season maize crop, which is characterized by intermittent drought events. This is novel because previous studies have been conducted during the main growing season under favorable environmental conditions. Moreover, there are few previous studies on the effect of biofertilizers on phytometric parameters of crops. The results indicate that the biofertilizer applied at adequate dosages to the soil around the plants may totally (100%) substitute synthetic N fertilizers without crop yield losses. Partial replacement (50%) of synthetic N with the biofertilizer tended to yield data inferior to those obtained with synthetic fertilizers. The mode of application of the biofertilizer (basal versus foliar) did not have any significant effect on either growth parameters or leaf nutrients. In future studies, more doses of biofertilizers, maize varieties, and timing of applications should be evaluated before reaching a final conclusion. Organic fertilizers often act as a long-term carbon sink and a slow-release pool for nutrients. Thus, multi-year experiments are needed in order to gain a deeper understanding of the action of the swine wastewater-based digestate used in the present study.

Author Contributions: Conceptualization, J.d.L.J. and M.S.S.M.C.; methodology, E.L.B., L.A.M.C., J.d.L.J., F.T.S., P.G. and M.S.S.M.C.; software, M.S.S.M.C.; validation, L.A.M.C., P.G. and M.S.S.M.C.; formal analysis, E.L.B. and F.T.S.; investigation, E.L.B., J.d.L.J. and F.T.S.; resources, L.A.M.C. and M.S.S.M.C.; data curation, P.G.; writing—original draft preparation, E.L.B.; writing—review and editing, E.L.B., L.A.M.C., J.d.L.J., F.T.S., P.G. and M.S.S.M.C.; supervision, M.S.S.M.C. and P.G.; project administration, M.S.S.M.C.; funding acquisition, M.S.S.M.C. All authors have read and agreed to the published version of the manuscript.

Funding: This research was funded by the Coordination for the Improvement of Higher Education Personnel (CAPES) who provided the scholarship to the first author, and the the Fundação para a Ciência e a Tecnologia (FCT grant number UIDB/04033/2020).

Institutional Review Board Statement: Not applicable.

Informed Consent Statement: Not applicable.

Data Availability Statement: The data that support the findings of this study are available within the article.

Acknowledgments: The authors are grateful to the Coordination for the Improvement of Higher Education Personnel (CAPES) and to the National Council for Scientific and Technological Development (CNPq).

Conflicts of Interest: The authors declare no conflict of interest. The funders had no role in the design of the study; in the collection, analyses, or interpretation of data; in the writing of the manuscript; or in the decision to publish the results.

References

1. Aragão, A.; Contini, E.O. *Agriculture in Brazil and in the World: A Synthesis of the 2000–2020 Period*, 1st ed.; SIRE/EMBRAPA: Brasília, Brazil, 2021; pp. 1–68. Available online: https://www.embrapa.br/documents/10180/62618376/O+AGRO+NO+BRASIL+E+NO+MUNDO.pdf/41e20155-5cd9-f4ad-7119-945e147396cb (accessed on 8 December 2022).
2. Asencios, Y.J.O. The Importance of hydrogen for Brazil: A source of clean energy and a path to the production of nitrogen fertilizers. *SciELO Prepr.* **2022**, 4408. [CrossRef]
3. Fioroni, B. *Deposits and Bioinputs Minimize the Lack of Foreign Fertilizer*; FAEP: Curitiba, Brazil, 2022; Volume 1558, pp. 18–27. Available online: https://www.sistemafaep.org.br/wp-content/uploads/2022/03/BI_1558_web.pdf (accessed on 8 December 2022).
4. Costa, M.S.S.M.; Lorin, H.E.F.; Costa, L.A.M.; Cestanaro, T.; Pereira, D.C.; Bernardi, F.H. Performance of four stabilization bioprocesses of beef cattle feedlot manure. *J. Environ. Manag.* **2016**, *181*, 443–448. [CrossRef] [PubMed]
5. Santos, F.T.; Costa, M.S.S.M.; Costa, L.A.M.; Trindade, H.; Tonial, L.M.S.; Lorin, H.E.F.; Goufo, P. Spectroscopic and physicochemical characterization of poultry waste-based composts and charcoal–compost mixtures for the prediction of dry matter yield of giant of Italy parsley. *Agronomy* **2021**, *12*, 256. [CrossRef]
6. Lamolinara, B.; Pérez- Martínez, A.; Guardado-Yordi, E.; Fiallos, C.G.; Diéguez- Santana, K.; Ruiz-Mercado, G.J. Anaerobic digestate management, environmental impacts, and techno-economic challenges. *Waste Manag.* **2022**, *140*, 14–30. [CrossRef] [PubMed]
7. Santos, C.; Goufo, P.; Fonseca, J.; Pereira, J.; Ferreira, L.; Coutinho, J.; Trindade, H. Effect of lignocellulosic and phenolic compounds on ammonia, nitric oxide and greenhouse gas emissions during composting. *J. Clean. Prod.* **2018**, *171*, 548–556. [CrossRef]
8. Santos, F.T.; Goufo, P.; Santos, C.; Botelho, D.; Fonseca, J.; Queirós, A.; Costa, M.S.S.M.; Trindade, H. Comparison of five agro-industrial waste–based composts as growing media for lettuce: Effect on yield, phenolic compounds and vitamin C. *Food Chem.* **2016**, *209*, 293–301. [CrossRef]

9. Novotny, E.H.; Balieiro, F.C.; Auccaise, R.; Benites, V.M.; Coutinho, H.L.C. Spectroscopic investigation on the effects of biochar and soluble phosphorus on grass clipping vermicomposting. *Agriculture* **2022**, *12*, 1011. [CrossRef]
10. Santos, C.; Monte, J.; Vilaça, N.; Fonseca, J.; Trindade, H.; Cortez, I.; Goufo, P. Evaluation of the potential of agro-industrial waste-based composts to control Botrytis gray mold and soilborne fungal diseases in lettuce. *Processes* **2021**, *9*, 2231. [CrossRef]
11. Santos, F.T.; Trindade, H.; Costa, M.S.S.M.; Costa, L.A.M.; Goufo, P. Effects of composts made from broiler chicken residues and blended with biochar on the minerals and phenolic compounds in parsley (*Petroselinum crispum* Mill.). *Agriculture* **2021**, *11*, 1168. [CrossRef]
12. Tambone, F.; Scaglia, B.; D'imporzano, G.; Schievano, A.; Orzi, V.; Salati, S.; Adani, F. Assessing amendment and fertilizing properties of digestates from anaerobic digestion through a comparative study with digested sludge and compost. *Chemosphere* **2010**, *81*, 577–583. [CrossRef]
13. Samoraj, M.; Mironiuk, M.; Izydorczyk, G.; Witek-Krowiak, A.; Szopa, D.; Moustakas, K.; Chojnacka, K. The challenges and perspectives for anaerobic digestion of animal waste and fertilizer application of the digestate. *Chemosphere* **2022**, *295*, 133799. [CrossRef]
14. Koszel, M.; Lorencowicz, E. Agricultural use of biogas digestate as a replacement fertilizer. *Agric. Sci. Proc.* **2015**, *7*, 119–124. [CrossRef]
15. Locoli, G.A.; Zabaloy, M.C.; Pasdevicelli, G.; Gómez, M.A. Use of biogas digestates obtained by anaerobic digestion and co-digestion as fertilizers: Characterization, soil biological activity and growth dynamic of *Lactuca sativa* L. *Sci. Total Environ.* **2019**, *647*, 11–19. [CrossRef]
16. Sigurnjak, I.; Vaneeckhaute, H.; Michels, E.; Ryckaert, B.; Ghkiere, G.; Tack, F.M.G.; Meers, E. Fertilizer performance of liquid fraction of digestate as synthetic nitrogen substitute in silage maize cultivation for three consecutive years. *Sci. Total Environ.* **2017**, *599–600*, 1885–1894. [CrossRef] [PubMed]
17. Tsachidou, B.; Scheuren, M.; Gennen, J.; Debbaut, V.; Toussaint, B.; Hissler, C.; George, I.; Delfosse, P. Biogas residues in substitution for chemical fertilizers: A comparative study on a grassland in the Walloon Region. *Sci. Total Environ.* **2019**, *166*, 212–225. [CrossRef]
18. Zilio, M.; Pigoli, A.; Rizzi, B.; Herrera, A.; Tambone, F.; Geromel, G.; Meers, E.; Schoumans, O.; Giordano, A.; Adani, F. Using highly stabilized digestate and digestate-derived ammonium sulphate to replace synthetic fertilizers: The effects on soil, environment, and crop production. *Sci. Total Environ.* **2022**, *815*, 152919. [CrossRef] [PubMed]
19. Nitsche, P.R.; Caramori, P.H.; Ricce, W.S.; Pinto, L.F.D. *Climate Atlas of the State of Paraná*, 1st ed.; IAPAR: Londrina, Brazil, 2019; pp. 1–210. Available online: https://www.idrparana.pr.gov.br/system/files/publico/agrometeorologia/atlas-climatico/atlas-climatico-do-parana-2019.pdf (accessed on 8 December 2022).
20. Pavinato, P.S.; Pauletti, V.; Motta, A.C.V.; Adônis, M. *Fertilization and Liming Manual for the State of Paraná in Brazil*, 2nd ed.; SBCS/NEPAR: Curitiba, Brazil, 2017; pp. 1–482. Available online: https://issuu.com/editoracubo/docs/issuu-nepar-calagem (accessed on 8 December 2022).
21. Malavolta, E.; Vitti, G.C.; Oliveira, S.A. *Evaluation of the Nutritional Status of Plants: Principles and Applications*, 2nd ed.; POTAFOS: Piracicaba, Brazil, 1997; pp. 1–319. Available online: https://repositorio.usp.br/item/001070906 (accessed on 8 December 2022).
22. Silva, F.C. *Manual of Chemical Analysis of Soils, Plants and Fertilizers*, 2nd ed.; EMBRAPA/IT: Brasília, Brazil, 2009; pp. 1–319. Available online: https://www.infoteca.cnptia.embrapa.br/handle/doc/330496 (accessed on 8 December 2022).
23. Guimarães, D.P.; Sans, L.M.A.; Moraes, A.V.C. Estimativa da Área Foliar de Cultivares de Milho. In Proceedings of the XXIV Congresso Nacional de Milho e Sorgo, Florianópolis, Brazil, 1–5 September 2002; Available online: https://ainfo.cnptia.embrapa.br/digital/bitstream/item/34888/1/Estimativa-area.pdf (accessed on 8 December 2022).
24. Benincasa, M.M.P. *Plant Growth Analysis, Basic Notions*, 1st ed.; FUNEP: Jaboticabal, Brazil, 1988; pp. 1–42. Available online: https://www.bdpa.cnptia.embrapa.br/consulta/busca?b=ad&id=418424&biblioteca=vazio&busca=autoria:"M"&qFacets=autoria:"M"&sort=&paginacao=t&paginaAtual=3 (accessed on 8 December 2022).
25. Weismann, M. Fases de desenvolvimento da cultura do milho. In *Technology and Production: Off-season Corn and Winter Crops*, 2008th ed.; Roscoe, R., Lourenção, A.L.F., Grigolli, J.F.J., Melotto, A.M., Pitol, C., Miranda, R.A.S., Eds.; Fundação MS: Maracaju, Brazil, 2008; pp. 4–7. Available online: https://www.fundacaoms.org.br/tipo_da_publicacao/conteudo-safrinha/ (accessed on 8 December 2022).
26. Bull, L.T.; Cantarella, H. *Maize Cultivation: Factors Affecting Productivity*, 1st ed.; POTAFOS: Piracicaba, Brazil, 1993; pp. 63–145. Available online: https://megaleitores.com.br/livro/estante/11993301-cultura-do-milho-fatores-que-afetam-a-produtividade-leonardo-theodoro-bull-e-heitor-cantarella/59797/4433/50300 (accessed on 8 December 2022).
27. Martinez, H.E.P.; Carvalho, J.G.; Souza, R.B. Diagnose foliar. In *Recommendations for the Use of Correctives and Fertilizers in Minas Gerais*, 5th ed.; Ribeiro, A.C., Guimarães, P.T.G., Alvarez, V.V.H., Eds.; CFSEMG/UFV: Viçosa, Brazil, 1999; pp. 143–168. Available online: https://www.sbcs.org.br/loja/index.php?route=product/product&product_id=133 (accessed on 8 December 2022).
28. Ferreira, D.F. SISVAR: A computer analysis system to fixed effects split plot type designs. *Rev. Bras. De Biom.* **2019**, *37*, 529–535. [CrossRef]
29. CONAB. Boletin da Safa de Grãos 2018/19, 9° Levantamento Junho 2019. In *Observatório Agrícola Acompanhamento da Safra Brasileira Grãos*, 1st ed.; Santos, C.M.R., Ruas, J.F., Salvador, M.A., Amazonas, L., Guth, T.L.F., Nogueira, B.P., Júnior, S.R.G.S., Soares, F.M.S., Campos, P.M., Eds.; CONAB: Brasília, Brazil, 2019; Volume 6, pp. 1–113. Available online: https://www.conab.gov.br/info-agro/safras/graos/boletim-da-safra-de-graos?start=4 (accessed on 8 December 2022).

30. Tejada, M.; Rodríguez-Morgado, B.; Gómez, I.; Franco-Andreua, L.; Benítezc, C.; Parrado, J. Use of biofertilizers obtained from sewage sludges on maize yield. *Eur. J. Agron.* **2016**, *78*, 3–19. [CrossRef]
31. Schulten, A.; Krämer, U. Interactions between copper homeostasis and metabolism in plants. *Prog. Bot.* **2017**, *79*, 111–146. [CrossRef]
32. Gao, F.; Robe, K.; Gaymard, F.; Izquierdo, E.; Dubos, C. The transcriptional control of iron homeostasis in plants: A tale of bHLH transcription factors? *Front. Plant Sci.* **2019**, *10*, 6. [CrossRef]
33. Fischer, W.W.; Hemp, J.; Johnson, J.E. Manganese and the evolution of Photosynthesis. *Orig. Life Evol. Biosph.* **2015**, *45*, 351–357. [CrossRef] [PubMed]
34. Hafeez, B.; Khanif, Y.M.; Saleem, M. Role of zinc in plant nutrition—A review. *Am. J. Exp. Agric.* **2013**, *3*, 374–391. [CrossRef]
35. Santos, H.G.; Jacomine, P.K.T.; Anjos, L.H.C.; Oliveira, V.A.; Lumbreras, J.F.; Coelho, M.R.; Almeida, J.A.; Araujo Filho, J.C.; Oliveira, J.B.; Cunha, T.J.F. *Brazilian System of Soil Classification*, 5th ed.; EMBRAPA: Brasília, Brazil, 2018; pp. 1–356. Available online: https://www.embrapa.br/busca-de-publicacoes/-/publicacao/1094003/sistema-brasileiro-de-classificacao-de-solos (accessed on 8 December 2022).
36. Ambrosini, V.G.; Rosa, D.J.; Basso, A.; Borghezan, M.; Pescador, R.; Miotto, A.; Melo, G.W.B.; Soares, C.R.F.S.; Comin, J.J.; Brunetto, G. Liming as an ameliorator of copper toxicity in black oat (Avena strigosa Schreb.). *J. Plant Nutr.* **2017**, *40*, 404–416. [CrossRef]
37. Casarin, V. Dinâmica de nutrientes no sistema solo-planta visando BPUFs. In Proceedings of the IX Regional Symposium of the International Plant Nutrition Institute, Paragominas, Brazil, 30–31 August 2016. Available online: https://www.npct.com.br/npctweb/npct.nsf/article/BRS-3424/$File/Palestra%20Valter%20Casarin.pdf (accessed on 8 December 2022).
38. Tejada, M.; Rodríguez-Morgado, B.; Gómez, I.; Parrado, J. Degradation of chlorpyrifos using different biostimulants/biofertilizers: Effects on soil biochemical properties and microbial community. *Appl. Soil Ecol.* **2014**, *84*, 158–165. [CrossRef]
39. Nascimento, R.C. O uso do biofertilizante em solos agrícolas do cerrado da região do Alto Paranaíba (MG). *Bol. Goiano Geogr.* **2010**, *30*, 55–56. [CrossRef]
40. Lima, J.F.; Peixoto, C.P.; Ledo, C.A.S. Physiological indexes and initial growth of papaya plants (*Carica papaya* L.) under greenhouse conditions. *Ciênc. Agrotec.* **2007**, *31*, 1358–1363. [CrossRef]
41. Urchei, M.A.; Rodrigues, J.D.; Stone, L.F. Growth analysis of two bean cultivars under irrigation in no tillage and the conventional tillage. *Pesqui. Agropecuária Bras.* **2000**, *35*, 497–506. [CrossRef]
42. Najafi, S.; Nazari Nasi, H.; Tuncturk, R.; Tuncturk, M.; Sayyed, R.Z.; Amirnia, R. Biofertilizer application enhances drought stress tolerance and alters the antioxidant enzymes in medicinal pumpkin (*Cucurbita pepo* convar. pepo var. *Styriaca*). *Horticulturae* **2021**, *7*, 588. [CrossRef]

agriculture

MDPI

Article

Urban Agriculture as an Alternative for the Sustainable Production of Maize and Peanut

José Luis Villalpando-Aguilar [1,*], Daniel Francisco Chi-Maas [2], Itzel López-Rosas [3,4], Victor Ángel Aquino-Luna [4], Jesús Arreola-Enríquez [4], Julia Cristel Alcudia-Pérez [4], Gilberto Matos-Pech [1], Roberto Carlos Gómez-García [1], Jesús Froylán Martínez-Puc [1] and William Cetzal-Ix [1]

1 Tecnológico Nacional de México, Instituto Tecnológico de Chiná, Calle 11 entre 22 y 28 Colonia Centro Chiná, Campeche 24050, Mexico
2 Tecnológico Nacional de México, Instituto Tecnológico Superior de Hopelchén, Carretera Federal, Campeche-Hopelchen Km. 83, Campeche 24600, Mexico
3 Consejo Nacional de Ciencia y Tecnología, Benito Juárez, Ciudad de Mexico 03940, Mexico
4 Colegio de Postgraduados Campus Campeche, Champotón 24450, Mexico
* Correspondence: jose.va@china.tecnm.mx

Abstract: Currently agriculture has difficulty keeping up with the demand for food around the world, which has generated a boom in the development of sustainable alternatives for producing food and caring for the environment. Therefore, the present study aims to show a backyard system comprising 50 cm × 50 cm pinewood boxes where maize and peanut were tested under control and compost conditions. The experiments were carried out for nine months starting from compost production and the sowing of the crops, which were irrigated with temporary rain. The compost was produced by converting ~213 kg of organic residues into ~300 kg of mature compost. The fertilizer treatment consisted of two doses of compost (1 kg doses). The developing plants were compared between conditions in both crops. In addition, the nutritional values of the compost and compost and soil were evaluated. Interestingly, the correlation analyses of the morphological properties of the soils showed that the effects of the nutrients were positively associated with the morphology of the crops studied. Finally, the yield produced for maize was 9 kg/m^2 and 6.6 kg/m^2 and that for peanuts was 184 g/m^2 and 73 g/m^2 under compost and control conditions, respectively. We consider that the development of new alternatives for producing food in times of crisis or situations of limited resources is necessary for the development of humanity and the care of the environment.

Keywords: compost; sustainable; backyard; urban agriculture; yield; nutrient mobility

Citation: Villalpando-Aguilar, J.L.; Chi-Maas, D.F.; López-Rosas, I.; Aquino-Luna, V.Á.; Arreola-Enríquez, J.; Alcudia-Pérez, J.C.; Matos-Pech, G.; Gómez-García, R.C.; Martínez-Puc, J.F.; Cetzal-Ix, W. Urban Agriculture as an Alternative for the Sustainable Production of Maize and Peanut. *Agriculture* **2023**, *13*, 59. https://doi.org/10.3390/agriculture13010059

Academic Editor: Maria Roulia

Received: 22 November 2022
Revised: 16 December 2022
Accepted: 20 December 2022
Published: 25 December 2022

1. Introduction

According to the FAO, climatic change, the demand for food, and pollution increases each year, and the environment and the cities suffer the consequences of using indiscriminate resources that affect nature, which generate crisis and pandemic situations that affect the quality of life for humanity. In times of crisis, such as the COVID-19 pandemic, the healthy and economic consequences are shown to the present day [1].

Agriculture has allowed food production and promotes humanity's development; however, extensive technical crop production has degraded the environment and reduced biodiversity and ecosystem services [2].

Currently, fulfilling food demand is a global challenge [3]. Crop cultivation faces the challenges of worn-out soils, infections produced by phytopathogen-resistant antibiotics, pesticides, or metal salts, and the limited availability of water [3,4]. In addition, another aspect that limits food production is the area for production from the origin of agriculture until the present day; principally the soil is exploited and requires steps of restoration to avoid process deforestation [5].

Now, agricultural food production should feed back to nature, understood as sustainable circular and urban agriculture, allowing the development and application of the techniques of friendly agriculture with the environment to decrease pollution during production and promote soil maintenance and restoration mediated by producing compost or vermicompost using organic waste [6,7].

Urban agriculture is an alternative during a crisis of food or sanitary emergencies (COVID-19 pandemic) and it promotes the efficient use of resources: space, waste management, use of water, and the growing of crops in the city to generate a positive impact to the planet, economy, and society [7]. These aspects are relevant in a changing world and an unstable economy, which cannot predict new catastrophes [3,8].

The present study aims to implement the production of maize and peanut in a backyard system with biofertilizer with compost obtained from organic waste from neighborhood locals in Campeche City, México. The experiments were performed in a control condition (absence of compost) and a compost condition (1 kg/box of compost), and the irrigation was performed during the temporal season (June–October 2021) in a backyard system consisting of 50 cm × 50 cm of pinewood boxes.

Our results showed significant differences in the physiological properties in both the crops and the soil nutritional values and showed the correlation between soil properties and morphological characteristics, highlighting the positive effect of nutrients in the soil, which are reflected in the plant's morphology. This research showed the relevance of using organic waste to produce compost to reduce city pollution. It also shows a way to produce maize in a place with reduced space such as a city and with reduced resources, which could be considered a valuable alternative tool for society and research in times of crisis situations with limited resources.

2. Materials and Methods

2.1. Site Description

This study was carried out in Campeche City, Campeche, Mexico. The global coordinates are 19°43′14″ N–90°24′59″ W. Campeche is a lower montane forest with an average ambient temperature of 28 to 38 °C (Figure 1).

Figure 1. Map of Campeche province showing the experimental location in San Rafael in Campeche City, México.

2.2. Seeds

The native maize, named "Na'al Te'el Rojo", was obtained from the village Suctue in Hopelchén province, and the peanuts were from Kesté village, Campeche State, Mexico.

Ten seeds of maize and peanut were considered as experimental units in each experimental box and were used in triplicate.

2.3. Organic Composting

The method applied was layered stacked composting that reached five layers (Supplementary Figure S1A,B). The organic residues (vegetables and fruits) recollected each week from greengrocers were cut into smaller pieces for composting (Supplementary Figure S1C,D). Each layer was composed of 10 cm of dry material (sheet, stem, and dry waste garden) and 10–20 cm of green material; each layer was hydrated with 10 L of water and the subsequent layers were collocated until five complete layers formed the heap (Supplementary Figure S1A,B).

Finally, a ground cover was added, the heap was incubated for five days, and it was oxygenated with mediated mechanical homogenization, followed by the addition of more residues for new layers and the building of the heap anew. This process was performed each week for five months, finalized with one month of final compost maturation (Supplementary Figure S1A). During the process, the compost was monitored along with the environmental temperature.

2.4. Analyzed Soil and Compost Samples

For each box (experimental unit) a composite soil sample of 1 kg was taken prior to harvest for control and compost conditions. Three samples of 1 kg of mature compost and six composite soil samples were sent for nutritional analysis to the Fertilab laboratory (https://www.fertilab.com.mx/, accessed on 1 August 2022).

2.5. Experimental Design

The experiment consisted of three experimental units (each box has dimensions of 50 cm × 50 cm of pine wood) by condition. Each experimental unit contains 10 seeds of maize and peanut under the same conditions correspondingly (control and compost). For compost conditions, two doses were used: first, when seeds were sowed and the second when plants started flowering; each dose corresponds to 1 kg of compost (Supplementary Figure S2).

2.6. Morphological Characteristics

The evaluation of the characteristics of maize measured was the height (cm), stem width (cm), number of leaves (cm), root length (cm), and dry weight of the maize plants (kg).

In the case of peanuts, the plant height, root length, number of spikes, dry weight, and leaf area index (LAI) were calculated using the following equations:

2.6.1. Leaf Area (la)

First, leaf area should be calculated with the following equation:
la = leaf area (average obtained from la calculated from all leaves of two plants);
l = length of leaf (corresponding to two plants);
w = width of leaf (corresponding to two plants);
c = 0.75 (constant).

$$\mathrm{la} = (\mathrm{l} \times \mathrm{w}) \times \mathrm{c}$$

After each leaf was calculated the overall average was used.

2.6.2. Leaf Area Index (LAI)

LAI = leaf area index;
P = population density (m^2);
A = sown area (m^2).

$$\mathrm{LAI} = \frac{\mathrm{L} \times \mathrm{P}}{\mathrm{A}}$$

2.7. Produce Obtained from Crops

The length (cm) and weight (kg) under each condition were measured for the maize cob. The total weight (g) of the peanuts was measured in a granary balance.

2.8. Maize and Peanut Yield

The yield was calculated for both maize and peanut using the following equation:
x = yield reported kg/m^2;
tw = total weight of grains;
A = sown area (in this study 0.25 m^2).

$$x = \frac{tw}{A}$$

2.9. Statistical Analyses

The nutrition provided by the soil and compost samples was analyzed and the values obtained from the morphology of the plants are presented in results section; the average and standard error (SE) were obtained. The significance of the difference was evaluated through the mediated application of the Student's *t*-test using Microsoft Excel compared to the control and compost conditions. The correlation analyses were realized using the Spearman coefficient and were performed with R version 4.1.1.1 and R studio.

3. Results

3.1. Compost Production and Yield

Previously, in order to sow the crops, we performed compost production and our results show the importance of this process; we monitored the temperature during the composting process. The compost showed an average temperature of 48 °C compared to the registered environmental temperature of ~30 °C (Figure 2A).

The quantity (kg) of organic residues was collected and added for composting over 16 weeks, which revealed an average of 14 kg of organic waste per week (Figure 2B). Then, we reduced and transformed a total of ~200 kg of organic waste that was transformed into ~300 kg as the yield of mature compost (Figure 2C). In addition, we complemented the study with nutritional evaluation analyses (Table 1). Despite our results, the compost showed compost enrichment in Fe, Zn, B, and Mn. The analyses revealed acceptable pH levels, electrical conductivity, N, P, K, Mg, Na, S, Cu, and Ashes (Table 1).

Table 1. Nutritional evaluation of compost sample analyses was performed in triplicate and the standard error is shown.

Determination	Results
pH	8.49 ± 0.00
Electric conductivity (d-S1 m)	1.90 ± 0.04
Total Nitrogen (%)	1.06 ± 0.00
Phosphorus (P) (%)	0.49 ± 0.01
Potassium (K) (%)	0.66 ± 0.00
Calcium (Ca) (%)	17.16 ± 0.39
Magnesium (Mg) (%)	0.27 ± 0.00
Sodium (Na) (%)	0.08 ± 0.00
Sulfur(S) (%)	0.27 ± 0.00
Iron (Fe) (%)	5249.33 ± 37.26
Cooper (Cu) (ppm)	10.76 ± 0.37
Manganese (Mn) (ppm)	275.33 ± 2.59
Zinc (Zn) (ppm)	113.36 ± 26.13
Boron (B) (ppm)	29.40 ± 21.92
Ashes (%)	81.06 ± 1.73

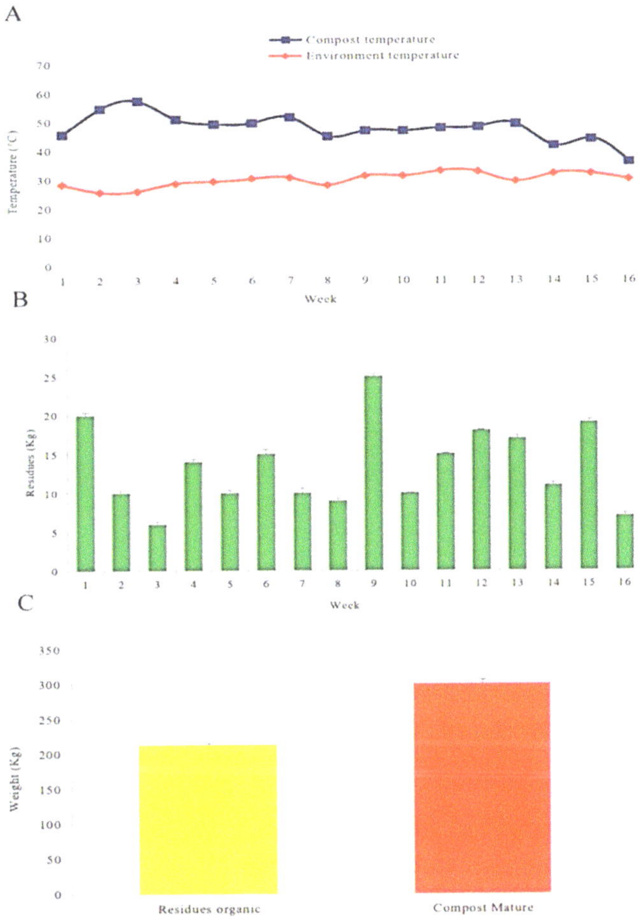

Figure 2. Production of compost analyses. (**A**) Environmental (line red) and compost temperature (line blue) were evaluated for 16 weeks. (**B**) Organic residues obtained during 16 weeks correspond to the processed compost; the error bars describe the measurements in triplicate. (**C**) Yield compost obtained after 16 weeks of composting organic residues (yellow bar) used and the total quantity of obtained compost (orange bar). The error bars describe the assay in triplicate.

3.2. *Experimental Crops*

Then, to assess the composting effect, we selected a model study of the crops: maize and peanut, and the experiment was performed in San Rafael, Campeche City, Mexico, during the COVID-19 pandemic in 2021. The design involved three experimental units (a box) for the control condition and three for the compost condition for each crop (Figure 3; Figure S2). Then, in the first stage, the germination of the seeds was evaluated (Figure 3); in the case of the maize, the germination of seven seeds under the control condition (control) and 10 seeds under the compost condition (compost–maize) was observed, respectively (Figure 3, orange bars). For peanuts, the germination of eight seeds under the control condition (control) and seven seeds under the compost condition (compost–peanut) was observed, respectively (Figure 3, green bars).

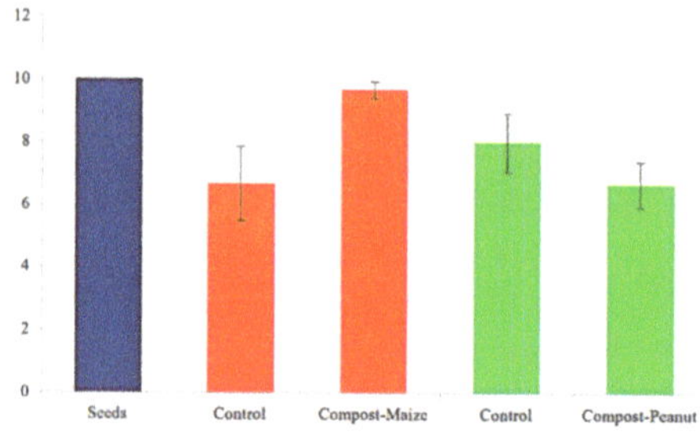

Figure 3. Germination analysis for compost–maize (orange bar) and control condition (orange bar). For peanut compost–maize (green bar) and control (green bar) conditions, both conditions and crops were compared with whole seeds sewn (blue bar). The error bar represents the reproducibility obtained from the triplicate assay.

Then, we illustrate the process of growing the crops; our experimental proposal consists of a design using box pine wood (Figure 4A). The development of the crops in the box system was monitored (Figure 4B). The first growth stage under both conditions (Figure 4C–F), present for maize control and compost conditions (Figure 4C,D), and peanuts under the same conditions (Figure 4E,F), and the final growth stage under the control and compost conditions for the maize (Figure 4G) and peanut (Figure 4H).

Figure 4. Experimental design assay. (**A**) Organization and built wooden boxes for control and compost conditions for maize and peanut. (**B**) Initial stage of grown maize and peanuts. (**C**) Triplicate of the experimental unit for the maize control condition. (**D**) Triplicate of the experimental unit for the maize compost condition. (**E**) Triplicate of the experimental unit for the peanut control condition. (**F**) Triplicate of the experimental unit for the peanut compost condition. (**G**) Final stage development of maize in control and compost conditions. (**H**) Final stage development of peanut in control and compost conditions.

Consequently, we evaluated the morphological characteristics (sheet number, stem width, height, and dry weight). The cob size and weight were evaluated for the maize.

In peanut plants, the height, root length, pikes number, dry weight, leaf area index, and fruit (peanuts weight) in both conditions were evaluated (Table 2). In the case of maize, sheets and plant height under the compost condition (14.15 $\pm$ 0.19 and 2.98 $\pm$ 0.07, respectively) showed a significant difference compared with the control condition (Table 2). In the case of peanuts, we found a significant difference in spikes 11.00 ± 0.81, dry weight: 49.11 ± 2.14, and LAI: 1.86 ± 0.00 cm (Table 2).

Table 2. Morphological measure for maize and peanut for control and compost conditions. The standard error was calculated for each morphological analysis for both conditions and the statistically significant difference (*) was tested by a Student's *t*-test with * $p < 0.05$, no significant difference was highlighted with ns.

Maize	**Leaves/Plant**	**Stem Width (cm)**	**Plant Height (cm)**	**Cob Size (cm)**	**Cob Weight (g)**	**Dry Weight (kg)**
Control	13.94 ± 0.12	2.60 ± 0.08	2.18 ± 0.04	28.30 ± 0.04	142.60 ± 4.77	4.61 ± 0.10
Compost	14.15 ± 0.19	2.73 ± 0.09	2.98 ± 0.07	28.21 ± 0.63	127.35 ± 4.05	4.83 ± 0.03
Significant *** $p < 0.05$**	*	ns	*	ns	ns	ns
Peanut	**Plant Height (cm)**	**Root Length (cm)**	**Spikes Number**	**Peanut Weight (g)**	**Dry Weight (kg)**	**Leaf Area Index**
Control	48.40 ± 2.15	8.96 ± 0.45	17.38 ± 1.78*	48.87 ± 1.12	227.22 ± 48.18*	2.27 ± 0.18
Compost	44.40 ± 1.57	8.01 ± 0.34	11.00 ± 0.81*	47.65 ± 0.70	49.11 ± 2.14	1.86 ± 0.00
Significant *** $p < 0.05$**	ns	ns	*	ns	*	*

Accordingly, we analyzed the nutritional properties of the sample soil used in the experiment where grown maize and peanut were under control and compost conditions (Table 3).

Table 3. Nutritional properties of soil analyses of maize and peanut in control and compost conditions were evaluated in the following aspects: properties of soil, relations between cations, cation exchange capacity, and soil fertility. The standard error was calculated for each measure analyzed for both conditions and the statistically significant difference (*) was performed using a Student's *t*-test with a * $p < 0.05$; no significant difference was highlighted with ns.

Properties	**Determination**	**Control**	**Compost**			
			Maize	*** $p < 0.05$**	**Peanut**	*** $p < 0.05$**
Soil properties	Class	Loamy	Loamy	ns	Loamy	ns
	Saturation point (%)	48.5 ± 1.76	55 ± 2.12	ns	55.1 ± 2.19	ns
	Field capacity (%)	25.85 ± 0.95	29.40 ± 1.13	ns	29.50 ± 1.20	ns
	Permanent Wilting Point (%)	15.40 ± 0.56	17.45 ± 0.67	ns	17.55 ± 0.74	ns
	Electric conductivity (cm/hr)	3.15 ± 0.17	2.05 ± 0.60	ns	2.05 ± 0.60	ns
	Apparent density (g/cm^3)	1.19 ± 0.00	1.10 ± 0.02	ns	1.05 ± 0.01	ns
	pH (1.2 water)	8.08 ± 0.05	8.40 ± 0.00	ns	8.38 ± 0.17	ns
	Total Carbonates (%)	36.95 ± 4.00	41.45 ± 3.57	ns	33.8 ± 1.76	ns
	Salinity (Extract CE) (dS/m)	1.38 ± 0.02	0.65 ± 0.04	*	0.78 ± 0.01	*
Relation between Cations	Ca/K	23.60 ± 0.07	14.55 ± 0.03	ns	13.85 ± 0.03	*
	Mg/K	1.90 ± 0.01	1.41 ± 0.00	ns	1.53 ± 0.00	*
	Ca/Mg/K	25.50 ± 0.07	16.0 ± 0.07	ns	15.4 ± 0.07	*
	Ca/Mg	12.55 ± 0.03	10.4 ± 0.00	*	9.04 ± 0.01	*
Cation Exchange Capacity	Calcium (Ca)	87.55 ± 0.03	84.40 ± 0.14	*	83.00 ± 0.14	*
	Magnesium (Mg)	6.06 ± 0.01	8.09 ± 0.00	*	9.18 ± 0.00	*
	Potassium (K)	3.70 ± 0.01	5.81 ± 0.03	*	6.00 ± 0.14	*
	Sodium (Na)	1.75 ± 0.01	1.71 ± 0.14	ns	1.78 ± 0.42	ns
Soil fertilization	Organic material (MO) (%)	3.95 ± 0.03	4.91 ± 0.17	ns	5.72 ± 0.17	*
	P-OLSEN (ppm)	144.50 ± 0.03	182.50 ± 3.18	*	193.50 ± 0.17	*
	K (ppm)	388.50 ± 0.03	582.50 ± 3.88	*	600.50 ± 0.17	*
	Ca (ppm)	4710.50 ± 0.03	4339.00 ± 31.82	*	4259.50 ± 0.17	ns
	Mg (ppm)	228.00 ± 0.03	252.50 ± 2.47	ns	285.50 ± 0.17	ns
	Na (ppm)	107.00 ± 0.03	100.80 ± 10.04	ns	104.00 ± 0.17	ns
	Fe (ppm)	6.63 ± 0.03	8.76 ± 0.01	*	9.63 ± 0.17	*
	Zn (ppm)	7.75 ± 0.03	7.99 ± 0.14	*	8.05 ± 0.17	ns
	Mn (ppm)	2.62 ± 0.03	3.89 ± 0.02	ns	3.38 ± 0.17	ns
	Cu (ppm)	1.17 ± 0.03	1.13 ± 0.02	ns	0.99 ± 0.17	ns
	B (ppm)	0.79 ± 0.03	1.01 ± 0.01	ns	1.04 ± 0.17	*
	S (ppm)	16.10 ± 0.03	1.40 ± 0.00	ns	2.09 ± 0.17	*
	N-NO_3 (ppm)	95.35 ± 0.03	14.35 ± 0.67	*	22.95 ± 0.17	*

The results showed that the soil used in the experiments for both crops was loamy (Table 3). Concerning the soil fertilization, P-OLSEN (ppm) of 182.50 ± 3.18, K (ppm) 582.50 ± 3.88, Fe (ppm) 8.76 ± 0.01, Zn (ppm) 7.79 ± 0.14, and N-NO_3 (ppm) 14.35 ± 0.67 was found (Table 3). Then, we found a statistically significant difference for maize in compost conditions regarding the aspect of the salinity, which was 0.65 ± 0.04. Another aspect evaluated was the relationship between cations, which showed that Ca/Mg was 10.40 ± 0.00. Regarding the cation exchange capacity, we found that Ca was 84.40 ± 0.14, Mg 8.09 ± 0.00, and K 5.81 ± 0.03.

In the case of peanuts, the compost condition for salinity was 0.78 ± 0.01 (Table 3). The relation between cations showed that Ca/K was 13.85 ± 0.03, Mg/K 1.53 ± 0.00, Ca/Mg/K 15.40 ± 0.07, and Ca/Mg 9.04 ± 0.01. Regarding the cation exchange capacity, Ca was 83.00 ± 0.14, Mg 9.18 ± 0.00, and K 6.00 ± 0.14.

For soil fertilization, it was found that MO(%) was 5.72 ± 0.17, P-OLSEN 193.50 ± 0.17 ppm, K 600.50 ± 0.17 ppm, Mg 285.50 ± 0.17 ppm, Fe 9.63 ± 0.17 ppm, B 1.04 ± 0.17 ppm, S 2.09 ± 0.17 ppm, and N-NO3 22.95 ± 0.17 ppm, which showed a statistical significance (Table 3). The results for maize and peanut under the compost condition compared with the control condition highlighted the statistically significant differences, with $p < 0.05$ (*) and $p > 0.05$ (ns) (Table 3).

Consequently, we performed the analyses of Pearson's correlation of variables, which produced statistically significant differences (Figure 5).

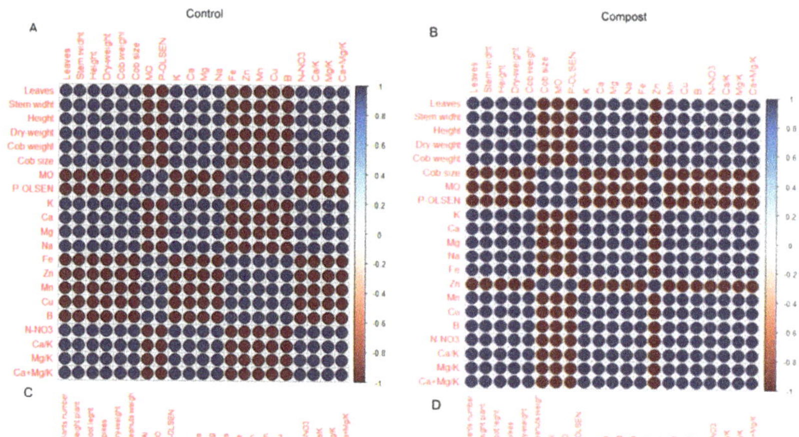

Figure 5. Correlation analyses of morphological characteristics and nutritional properties of soil used in the experiment for growing maize and peanuts in a backyard system in Campeche City, Mexico. (**A**) Plot correlation of maize control condition (**B**) Plot correlation of maize compost condition (**C**) Plot correlation of peanut control condition and (**D**) Plot correlation of peanut compost condition.

Our results are presented in the correlation plot, which showed the scale Rho of Spearman values between the variables. This analysis compared the variables obtained from the morphological characteristics of the plants of both crops evaluated with the nutritional properties of soil samples analyzed. For the case of the maize, both conditions were evaluated using the number leaves, stem width, height, cob weight, and size analyzed in correlation with MO, P-OLSEN, K, Ca, Mg, Na, Fe, Zn, Mn, Cu, B, N-NO_3, and the relationship between metals Ca/K, Mg/K, and Ca/Mg/K (Figure 5A,B). The case of the maize under control conditions showed that the K, Ca, Mg, Na, N-NO_3, and a combination of metals were positively related to the effect on the morphological characteristics of the maize plants (Figure 5A). The compost–maize condition found that the MO, P-OLSEN, and Zn affected the plants' properties negatively, showing that the proportion of the other properties in the soil with the treatment of compost was positively related to the morphological plants and cob (Figure 5B).

In the case of the peanuts used as variables, morphological characteristics such as plant number, height, root length, spikes, dry weight, and LAI on peanut plants and the total weight of peanuts were obtained; these were correlated with MO, P-OLSEN, K, Ca, Mg, Na, Fe, Zn, Mn, Cu, B, N-NO_3, and the relationship between metals Ca/K, MG/K, and Ca/Mg/K (Figure 5C,D). For the control condition, Ca, Mg, B, and N-NO_3 improved a plant's number, height, and the number of spikes, but the dry weight and peanuts' weight were positively correlated with almost all the properties except Ca, Mg, B, and N-NO_3 (Figure 5C). The peanut compost conditions showed that a plant's number, height, spikes, dry weight, and LAI correlated positively with MO, P-OLSEN, Fe, Zn, Mn, Cu, and the relationship between metals, Mg/K, and Ca/Mg/K (Figure 5D). Only the root length was positively correlated with K, Ca, Mg, Na, N-NO_3, and Ca/K (Figure 5D).

Finally, we calculated the yield for the maize and peanuts produced from the proposed box system; maize was obtained at 6.6 kg/m^2 and 9 kg/m^2 for control and compost conditions, respectively (Figure 6). A representation of the cob produced is shown in Supplementary Figure S3. In the case of the peanuts, they were obtained at 184 g/m^2 for the control and x at 73 g/m^2 for the compost condition (Figure 6).

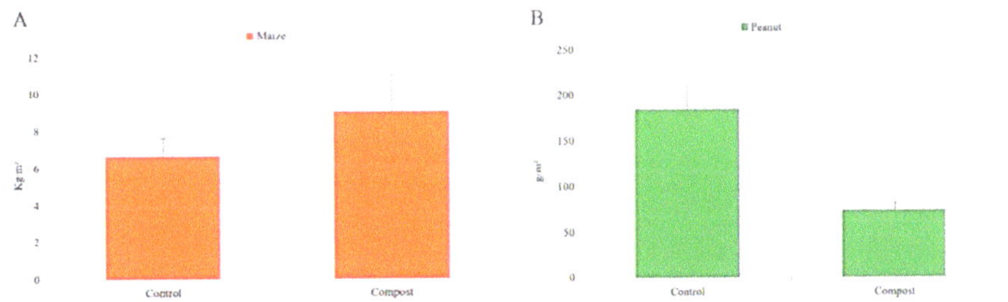

Figure 6. Yield of maize grains and peanut seeds determination for control and compost conditions were calculated for maize in control (orange bar), and compost–maize (orange bar) shown in kg/m^2. For peanuts, the yield obtained for control (green bar) and compost–peanut (green bar) was presented in g/m^2. The error bar represents the reproducibility obtained from the triplicate assay.

4. Discussion

From its origin to the present day, agriculture has been responsible for providing food for humans and animals for livestock production [5]. The techniques of traditional and extensive production are not enough to supply the demand for food around the world [9]. In addition, production is lost by limited water, the presence of disease resistance, high fertilizer prices, and excessively used soil creating degradation and malnourished soil [10]. These difficulties have promoted the emergence of new techniques and new places where agriculture is performed [6]. Currently, there is a boom in developing techniques that make efficient use of resources and that are friendly to the environment [6,11]. Sustainable urban agriculture is an alternative for producing more food with high quality in the city [11]. This kind of agriculture allows for the approach of using resources from the city (space, organic waste, and water) for promoting sustainability, producing food innocuously and with high nutritional value. However, other reasons have promoted the practices of alternatives to food production, one such case more recently being the COVID-19 pandemic, which stopped the development of humanity and produced times of crisis where it was difficult to perform activities such as agriculture [12].

This project aimed to promote the use of organic waste and implement a system in the city to produce food in limited spaces and used models of maize and peanut, which are important food for the human diet [13,14].

Our research allowed us to reduce organic waste and transform it into compost that was used as a biofertilizer for the crops tested; organic compost is a natural tool that returns macro and micronutrients, minerals, and microorganisms to the environment that help the soil to produce Nitrogen and solubility and improves the adsorption of nutrients by plants [15]. Our results showed the achievable production of compost in six months in a tropical zone (Campeche, México) with an average temperature of 28–30 °C, and each week we added mixed and new organic waste, which allowed us to improve the process and obtain compost [14–16].

Compost is an organic material with concentrated nutrients such as Nitrogen, phosphorus, potassium, and minerals, which can be solubilized easily and absorbed by radicle system plants. It is a technique that uses natural materials to maintain fertilized soils [16,17].

This research proposes a new method of producing maize and peanut in the backyard using pinewood boxes. In addition, we used what we had produced as fertilizer compost [16]. The compost quality is determined by a combination of different aspects: the carbon: Nitrogen (C/N) ratio of 30:1 and a moisture content of about 60% are recommended. Temperature is another crucial aspect of the production of compost-free pathogens [18]. The temperature during the thermophilic phase should be 54–55 °C (above 40 °C), which would guarantee that the compost would inactivate pathogen microorganisms [18]. Our results coincide with the quality of compost produced, which is interesting because it was produced in a short time, and still reached the quality parameters. An additional interesting aspect was the positive impact that the reduction of organic waste has on society.

Our results coincided with those reported by Shrestha Paliza et al., who described a two-year compost maturity experiment and compared it with synthetic phosphorus and Nitrogen; they showed that the compost contained a high amount of available phosphorus (P) and Nitrogen (N) using cultures compared to synthetic treatment [19]. Interestingly, the compost that we tested coincided with the reported compost quality and nutritional values, and producing compost to convert organic residues into biofertilizers is useful [18].

Then, the idea was to test the compost, for which we designed a box for growing maize and peanut as test crops. A limiting aspect was the space. Each box had dimensions of 50 cm × 50 cm, which could affect the development of the plant. In the case of maize, the space tested as normal for grown maize is described by Wenshun et al., who showed that a space of 0.54 m × 0.27m × 1.00 m obtained a good development but that if this is reduced to 0.20 m then this obtained a negative effect in the maize plants [20]. Our system allowed us to obtain maize plants of >2.90 m, which could explain why the space did not affect the development of the plants; evidence of root development on the maize is shown (Supplementary Figure S3A). In addition, the seeds used were of native crops of Campeche; the maize and peanut may adapt to the box system because they are not extensive crops.

In fact, the morphology was not affected in the case of the maize in both conditions; however, we found significant differences in morphological properties associated with the impact of the compost fertilizer [21].

Our results coincided with those described by Kandil et al., who depicted the potential of organic manure to improve the production of *Zea mays*. The study revealed that the compost treatment described (0, 5, and 10 tons/ha) positively affected the plant height, ear length, number of grains/rows, number of grains/ears, and the weight of 100 grains and improved the yield [22].

In the case of peanuts, our results showed that the compost had a negative effect because major development was observed in the control condition of the peanuts. This was a mistake because space is important for the development of this plant [23]. The recommendation is 10 to 12 plants distributed in a groove; however, our crop was observed to have a more negative effect in compost conditions (Supplementary Figure S3B,C) [20,23]. Another aspect that could explain the negative effect on peanut plants was that the compost had the minimum quantity acceptable of Phosphorus, Boron, and Nitrogen, deficiencies of which affect growth, and Nitrogen deficiency symptoms include stunted plants with yellow and small leaves, which affects flowering and nodule formation [24].

Agegnehu et al. reported that the application of biochar and compost, or a combination of them, improved the development of peanut plants and enhanced the organic carbon present in the soil, the nutrients available for plants, and soil water retention, improving growth and crop yield [25]. However, our results for peanuts contradicted those described by Agegnehu et al., in addition, space was a factor that was even more important for the development of these crops [25].

Our study provided an alternative technique for producing maize and peanuts in the city. The availability of nutrients in the soil depends on the pH condition, quality, properties, and structure of the soil, and the availability of the principal nutrients helps in the growth of healthy plants [16,18].

The correlation analyses between the variables evaluated for each condition of the crops demonstrated the positive effect shown by the crops for the respective nutrients coinciding with those described for the function of each nutrient for maize and peanut, respectively.

Then, the nutrients evaluated in the soil were correlated with the morphological characteristics of the crops showing their effect on the plant's development, the soil properties, the relations between cations, and that the cation exchange favors the availability of micronutrients (Ca, Mg, Na, Fe, Cu, B, and S), which showed significant differences between the compared conditions in both crops. This explained the positive results observed in the development of the plants (Table 2), as these micronutrients are responsible for activating metabolism by developing plant protein synthesis (N, S), photosynthesis (Fe, Cl), DNA and RNA synthesis (P), enzyme activation (Zn, Cu, Ca, Mn, Ni), movement sugar (K), carbohydrate metabolism (B), respiration, and nitrogen fixation (Fe, Mo) [26].

The high pH in the compost condition could explain the low germination in the seed of peanuts compared to the control condition, which reduced the availability of zinc, iron, manganese, copper, cobalt, calcium, and magnesium, affecting the development of the plants in the case of the peanut [27].

Finally, the tested conditions showed that maize could be produced in a limited space. However, we think that the backyard system affected the development of peanuts. Since peanuts require a significant extension of soil in the pods when introduced in the soil to absorb nutrients and develop healthily, we believe that the compost and box system limited the development of the peanut plants, reflected in the low yield of the peanuts.

Our success story is that maize was produced with tall plants and good numbers of germinated seeds. Even though we did not find a significant difference in the yield produced in maize, the results provide an alternative for the sustainable production of this crop that is an essential part of the food base of the human population. The application of urban agriculture was described in a study in Sidney, Australia. Small-scale urban agriculture has a high yield but requires judicious input management to achieve sustainability [28]. In addition, the backyard production system was found to be related to the initiation of socioeconomic improvement in rural towns in Sinaloa, Mexico [29].

5. Conclusions

This study proposed a system for producing compost and using it as a fertilizer to produce, in this case, maize in the city; our research showed that the box system is an alternative for producing food that could be recommended for crops of the family *Poaceae*, which is a main principal ingredient in cereals; in contrast, this system is not recommended for legumes such as peanut. Finally, we consider that the development of new alternatives for producing food in times of crisis or situations of limited resources is necessary for the development of humanity and the care of the environment.

Supplementary Materials: The following supporting information can be downloaded at: https://www.mdpi.com/article/10.3390/agriculture13010059/s1, Figure S1: Compost obtained from organic waste: (A) Pile of mature compost; (B) Evidence of high temperature (~55 °C) in the composting process; (C) Organic waste used for producing compost; and (D) An example of a layer of the compost pile method. Figure S2: Design of a backyard system comprising two crops for testing, two conditions for each, and our proposal using ten seeds per crop per condition. The point represents where two seeds were placed in each experimental unit (a box) by crop. Figure S3: Morphological characteristics of crops and cob. (A) Root developed from maize in the proposed backyard system; (B) Peanut plant grown in the absence of compost; (C) Peanut grown in the presence of compost; (D) Example cobs obtained from experiments performed in the backyard system.

Author Contributions: J.L.V.-A., I.L.-R. and D.F.C.-M. designed the experiment; J.L.V.-A., I.L.-R., R.C.G.-G., D.F.C.-M. and G.M.-P. carried out the experiment and analyzed the data; J.L.V.-A., V.Á.A.-L., J.C.A.-P. and J.F.M.-P. drafted and revised the manuscript; J.L.V.-A., J.A.-E., I.L.-R. and W.C.-I., critically revised and edited the manuscript; J.F.M.-P. was responsible for project administration. All authors have read and agreed to the published version of the manuscript.

Funding: The authors acknowledge the funding provided by the Tecnológico Nacional de México (TecNM) for the Project "El Compostaje como alternativa de restauración de suelos agrícolas de Campeche" (10996.21-P).

Data Availability Statement: The datasets generated and/or analyzed during the current study are available from the corresponding author.

Acknowledgments: We thank Fertilab for processing the samples of compost and soil. We thank the farmer, Santos of Tecnológico Nacional de Mexico (IT-Chiná), for his support in the field establishment and management of crops during the trials.

Conflicts of Interest: The authors declare no competing interest.

References

1. Haque, M.; Zaman, M.; Uz, R.; Rahman, M.; Hossain, M.; Shurid, T.I.; Rimi, T.A.; Arby, H.; Rabbany, M. A Review on Impacts of COVID-19 on Global Agricultural System and Scope for Bangladesh after Pandemic. *Environ. Sci. Pollut. Res. Int.* **2022**, *29*, 54060–54071. [CrossRef] [PubMed]
2. Rehman, A.; Farooq, M.; Lee, D.-J.; Siddique, K.H.M. Sustainable Agricultural Practices for Food Security and Ecosystem Services. *Environ. Sci. Pollut. Res.* **2022**, *29*, 84076–84095. [CrossRef] [PubMed]
3. Spiertz, H. Challenges for Crop Production Research in Improving Land Use, Productivity and Sustainability. *Sustainability* **2013**, *5*, 1632–1644. [CrossRef]
4. Erenstein, O.; Jaleta, M.; Sonder, K.; Mottaleb, K.; Prasanna, B.M. Global Maize Production, Consumption and Trade: Trends and R&D Implications. *Food Sec.* **2022**, *14*, 1–25. [CrossRef]
5. Wells, J.C.K.; Stock, J.T. Life History Transitions at the Origins of Agriculture: A Model for Understanding How Niche Construction Impacts Human Growth, Demography and Health. *Front. Endocrinol.* **2020**, *11*, 325. [CrossRef]
6. Rahmann, G.; Azim, K.; Brányiková, I.; Chander, M.; David, W.; Erisman, J.W.; Grimm, D.; Hammermeister, A.; Ji, L.; Kuenz, A.; et al. Innovative, Sustainable, and Circular Agricultural Systems for the Future. *Org. Agric.* **2021**, *11*, 179–185. [CrossRef]
7. Nowysz, A.; Mazur, Ł.; Vaverková, M.D.; Koda, E.; Winkler, J. Urban Agriculture as an Alternative Source of Food and Water Security in Today's Sustainable Cities. *Int. J. Environ. Res. Public Health* **2022**, *19*, 15597. [CrossRef]
8. Myers, S.; Fanzo, J.; Wiebe, K.; Huybers, P.; Smith, M. Current Guidance Underestimates Risk of Global Environmental Change to Food Security. *BMJ* **2022**, *378*, e071533. [CrossRef]
9. McColl, K. Can We Feed the World? *BMJ* **2008**, *336*, 1336–1338. [CrossRef]
10. Giller, K.E.; Hijbeek, R.; Andersson, J.A.; Sumberg, J. Regenerative Agriculture: An Agronomic Perspective. *Outlook Agric.* **2021**, *50*, 13–25. [CrossRef]
11. Schneider, P.; Rochell, V.; Plat, K.; Jaworski, A. Circular Approaches in Small-Scale Food Production. *Circ. Econ. Sustain.* **2021**, *1*, 1231–1255. [CrossRef]
12. Mead, B.R.; Davies, J.A.C.; Falagán, N.; Kourmpetli, S.; Liu, L.; Hardman, C.A. Urban Agriculture in Times of Crisis: The Role of Home Food Growing in Perceived Food Insecurity and Well-Being during the Early COVID-19 Lockdown. *Emerald. Open Res.* **2021**, *3*, 7. [CrossRef]

13. Menyuka, N.N.; Sibanda, M.; Bob, U. Perceptions of the Challenges and Opportunities of Utilising Organic Waste through Urban Agriculture in the Durban South Basin. *Int. J. Environ. Res. Public Health* **2020**, *17*, 1158. [CrossRef]
14. Brock, C.; Oltmanns, M.; Matthes, C.; Schmehe, B.; Schaaf, H.; Burghardt, D.; Horst, H.; Spieß, H. Compost as an Option for Sustainable Crop Production at Low Stocking Rates in Organic Farming. *Agronomy* **2021**, *11*, 1078. [CrossRef]
15. Bernal, M.P.; Sommer, S.G.; Chadwick, D.; Qing, C.; Guoxue, L.; Michel, F.C. Current Approaches and Future Trends in Compost Quality Criteria for Agronomic, Environmental, and Human Health Benefits. In *Advances in Agronomy*; Elsevier: Amsterdam, The Netherlands, 2017; Volume 144, pp. 143–233. ISBN 978-0-12-812419-2.
16. Azim, K.; Soudi, B.; Boukhari, S.; Perissol, C.; Roussos, S.; Thami Alami, I. Composting Parameters and Compost Quality: A Literature Review. *Org. Agric.* **2018**, *8*, 141–158. [CrossRef]
17. Yang, X.; Liu, E.; Zhu, X.; Wang, H.; Liu, H.; Liu, X.; Dong, W. Impact of Composting Methods on Nitrogen Retention and Losses during Dairy Manure Composting. *Int. J. Environ. Res. Public Health* **2019**, *16*, 3324. [CrossRef]
18. Peña, H.; Mendoza, H.; Diánez, F.; Santos, M. Parameter Selection for the Evaluation of Compost Quality. *Agronomy* **2020**, *10*, 1567. [CrossRef]
19. Shrestha, P.; Small, G.E.; Kay, A. Quantifying Nutrient Recovery Efficiency and Loss from Compost-Based Urban Agriculture. *PLoS ONE* **2020**, *15*, e0230996. [CrossRef]
20. Jiang, W.; Wang, K.; Wu, Q.; Dong, S.; Liu, P.; Zhang, J. Effects of Narrow Plant Spacing on Root Distribution and Physiological Nitrogen Use Efficiency in Summer Maize. *Crop J.* **2013**, *1*, 77–83. [CrossRef]
21. Shareef, R.; Mamat, A.; Al-Shaheen, M.R. The Effect of Soil PH, High-Calcium Compost and Cadmium on Some of Growth Characters in Corn (Zea Maysl.). *J. Pharm. Sci. Res.* **2019**, *4*, 16–27. [CrossRef]
22. Kandil, E.E.; Abdelsalam, N.R.; Mansour, M.A.; Ali, H.M.; Siddiqui, M.H. Potentials of Organic Manure and Potassium Forms on Maize (Zea Mays L.) Growth and Production. *Sci. Rep.* **2020**, *10*, 8752. [CrossRef] [PubMed]
23. Sharma, K.K.; Bhatnagar-Mathur, P. Peanut (*Arachis hypogaea* L.). In *Agrobacterium Protocols*; Wang, K., Ed.; Methods in Molecular Biology; Humana Press: Totowa, NJ, USA, 2006; pp. 347–358. ISBN 978-1-59745-130-7.
24. Li, L.; Li, Q.; Davis, K.E.; Patterson, C.; Oo, S.; Liu, W.; Liu, J.; Wang, G.; Fontana, J.E.; Thornburg, T.E.; et al. Response of Root Growth and Development to Nitrogen and Potassium Deficiency as Well as MicroRNA-Mediated Mechanism in Peanut (*Arachis hypogaea* L.). *Front. Plant Sci.* **2021**, *12*, 695234. [CrossRef] [PubMed]
25. Agegnehu, G.; Bass, A.M.; Nelson, P.N.; Muirhead, B.; Wright, G.; Bird, M.I. Biochar and Biochar-Compost as Soil Amendments: Effects on Peanut Yield, Soil Properties and Greenhouse Gas Emissions in Tropical North Queensland, Australia. *Agric. Ecosyst. Environ.* **2015**, *213*, 72–85. [CrossRef]
26. Liu, Y.Y.; Ukita, M.; Imai, T.; Higuchi, T. Recycling Mineral Nutrients to Farmland via Compost Application. *Water Sci. Technol.* **2006**, *53*, 111–118. [CrossRef]
27. Gondek, M.; Weindorf, D.C.; Thiel, C.; Kleinheinz, G. Soluble Salts in Compost and Their Effects on Soil and Plants: A Review. *Compost. Sci. Util.* **2020**, *28*, 59–75. [CrossRef]
28. McDougall, R.; Kristiansen, P.; Rader, R. Small-Scale Urban Agriculture Results in High Yields but Requires Judicious Management of Inputs to Achieve Sustainability. *Proc. Natl. Acad. Sci. USA* **2019**, *116*, 129–134. [CrossRef]
29. González-Félix, G.K.; Peinado-Guevara, V.M.; Peinado-Guevara, H.J.; Cuadras-Berrelleza, A.A.; Herrera-Barrientos, J.; de López-López, J.J.; Zúñiga-Espinoza, N.G. Backyard Agricultural and Farm Activity as an Option of Socioeconomic and Food Improvement in the Rural Towns of the Municipality of Guasave, Sinaloa. *Sustainability* **2021**, *13*, 3606. [CrossRef]

agriculture

MDPI

Article

Response of Maize Yield and Nutrient Uptake to Indigenous Organic Fertilizer from Corn Cobs

Maria Theresia Sri Budiastuti [1,*], Djoko Purnomo [1], Bambang Pujiasmanto [1] and Desy Setyaningrum [2]

[1] Department of Agrotechnology, Faculty of Agriculture, Sebelas Maret University, Surakarta 57126, Indonesia
[2] Diploma in Agribusiness, Vocational School, Sebelas Maret University, Surakarta 57126, Indonesia
* Correspondence: mariatheresia@staff.uns.ac.id; Tel.: +62 812-2592-692

Abstract: Indonesia's corn harvest area is decreasing so that corn production is also decreasing. The use of suboptimal land can be done to increase the harvested corn area by adding nutrients with organic fertilizers. One of the organic fertilizer ingredients is corn cob waste. The aim of the study was to examine the role of corn cob fertilizer on the growth, yield and nutrient uptake of corn. The study used a completely randomized block design with one fertilization factor with six levels, namely chemical fertilizers and corn cob organic fertilizer at a dose of 2.5, 5, 7.5, 10 or 12.5 tons/ha. Corn cob organic fertilizer has met the standard as an organic fertilizer with an organic C content of 62.21% and organic matter of 85.71%, ranking it in the high category. The total nitrogen is 1.44%, total phosphate is 1.43% and total potassium is 2.17%. Corn cob organic fertilizer had an effect on the leaf area index, root length, levels of chlorophyll a and chlorophyll b, weight of 100 seeds, cob diameter and phosphate uptake. Doses of 12.5 tons/ha produced the highest changes in chlorophyll a and b, root length and phosphate uptake. Phosphate and potassium uptake correlated with plant biomass and root length. Therefore, the results of the present study suggest that corn cob organic fertilizer is able to support the growth, yield and nutrient uptake of corn in sub-optimum land. Several gaps and research priorities in soil fertility have been identified, which need to be addressed in the future.

Keywords: alfisol; phosphate uptake; potassium uptake; root length

Citation: Budiastuti, M.T.S.; Purnomo, D.; Pujiasmanto, B.; Setyaningrum, D. Response of Maize Yield and Nutrient Uptake to Indigenous Organic Fertilizer from Corn Cobs. *Agriculture* **2023**, *13*, 309. https://doi.org/10.3390/agriculture13020309

Academic Editors: Maria Roulia and William A. Payne

Received: 7 November 2022
Revised: 15 January 2023
Accepted: 24 January 2023
Published: 27 January 2023

1. Introduction

Corn (*Zea mays* L.) is used as food, animal feed, biofuel and industrial raw materials in Indonesia and throughout the world. In 2021, the harvested corn area will be 5.5 million hectares with a corn production of 23 million tons in Indonesia. The harvested corn area in 2021 was lower than in 2018 which was 5,734,326 ha with a production of 30,055.623 tons [1]. This shows that there is a decrease in the area of agricultural land, causing production to decrease. The United Nations Department of Economics and Social Affairs estimates that crop production will need to increase by 70–100% to feed the world's 9.3 billion people by 2050 [2]. However, the potential for increasing the planted area is limited because the area of agricultural land is decreasing every year. Suboptimal land use is the main strategy to increase the planting area. Suboptimal land with high availability in Indonesia is alfisol soil. Utilization of alfisol soil for plant cultivation requires high intensification due to the low organic matter, acidic pH and low cation exchange capacity. This soil often experiences soil degradation due to its low organic matter content [3–6].

Agricultural intensification that is often done by Indonesian farmers is the provision of high inputs such as fertilizing with chemical fertilizers [7] since nutrition is a determinant in plant cultivation activities. Based on [8], corn grain yield is strongly influenced by nutrition, especially nitrogen. Corn yields decreased with decreased nutrient application [9]. In addition, the increase in grain yield was more significant with high nitrogen and phosphate inputs [10]. This is because the element nitrogen plays a role in the preparation of amino acids so that it can increase protein and corn grain yields. The challenges in maize nutrient management are the spatial variability of soil nutrient supply and plant performance and yield response to soil properties; plant growth conditions vary between fields, growth stages and years [11–13]. However, the fulfillment of nutrients with chemical fertilizers in high doses for a long period of time can increase soil hardness, nutrient imbalances and decrease organic matter so that soil fertility decreases [14]. Organic fertilizer is one way to avoid soil and environmental damage due to the long-term use of chemical fertilizers [15].

Organic fertilizers are fertilizers that mostly or wholly consist of organic materials derived from plants and/or animals that have gone through an engineering process such as composting [16,17]. Organic fertilizer made through the composting process (which can occur aerobically or anaerobically) is called compost. In efforts to improve soil quality and the sustainability of agricultural systems, organic materials such as agricultural waste can be proposed as raw materials for organic fertilizers and compost to improve soil quality and productivity [18,19]. Organic materials that can be used as compost include corn cob waste. Corn cobs are agricultural waste that has not been utilized and its availability is high. In addition, corn cobs contain 20–30% lignin, 25–35% hemicellulose and 45–55% cellulose, each of which can be converted into other biological compounds [20]. Cellulose is a carbon source that can be used by microorganisms as a substrate in the fermentation process to produce organic fertilizers [21]. Based on previous research, organic waste from agricultural activities such as *Indigofera tinctoria* used as organic fertilizer through the composting process contains high nutrients [22]. The utilization of corn cobs as compost is an effort to recycle organic agricultural waste. In addition, compost is considered a fertilizer that can improve soil structure and fertility. Based on previous research, processing organic matter through composting can increase the availability of exchangeable nutrients, namely K^+, Ca^{2+} and Mg^{2+}, so corn biomass could increase by 243% compared to mineral NPK [23]. To our knowledge, research related to the use of corn cobs as organic fertilizer and its role in the growth of corn plants has never been carried out. Therefore, this research was conducted to study the chemical content of corn cob organic fertilizer. Furthermore, it examines the role of corn cob organic fertilizer on the growth and yield of corn plants.

2. Materials and Methods

The research was carried out in Sukosari Village, Jumantono District, Karanganyar Regency, Central Java, Indonesia, precisely at coordinates S 07°38′07.01″ E 110°57′00.0″ with an altitude of 198.7 masl. The study used a one-factor Complete Randomized Block Design, namely: chemical fertilizer (nitrogen fertilizer = 350 $kg.ha^{-1}$, phosphate fertilizer = 125 $kg.ha^{-1}$, potassium fertilizer = 125 $kg.ha^{-1}$) and six doses of organic corn cob fertilizer at 2.5, 5, 7.5, 10 or 12.5 $tons.ha^{-1}$. The treatment was repeated five times. The application of the fertilizer treatment was carried out two days before planting and follow-up fertilization was carried out according to the level of treatment twice, namely 14 and 28 days after planting. The corn seeds used for planting were the corn variety Pertiwi 3. The spacing used was 75 $\times$20 cm with a total of 16 plants per plot. The type of irrigation used was surface irrigation with a frequency of irrigation of twice a week. Provision of water was done by flowing water between the beds.

The materials used in the manufacture of organic fertilizer on corn cobs are corn cobs, decomposer, molasses, bran, and cow dung fertilizer. The composting process was carried out in accordance with [24], namely using cow manure (1:10 ratio = 14.5 kg cow manure: 145 kg corn cobs waste), bran (1:60 ratio = 2.41 kg bran: 145 kg waste), molasses (1:3 = 511 ml molasses: 145 kg waste), and decomposer 500 ml (1:3 = 511 ml decomposer: 145 kg waste). Furthermore, all the ingredients that have been mixed were put into a storage barrel by volume. The composting process was carried out anaerobically [25]. The composting process was carried out for 52 days. Corn cob organic fertilizer was ready to be used when it has reached the quality criteria for organic fertilizer, which has a color and aroma like soil, does not feel hot or is at room temperature, and when held it will slightly curdle but remain crumbly.

The observed variables were the chemical composition of corn cob organic fertilizer, growth and yield of corn plants, and the uptake of phosphate and potassium nutrients by corn plants. Variable observations of the chemical content of fertilizers include pH, water content, C-Organic, organic matter, total and available nitrogen, total and available phosphate, and total and available potassium. The variables of plant growth observation included: plant height, number of leaves, leaf area index, root length, and chlorophyll a, b, and total content. The variables for crop yields included fresh plant weight, plant biomass, cob diameter, cob weight with husks, cob weight without husks, and weight of 100 seeds. The observation of the yield variables was carried out at harvest, which was 103 days after planting. The leaf area index was carried out in the maximum vegetative phase, 75 days after planting, using the length x width method. Chlorophyll content was analyzed when six weeks after planting. The chlorophyll content test was carried out using a spectrophotometer with a modified Arnon method [26]. The analytical method uses 80% acetone and a 645 and 663 nm wavelength on a spectrophotometer to measure the value of chlorophyll absorption. Analysis of plant phosphate was carried out in the maximum vegetative phase using a spectrophotometer. The process of preparing plant destruction samples was carried out by modifying the procedure by Friel with wet ashing using HNO_3 and $HClO_4$. The prepared solution was analyzed using a spectrophotometer at a wavelength of 400–470 nm [27]. Potassium analysis was carried out using atomic absorption spectrophotometry (AAS); this method was chosen because it complies with the ISO/IEC 17025 standard and is simple, precise, accurate, and the easiest method [28]. The process of preparing plant destruction samples was carried out by modifying the procedure by Friel with wet ashing using HNO_3 and $HClO_4$. The sample solution was analyzed using AAS at a wavelength of 766.5 nm. The data were analyzed using SPSS 22, namely the analysis of variance (ANOVA) with the F-test at 5% level and it is significant, then continued with Duncan's Multiple Distance Test (DMRT) at 5% level.

3. Results

3.1. Chemical Content of Corn Cob Organic Fertilizer

The results showed that organic fertilizer from corn cobs complied with the Ministry of Agriculture Regulation No.70/Permentan/SR.140/10/2011 regarding organic fertilizers, biological fertilizers, and soil conditioners (Table 1). Total N + P_2O_5 + K_2O of 5.04% met the standard. Corn cob organic fertilizer contains 62.21% organic carbon in the high category and meets the standards. C-Organic content is an essential factor determining the quality of mineral soil. The higher the total C-Organic content, the better the quality of the mineral soil. Organic carbon can induce phytochemical and biological changes in the soil to restore soil properties and increase soil quality [29]. High C-organic content can increase plant growth and yield because plants can absorb high levels of nutrients for optimal growth processes. In addition, the content of organic matter in corn cob fertilizer was relatively high, namely 85.71%. Organic matter content is significant for structural stability and supports biodiversity [30]. Fertilization with a high organic matter content can increase soil carbon absorption to encourage the availability of other elements such as nitrogen, phosphate, and potassium [31,32]. This research showed that the availability of nitrogen, phosphate, and potassium has met the standard for organic fertilizer.

3.2. Corn Growth

Organic fertilizers with a high content of organic matter, nitrogen, phosphate, and potassium can increase the storage capacity of nitrogen, phosphate, calcium, sulfur, and soil enzymatic catalytic activity, further promoting plant growth and yield. However, in this study, the dose of corn cob organic fertilizer did not affect plant height and the number of leaves (Table 2). These results indicate that corn cob organic fertilizer has not been able to support the growth of corn plants. This is because organic fertilizers gradually release nutrients into the soil solution [33]. According to [34], nitrogen mineralization from organic fertilizers takes 10 days and immobilization takes 20 days. In addition, phosphate mineralization takes 150 days, with a total immobilization phase from 180 to 240 days. In this case, nitrogen and phosphate are essential nutrients in physiological functions, namely plant metabolism [35]. The results showed that corn cob organic fertilizer had an effect on leaf area and root length index (Table 2). The leaf area index was higher with higher dose treatments. The dose of corn cob fertilizer of 12.5 tons/ha was able to produce the highest leaf area index with no significant difference compared to chemical fertilizer treatment. The higher the leaf area index, the higher the sunlight absorbed by the leaves [36]. This will support the process of photosynthesis which is an essential factor for plant growth and development [37]. The leaf area index is one of the biophysical properties of plants used to estimate plant biomass [38]. The leaf area index correlates with maize's morphological structure and yield [39]. The corn cob fertilizer also affected root length. The highest root length was in the corn cob fertilizer treatment at a dose of 12.5 tons/ha, namely 111.06 cm, and the lowest was in the chemical fertilizer treatment, which was only 70.93 cm (Table 2). The high organic carbon and organic matter content of corn cob fertilizer supports this increase in root length (Table 1), so that the organic fertilizer can trigger microbial activity, act as a source of energy, and increase synergistic interactions in the soil microbiome in improving soil structure. This causes plant roots to penetrate deeper into the soil layer so that the length and reach of roots becomes wider [40], further supporting plant growth and yield [18,37].

Table 1. Chemical content of corn cob organic fertilizer.

Parameter	Unit	Test Result	Standard *	Information
Water content	%	20.6	15–25	Meets the standards
pH		7.34	4–9	Meets the standards
Organic C	%	62.21	Minimum 1515	Meets the standards
Organic ingredients	%	85.71	-	Meets the standards
Total nitrogen	%	1.44	Macro nutrients (N + P_2O_5 + K_2O) minimum 4%	Meets the standards
Total phosphate	%	1.43		Meets the standards
Total potassium	%	2.17		Meets the standards
Nitrogen available	%	2.10	-	Meets the standards
Phosphate available	%	0.98	-	Meets the standards
Potassium available	%	1.75	-	Meets the standards
Cation exchange Capacity	(cmol/kg)	65	-	Meets the standards

* Standard according to the Ministry of Agriculture of the Republic of Indonesia (2011).

Table 2. The role of various doses of corn cob organic fertilizer on corn growth.

Treatment	Plant Height (cm)	Number of Leaves (Strand)	Leaf Area Index (cm^2)	Root Length (cm)
Chemical fertilizer	122.67	13.00 b	3.14 ab	70.93 a
Organic corn cob fertilizer at dose of 2.5 tons/ha	118.43	8.33 a	2.64 a	90.67 ab
Organic corn cob fertilizer at dose of 5 tons/ha	108.166	8.67 a	2.94 ab	95.40 ab
Organic corn cob fertilizer at dose of 7.5 tons/ha	121.67	8.33 a	2.91 ab	87.00 ab
Organic corn cob fertilizer at dose of 10 tons/ha	116.53	9.57 a	3.01 ab	96.90 ab
Organic corn cob fertilizer at dose of 12.5 tons/ha	113.67	9.33 a	3.76 b	111.06 b

Note: Numbers followed by the same letter in the same column are not significantly different based on DMRT ($\alpha = 0.05$)

3.3. Corn Chlorophyll

The results showed that the dose of corn cob fertilizer affected chlorophyll a, b, and total chlorophyll. The 12.5 tons/ha dose showed the highest amount of chlorophyll a and b (Table 3). This result is due to corn cob organic fertilizer containing high nitrogen and phosphate levels. Nitrogen supports an increase in soluble sugar in the leaves to increase the chlorophyll content [41]. Increased photosynthetic pigments in cob fertilizer treatment could be caused by improved soil structure, increased soil moisture retention capacity, and nutrient supply [42]. Based on [43], the nitrogen and phosphate content of organic fertilizers correlates with the chlorophyll content of plants. High chlorophyll content will increase the potential capacity of plant photosynthesis to support plant growth and yield from plant metabolic processes [44]. Based on [45], the element phosphate in fertilizer can increase the chlorophyll a and b content of plants. The photosynthetic performance index is an indicator to determine the yield potential of corn seeds [46]. Chlorophyll a and b were positively correlated with the photosynthetic performance index. This is supported by the role of jasmonic acid on chlorophyll which is synthesized from linolenic acid via the octadecane pathway [47]. Chlorophyll b is converted to chlorophyll an under the catalysis of chlorophyllide b reductase encoded by the non-yellow color 1 (NYC1) gene [48,49]. Based on [50], organic fertilizers can increase corn plants' chlorophyll content and grain yield by between 10 and 29%. The increase in seed yield was positively correlated with leaf chlorophyll, plant nitrogen, and phosphate uptake. The results of this study indicate that the chemical fertilizer treatment showed the lowest amount of chlorophyll a and b. The results of the study by [51] showed that the chlorophyll content in chemical fertilizer treatment is lower than in organic fertilizers with high nitrogen and phosphate levels. This is because organic fertilizer treatment can improve soil's physical and chemical properties such as soil water content (SWC), total soil organic carbon (SOC), total nitrogen (N), available phosphorus (P), and nitrate–nitrogen (NO_3-N) and ammonium–nitrogen (NH_4^+-N) cation exchange capacity compared to chemical fertilizer treatment [52].

Table 3. The role of various doses of corn cob compost on chlorophyll content.

Treatments	Chlorophyll a (mg.g^{-1})	Chlorophyll b (mg.g^{-1})	Total Chlorophyll (mg.g^{-1})
Chemical fertilizer	0.4307 a	1.117 a	1.5763 b
Organic corn cob fertilizer at dose of 2.5 tons/ha	0.4406 ab	1.0841 a	1.5339 ab
Organic corn cob fertilizer at dose of 5 tons/ha	0.4477 b	1.1191 a	1.5746 b
Organic corn cob fertilizer at dose of 7.5 tons/ha	0.4499 b	1.1974 ab	1.5394 ab
Organic corn cob fertilizer at dose of 10 tons/ha	0.4535 b	1.2265 ab	1.5228 a
Organic corn cob fertilizer at dose of 12.5 tons/ha	0.4685 c	1.3889 b	1.5474 ab

Note: Numbers followed by the same letter in the same column are not significantly different based on DMRT ($\alpha = 0.05$)

3.4. Corn Yield

Corn cob fertilizer treatment affected the weight of 100 seeds and cob diameter. The 12.5 ton/ha dose showed the highest cob diameter and weight without husks (Table 4). The weight of 100 seeds in the chemical fertilizer treatment was not significantly different from the corn cob fertilizer treatment at a dose of 12.5 tons/ha. These results indicate that corn cob fertilizer at a dose of 12.5 tons/ha has been able to meet the nutrients for the growth and development of corn plants. This was supported by the nutritional content of corn cob fertilizer (Table 1). In addition, the solid organic fertilizer produced from the composting process contains a diversity of microbes and nutrients for the soil and, when applied to corn plants, showed an effect on increasing the number of leaves, leaf area, root biomass, root length, and weight of 100 corn seeds [53]. The highest cob diameter and cob weight without husks was in the treatment of corn cob fertilizer at a dose of 12.5 tons/ha. The organic matter content supports this result by the fertilizer increasing phosphate solubility. Applying fertilizers with high phosphate content to alfisol soils can increase soil phosphate availability and make it available for plant uptake [54].

Table 4. The role of various doses of corn cob compost on corn yield.

Treatments	Plant Fresh Weight (g)	Plant Biomass (g)	Weight of 100 Seeds (g)	Cob Diameter (mm)	Cob Weight with Cornhusk (g)
Chemical fertilizer	127.91	26.62	27.97 c	44.53 ab	174.55
Organic corn cob fertilizer at dose of 2.5 tons/ha	91.63	27.78	19.02 a	42.33 ab	92.93
Organic corn cob fertilizer at dose of 5 tons/ha	95.21	25.92	19.23 a	32.23 a	94.98
Organic corn cob fertilizer at dose of 7.5 tons/ha	141.86	41.43	21.77 ab	38.80 ab	91.38
Organic corn cob fertilizer at dose of 10 tons/ha	95.13	27.20	25.22 bc	37.40 ab	93.32
Organic corn cob fertilizer at dose of 12.5 tons/ha	112.28	31.91	26.89 c	47.40 b	162.92

Note: Numbers followed by the same letter in the same column are not significantly different based on DMRT ($\alpha = 0.05$)

Based on the correlation test, plant phosphate uptake correlated with the weight of 100 seeds (Table 4). Phosphate uptake efficiency can increase crop production and productivity [55]. Plant nutrient uptake positively correlated with plant root length. Root growth exhibited a high degree of plasticity in response to nutrient availability, and the depth of fertilizer placement influenced the development and distribution of plant roots. Fertilization with high nitrogen and phosphate content can promote root growth by reducing metabolism [56]. It has a more significant number of cortical root aerenchyma and larger cortical cell sizes and increases nitrogen and phosphate uptake [57]. Furthermore, roots will synergistically increase corn seed yield and efficient utilization of nutrients throughout the plant. The decomposition of organic matter helps increase organic acids in the soil. This supports plant development, such as increasing cob diameter and cob weight. Improved phosphate uptake results in increased phosphate utilization efficiency and greater maize grain yields [58]. Plants with intensive development and short cycles, such as maize plants, require a higher amount of phosphate in solution and more rapid replenishment of phosphate than annual plants [59].

3.5. Plant Nutrient Uptake

The results showed that the dose of corn cob fertilizer had a positive effect on phosphorus uptake and no effect on potassium uptake (Table 5). The highest phosphorus absorption was in the corn cob fertilizer treatment at a dose of 12.5 tons/ha. In contrast, the lowest phosphorus absorption was in the chemical fertilizer treatment at only 0.8528 (g.plant^{-1}). There is high phosphorus content in the fertilizer so the uptake of phosphorus nutrients from corn plants is high. Increased phosphorus uptake supports the rate of photosynthesis, thereby causing an increase in root length, contract surface area, and root volume [60]. The results of this study indicate that phosphorus uptake correlates with root length (Table 6). The ability of plants to absorb phosphorus from the soil depends on root morphology and root exudate release and can be modulated by beneficial soil microbes [61]. The results of the study by [62] showed that root morphology correlates with phosphate uptake, seed yield, and biomass. Similarly, increased grain yields and nutrient uptake are associated with improved soil properties supported by organic fertilizers [55].

3.6. Correlation

Chlorophyll a and b were positively correlated (Table 6). This is supported by the role of jasmonic acid on chlorophyll which is synthesized from linolenic acid via the octadecane pathway [47]. Chlorophyll b is converted to chlorophyll a under the catalysis of chlorophyllide b reductase encoded by the non-yellow gene color 1 (NYC1) [44,49]. The results of this study showed that phosphorus uptake was correlated with root length (Table 6). The ability of plants to absorb phosphorus from the soil depends on root morphology and the release of root exudates and can be modulated by beneficial soil microbes [61]. Based on the correlation test, plant phosphate uptake was correlated with the weight of 100 seeds (Table 6).

Table 5. The role of corn cob compost dose on phosphate and potassium uptake.

Treatments	Phosphate Uptake (g.plant^{-1})	Potassium Uptake (g.plant^{-1})
Chemical fertilizer	0.8528 a	3.5354
Organic corn cob fertilizer at dose of 2.5 tons/ha	1.1456 ab	4.8206
Organic corn cob fertilizer at dose of 5 tons/ha	1.0191 ab	3.4415
Organic corn cob fertilizer at dose of 7.5 tons/ha	1.0706 ab	4.6906
Organic corn cob fertilizer at dose of 10 tons/ha	1.0411 ab	4.5381
Organic corn cob fertilizer at dose of 12.5 tons/ha	1.4339 b	5.7712

Note: Numbers followed by the same letter in the same column are not significantly different based on DMRT (α = 0.05)

Table 6. Correlation between observation variables.

	Plant Biomass	Root Length	Weight of 100 Seeds	Phosphate Uptake	Potassium Uptake	Chlorophyll a	Chlorophyll b
Plant biomass	1	0.582 *	0.403	0.725 **	0.720 **	0.100	0.084
Root length	0.582 *	1	0.170	0.658 **	0.579 **	0.075	0.065
Weight of 100 seeds	0.403	0.170	1	−0.029	−0.106	0.050	0.046
Phosphate uptake	0.725 **	0.658 **	−0.029	1	0.832 **	0.113	0.109
Potassium uptake	0.720 **	0.579 **	−0.106	0.832 **	1	0.112	0.100
Chlorophyll a	0.100	0.075	0.050	0.113	0.112	1	0.724 **
Chlorophyll b	0.084	0.065	0.046	0.109	0.100	0.724 **	1

Note: ** Correlation is significant at the 0.01 level (2-tailed), * Correlation is significant at the 0.05 level (2-tailed)

4. Discussion

The results showed that organic fertilizer from corn cobs met the standards of the Ministry of Agriculture No. 70/Permentan/SR.140/10/2011 concerning organic fertilizers, biological fertilizers, and soil enhancers (Table 1). A total N + P_2O_5 + K_2O of 5.04% already meets the standard. Corn cob organic fertilizer contains 62.21% organic C in the high category and has met the standard. Organic carbon can induce phytochemical and biological changes in the soil so that it is able to restore soil properties, increasing soil quality [29]. The content of organic matter in corn cobs fertilizer is quite high, namely 85.71%. The organic matter content is very important because it is required for structural stability and supports biodiversity [30]. Organic matter is a reserve of nitrogen and other nutrients needed by plants, so the high organic matter is able to support the availability of other elements such as nitrogen, phosphate, and potassium [63]

Organic fertilizers with high organic matter content can increase soil microbial biomass, populations of bacteria, fungi, and actinomycetes, as well as soil enzyme activity. All enzymatic activities linearly correlated with soil organic matter content [64]. Soil enzyme activity is a catalyst for the decomposition of organic matter, which is correlated with soil's physical and chemical properties [65], so high soil enzyme activity can increase root growth and plant biomass [66]. These soil microorganisms can promote plant metabolism by increasing their photosynthetic, carbohydrate, and protein processes, thereby increasing plant yields [67]. Microorganisms can increase plant growth and yield by increasing nutrient uptake [68,69]. As in this research, applying organic fertilizers at higher doses increased the uptake of phosphate and potassium.

Furthermore, organic fertilizers indirectly increase plant growth and yield. This study shows that organic corn cob fertilizer can increase the growth of roots and leaves, and increase the yield of corn plants. The application of corn cob fertilizer also affects root length. The highest root length was 111.06 cm in the corn cob fertilizer treatment with a dose of 12.5 tons.ha^{-1}, and was the lowest in the chemical fertilizer treatment at only 70.93 cm (Table 2). The root length is supported by organic fertilizers, which effectively source energy from soil microbes to improve soil structure. This allows plant roots to penetrate deeper soil layers to support plant growth and yields. Based on this study's results, root length correlated with phosphate uptake. The results of the study [60] showed that root morphology is correlated with phosphate uptake, seed yield, and biomass. Similarly, increases in grain yield and nutrient uptake are associated with improved soil properties supported by organic fertilizers [53].

The higher the dose of organic corn cob fertilizer can increase the plant area index. The higher the leaf area index, the higher the sunlight absorbed by the leaves [70]. This will support the process of photosynthesis which is an essential factor for plant growth and development [39]. The leaf area index, one of the plants' biophysical properties, is used to estimate plant biomass [9]. The leaf area index correlates with maize's morphological structure and yield [39]. Organic corn cob fertilizer can also increase the content of chlorophyll a and b. This result is due to organic corn cob fertilizer containing high levels of nitrogen and phosphate so that it can increase the availability of soil nutrients, affecting plant nutrient uptake. This supports the increase in soluble sugar in the leaves to increase the chlorophyll content [41]. Increased photosynthetic pigments in cob fertilizer treatment could be caused by improved soil structure, increased soil moisture retention capacity, and nutrient supply [71]. Based on [72], the nitrogen content of organic fertilizers correlates with the chlorophyll content of plants. In addition, the induction of organic matter in the soil can help increase the soil's organic acids, which can support plant development.

5. Conclusions

This study shows that corn cob organic fertilizer has met the standard as organic fertilizer with an organic C content of 62.21% and organic matter content of 85.71% in the high category. The total nitrogen is 1.44%, total phosphate is 1.43%, and total potassium is 2.17% with a neutral pH of 7.34. Corn cob organic fertilizer had an effect on leaf area index, root length, chlorophyll a, chlorophyll b, weight of 100 seeds, cob diameter, and phosphate uptake. The higher doses of corn cob organic fertilizer affected the increase in leaf area index, root length, chlorophyll a, chlorophyll b, 100 seed weight, cob diameter, and phosphate uptake. The nutritional content of organic corn cob fertilizer supports this. Doses of 12.5 tons/ha increased chlorophyll a and b and root length and had the highest phosphate uptake. Phosphate and potassium uptake correlated with plant biomass and root length. Therefore, the results of the present study suggest that corn cob organic fertilizer is able to support the growth, yield, and nutrient uptake of corn in sub-optimum land. Several gaps and research priorities in soil fertility have been identified, which need to be addressed in the future.

Author Contributions: Conceptualization, M.T.S.B. and D.S.; methodology, M.T.S.B. and D.P.; software, D.S.; validation, B.P. and D.P.; formal analysis, D.P. and D.S.; investigation, M.T.S.B. and D.S.; resources, M.T.S.B. and D.S.; data curation, D.S. and D.P.; writing—original draft preparation, M.T.S.B. and D.S.; writing—review and editing, M.T.S.B. and D.S.; visualization, D.P. and B.P.; supervision, M.T.S.B.; project administration, M.T.S.B.; funding acquisition, M.T.S.B. All authors have read and agreed to the published version of the manuscript.

Funding: This research was funded by Sebelas Maret University with a grant of Non APBN 2022 Fundamental Research Scheme with Contract Number 254/UN27.22/PT.01.03/2022.

Informed Consent Statement: Not applicable.

Data Availability Statement: The authors declare that the data supporting the findings of this study are available within the article.

Acknowledgments: The authors would like to thank Sebelas Maret University which has provided funding for this research.

Conflicts of Interest: The authors declare no conflict of interest.

References

1. BPS. *Indonesian Corn Production*; Central Bureau of Statistics: Jakarta, Indonesia, 2021.
2. Tian, X.; Engel, B.A.; Qian, H.; Hua, E.; Sun, S.; Wang, Y. Will reaching the maximum achievable yield potential meet future global food demand? *J. Clean. Prod.* **2021**, *294*, 126285. [CrossRef]
3. Libutti, A.; Trotta, V.; Rivelli, A.R. Biochar, vermicompost, and compost as soil organic amendments: Influence on Growth Parameters, Nitrate and Chlorophyll Content of Swiss Chard (*Beta vulgaris* L. var. *cycla*). *Agronomy* **2020**, *10*, 346. [CrossRef]
4. Baghel, V.; Thakur, J.K.; Yadav, S.S.; Manna, M.C.; Mandal, A.; Shirale, A.O.; Sharma, P.; Sinha, N.K.; Mohanty, M.; Singh, A.B.; et al. Phosphorus and Potassium Solubilization From Rock Minerals by Endophytic *Burkholderia* sp. Strain FDN2-1 in Soil and Shift in Diversity of Bacterial Endophytes of Corn Root Tissue with Crop Growth Stage. *Geomicrobiol. J.* **2020**, *37*, 550–563. [CrossRef]
5. Joy, B.; Sudirja, R.; Sofyan, E.T.; Harriyanto, R.; Mulyani, O.; Herdiansyah, G. Evaluation of soil fertility status for rice, corn, soybean on suboptimal land in West Java Indonesia. *IOP Conf. Ser. Earth Environ. Sci.* **2019**, *393*, 012030. [CrossRef]
6. Budiastuti, M.; Purnomo, D.; Setyaningrum, D. Agroforestry System as the Best Vegetation Management to Face Forest Degradation in Indonesia. *Rev. Agric. Sci.* **2022**, *10*, 14–23. [CrossRef]
7. Kiloes, A.M.; Puspitasari; Mulyono, D.; Syah, M.J.A. Land resources management of shallots farming: A case study in the highlands of Solok Regency, West Sumatera. *IOP Conf. Ser. Earth Environ. Sci.* **2021**, *648*, 012097. [CrossRef]
8. Djaman, K.; Allen, S.; Djaman, D.S.; Koudahe, K.; Irmak, S.; Puppala, N.; Darapuneni, M.K.; Angadi, S.V. Planting date and plant density effects on maize growth, yield and water use efficiency. *Environ. Chall.* **2022**, *6*, 100417. [CrossRef]
9. Song, L.; Jin, J. Improving CERES-Maize for simulating maize growth and yield under water stress conditions. *Eur. J. Agron.* **2020**, *117*, 126072. [CrossRef]
10. Zhang, W.; Xiong, Y.; Li, Y.; Qiu, Y.; Huang, G. Effects of organic amendment incorporation on maize (*Zea mays* L.) growth, yield and water-fertilizer productivity under arid conditions. *Agric. Water Manag.* **2022**, *269*, 107663. [CrossRef]
11. Liu, J.; Shu, A.; Song, W.; Shi, W.; Li, M.; Zhang, W.; Li, Z.; Liu, G.; Yuan, F.; Zhang, S.; et al. Long-term organic fertilizer substitution increases rice yield by improving soil properties and regulating soil bacteria. *Geoderma* **2021**, *404*, 115287. [CrossRef]
12. Liu, B.; Gao, P.; Zhang, S. Photosynthesis and Fine Root Growth Dynamics of Soybean in Walnut-Soybean Agroforestry System. *Comput. Intell. Neurosci.* **2022**, *2022*, 2246824. [CrossRef] [PubMed]
13. Liu, Z.; Hao, Z.; Sha, Y.; Huang, Y.; Guo, W.; Ke, L.; Chen, F.; Yuan, L.; Mi, G. High responsiveness of maize grain yield to nitrogen supply is explained by high ear growth rate and efficient ear nitrogen allocation. *Field Crops Res.* **2022**, *286*, 108610. [CrossRef]
14. Yang, F.; Shen, S.; Gao, W.; Ma, Y.; Han, B.; Ding, Y.; Wang, X.; Zhang, K. Deciphering discriminative antibiotic resistance genes and pathogens in agricultural soil following chemical and organic fertilizer. *J. Environ. Manag.* **2022**, *322*, 116110. [CrossRef] [PubMed]
15. Arifah, S.M.; Sri Budiastuti, M.T.; Dewi, W.S. Supriyadi Vermicompost formulation based on soybean husk and cow manure on shallots. *Int. J. Des. Nat. Ecodynamics* **2021**, *16*, 327–333. [CrossRef]
16. Boldrin, A.; Andersen, J.K.; Møller, J.; Christensen, T.H.; Favoino, E. Composting and compost utilization: Accounting of greenhouse gases and global warming contributions. *Waste Manag. Res.* **2009**, *27*, 800–812. [CrossRef] [PubMed]
17. Wu, S.C.; Cao, Z.H.; Li, Z.G.; Cheung, K.C.; Wong, M.H. Effects of biofertilizer containing N-fixer, P and K solubilizers and AM fungi on maize growth: A greenhouse trial. *Geoderma* **2005**, *125*, 155–166. [CrossRef]
18. Jamison, J.; Khanal, S.K.; Nguyen, N.H.; Deenik, J.L. Assessing the effects of digestates and combinations of digestates and fertilizer on yield and nutrient use of *Brassica juncea* (Kai choy). *Agronomy* **2021**, *11*, 509. [CrossRef]
19. Galsim, F.; Golabi, M.H.; Kim, Y.S.; Iyekar, C. Comparative effects of composted organic waste and inorganic fertilizer on nitrate leachate from the farm soils of northern Guam. *Int. Soil Water Conserv. Res.* **2021**, *9*, 87–102. [CrossRef]
20. Castillo-González, E.; De Medina-Salas, L.; Giraldi-Díaz, M.R.; Sánchez-Noguez, C. Vermicomposting: A valorization alternative for corn cob waste. *Appl. Sci.* **2021**, *11*, 5692. [CrossRef]
21. Liew, J.X.; Loy, A.C.M.; Chin, B.L.F.; AlNouss, A.; Shahbaz, M.; Al-Ansari, T.; Govindan, R.; Chai, Y.H. Synergistic effects of catalytic co-pyrolysis of corn cob and HDPE waste mixtures using weight average global process model. *Renew. Energy* **2021**, *170*, 948–963. [CrossRef]
22. Budiastuti, M.T.S.; Thaidy, S.; Sulistyo, T.D.; Manurung, I.R.; Setyaningrum, D. The effectiveness of organic fertilizer from natural dyes waste on the growth of *Indigofera tinctoria* L. *IOP Conf. Ser. Earth Environ. Sci.* **2022**, *1016*, 012014. [CrossRef]
23. Pandit, N.R.; Schmidt, H.P.; Mulder, J.; Hale, S.E.; Husson, O.; Cornelissen, G. Nutrient effect of various composting methods with and without biochar on soil fertility and maize growth. *Arch. Agron. Soil Sci.* **2020**, *66*, 250–265. [CrossRef]
24. Budiastuti, M.; Pujiasmanto, B.; Sulistyo, T.D.; Ike, A.; Agroteknologi, P.S.; Pertanian, F.; Maret, U.S.; Agronomi, P.M.; Pertanian, F.; Maret, U.S. Pemanfaatan Limbah Ekstraksi *Indigofera tinctoria* L. sebagai Pupuk Organik pada Usaha Batik Pewarna Alami di Sukoharjo. *PRIMA J. Community Empower. Serv.* **2020**, *4*, 109–119. [CrossRef]
25. Cucina, M.; De Nisi, P.; Trombino, L.; Tambone, F.; Adani, F. Degradation of bioplastics in organic waste by mesophilic anaerobic digestion, composting and soil incubation. *Waste Manag.* **2021**, *134*, 67–77. [CrossRef]
26. Arnon, D.I. Copper Enzymes in Isolated Chloroplasts. Polyphenoloxidase in *Beta vulgaris*. *Plant Physiol.* **1949**, *24*, 1–15. [CrossRef]
27. Song, Y.; Shang, W.; Wang, Z.; He, S.; Sun, Y.; Shi, L.; Shen, Y.; He, D. Functional Analysis of Glucose-6-Phosphate Translocator PoGPT1 in Embryogenic Callus Formation of Tree Peony. *Horticulturae* **2022**, *8*, 957. [CrossRef]

28. Du, J.; Chang, Y.; Zhang, X.; Hu, C. Development of a method of analysis for profiling of the impurities in phenoxymethylpenicillin potassium based on the analytical quality by design concept combined with the degradation mechanism of penicillins. *J. Pharm. Biomed. Anal.* **2020**, *186*, 113309. [CrossRef]
29. Vida, C.; de Vicente, A.; Cazorla, F.M. The role of organic amendments to soil for crop protection: Induction of suppression of soilborne pathogens. *Ann. Appl. Biol.* **2020**, *176*, 1–15. [CrossRef]
30. Paterson, E.; Sim, A. Soil-specific response functions of organic matter mineralization to the availability of labile carbon. *Glob. Chang. Biol.* **2013**, *19*, 1562–1571. [CrossRef]
31. Soria, R.; Ortega, R.; Bastida, F.; Miralles, I. Role of organic amendment application on soil quality, functionality and greenhouse emission in a limestone quarry from semiarid ecosystems. *Appl. Soil Ecol.* **2021**, *164*, 103925. [CrossRef]
32. Cai, A.; Xu, M.; Wang, B.; Zhang, W.; Liang, G.; Hou, E.; Luo, Y. Manure acts as a better fertilizer for increasing crop yields than synthetic fertilizer does by improving soil fertility. *Soil Tillage Res.* **2019**, *189*, 168–175. [CrossRef]
33. Shaji, H.; Chandran, V.; Mathew, L. *Organic Fertilizers as a Route to Controlled Release of Nutrients*; Elsevier Inc.: Amsterdam, The Netherlands, 2021; ISBN 9780128195550.
34. Leno, N.; Sudharmaidevi, C.R. Physicochemical and Nutrient Release Characteristics of a Thermochemical Organic Fertilizer Produced from Degradable Solid Waste and Its Effect on Productivity of Banana. *Commun. Soil Sci. Plant Anal.* **2021**, *52*, 2562–2577. [CrossRef]
35. Naumann, M.; Koch, M.; Thiel, H.; Gransee, A.; Pawelzik, E. The Importance of Nutrient Management for Potato Production Part II: Plant Nutrition and Tuber Quality. *Potato Res.* **2020**, *63*, 121–137. [CrossRef]
36. Hosseini, M.; McNairn, H.; Mitchell, S.; Robertson, L.D.; Davidson, A.; Ahmadian, N.; Bhattacharya, A.; Borg, E.; Conrad, C.; Dabrowska-Zielinska, K.; et al. A comparison between support vector machine and water cloud model for estimating crop leaf area index. *Remote Sens.* **2021**, *13*, 1348. [CrossRef]
37. Zhang, Y.; Xia, C.; Zhang, X.; Sha, Y.; Feng, G.; Gao, Q. Quantifying the relationships of soil properties and crop growth with yield in a NPK fertilizer application maize field. *Comput. Electron. Agric.* **2022**, *198*, 107011. [CrossRef]
38. Raj, R.; Walker, J.P.; Pingale, R.; Nandan, R.; Naik, B.; Jagarlapudi, A. Leaf area index estimation using top-of-canopy airborne RGB images. *Int. J. Appl. Earth Obs. Geoinf.* **2021**, *96*, 102282. [CrossRef]
39. Shao, G.; Han, W.; Zhang, H.; Liu, S.; Wang, Y.; Zhang, L.; Cui, X. Mapping maize crop coefficient Kc using random forest algorithm based on leaf area index and UAV-based multispectral vegetation indices. *Agric. Water Manag.* **2021**, *252*, 106906. [CrossRef]
40. Zhang, J.; Bei, S.; Li, B.; Zhang, J.; Christie, P.; Li, X. Organic fertilizer, but not heavy liming, enhances banana biomass, increases soil organic carbon and modifies soil microbiota. *Appl. Soil Ecol.* **2019**, *136*, 67–79. [CrossRef]
41. Salehi, A.; Tasdighi, H.; Gholamhoseini, M. Evaluation of proline, chlorophyll, soluble sugar content and uptake of nutrients in the German chamomile (*Matricaria chamomilla* L.) under drought stress and organic fertilizer treatments. *Asian Pac. J. Trop. Biomed.* **2016**, *6*, 886–891. [CrossRef]
42. Sorkhi, F. Effect of Vermicompost Fertilizer on Antioxidant Enzymes and Chlorophyll Contents of Borago Officinalis Under Salinity Stress. *Iran. J. Plant Physiol.* **2021**, *11*, 3559–3598. [CrossRef]
43. Zafar-ul-hye, M.; Naeem, M.; Danish, S.; Khan, M.J.; Fahad, S.; Datta, R.; Brtnicky, M.; Kintl, A.; Hussain, M.S.; El-esawi, M.A. Effect of cadmium-tolerant rhizobacteria on growth attributes and chlorophyll contents of bitter gourd under cadmium toxicity. *Plants* **2020**, *9*, 1386. [CrossRef]
44. Li, Y.; Sun, Y.; Jiang, J.; Liu, J. Spectroscopic determination of leaf chlorophyll content and color for genetic selection on *Sassafras tzumu*. *Plant Methods* **2019**, *15*, 73. [CrossRef] [PubMed]
45. Jama-Rodzeńska, A.; Chohura, P.; Gałka, B.; Szuba-Trznadel, A.; Falkiewicz, A.; Białkowska, M. Effect of Different Doses of Phosgreen Fertilization on Chlorophyll, K, and Ca Content in Butterhead Lettuce (*Lactuca sativa* L.) Grown in Peat Substrate. *Agriculture* **2022**, *12*, 788. [CrossRef]
46. Ghorbanzadeh, P.; Aliniaeifard, S.; Esmaeili, M.; Mashal, M.; Azadegan, B.; Seif, M. Dependency of Growth, Water Use Efficiency, Chlorophyll Fluorescence, and Stomatal Characteristics of Lettuce Plants to Light Intensity. *J. Plant Growth Regul.* **2021**, *40*, 2191–2207. [CrossRef]
47. Fang, H.; Luo, F.; Li, P.; Zhou, Q.; Zhou, X.; Wei, B.; Cheng, S.; Zhou, H.; Ji, S. Potential of jasmonic acid (JA) in accelerating postharvest yellowing of broccoli by promoting its chlorophyll degradation. *Food Chem.* **2020**, *309*, 125737. [CrossRef]
48. Li, Y.; Song, H.; Zhou, L.; Xu, Z.; Zhou, G. Tracking chlorophyll fluorescence as an indicator of drought and rewatering across the entire leaf lifespan in a maize field. *Agric. Water Manag.* **2019**, *211*, 190–201. [CrossRef]
49. Ruan, J.; Zhou, Y.; Zhou, M.; Yan, J.; Khurshid, M.; Weng, W.; Cheng, J.; Zhang, K. Jasmonic acid signaling pathway in plants. *Int. J. Mol. Sci.* **2019**, *20*, 2479. [CrossRef] [PubMed]
50. Agegnehu, G.; Bass, A.M.; Nelson, P.N.; Bird, M.I. Benefits of biochar, compost and biochar-compost for soil quality, maize yield and greenhouse gas emissions in a tropical agricultural soil. *Sci. Total Environ.* **2016**, *543*, 295–306. [CrossRef]
51. Zandvakili, O.R.; Barker, A.V.; Hashemi, M.; Etemadi, F. Biomass and nutrient concentration of lettuce grown with organic fertilizers. *J. Plant Nutr.* **2019**, *42*, 444–457. [CrossRef]
52. Urra, J.; Alkorta, I.; Lanzén, A.; Mijangos, I.; Garbisu, C. The application of fresh and composted horse and chicken manure affects soil quality, microbial composition and antibiotic resistance. *Appl. Soil Ecol.* **2019**, *135*, 73–84. [CrossRef]

53. Dahunsi, S.O.; Oranusi, S.; Efeovbokhan, V.E.; Adesulu-Dahunsi, A.T.; Ogunwole, J.O. Crop performance and soil fertility improvement using organic fertilizer produced from valorization of *Carica papaya* fruit peel. *Sci. Rep.* **2021**, *11*, 4696. [CrossRef]
54. Qiu, Y.; Fall, T.; Su, Z.; Bortolozo, F.; Mussoline, W.; England, G.; Dinkins, D.; Morgan, K.; Clark, M.; Liu, G. Effect of Phosphorus Fertilization on Yield of Chipping Potato Grown on High Legacy Phosphorus Soil. *Agronomy* **2022**, *12*, 812. [CrossRef]
55. Nunes, R.D.S.; de Sousa, D.M.G.; Goedert, W.J.; de Oliveira, L.E.Z.; Pavinato, P.S.; Pinheiro, T.D. Distribution of Soil Phosphorus Fractions as a Function of Long-Term Soil Tillage and Phosphate Fertilization Management. *Front. Earth Sci.* **2020**, *8*, 350. [CrossRef]
56. Wang, C.; Wang, Y.; Yan, S.; Li, Y.; Zhang, P. Biochar-amended composting of lincomycin fermentation dregs promoted microbial metabolism and reduced antibiotic. *Bioresour. Technol.* **2022**, *367*, 128253. [CrossRef] [PubMed]
57. Chen, X.; Liu, P.; Zhao, B.; Zhang, J.; Ren, B.; Li, Z.; Wang, Z. Root physiological adaptations that enhance the grain yield and nutrient use efficiency of maize (*Zea mays* L.) and their dependency on phosphorus placement depth. *Field Crops Res.* **2022**, *276*, 108378. [CrossRef]
58. Pereira, S.I.A.; Abreu, D.; Moreira, H.; Vega, A.; Castro, P.M.L. Plant growth-promoting rhizobacteria (PGPR) improve the growth and nutrient use efficiency in maize (*Zea mays* L.) under water deficit conditions. *Heliyon* **2020**, *6*, e05106. [CrossRef] [PubMed]
59. Saia, S.; Aissa, E.; Luziatelli, F.; Ruzzi, M.; Colla, G.; Ficca, A.G.; Cardarelli, M.; Rouphael, Y. Growth-promoting bacteria and arbuscular mycorrhizal fungi differentially benefit tomato and corn depending upon the supplied form of phosphorus. *Mycorrhiza* **2020**, *30*, 133–147. [CrossRef]
60. Iqbal, A.; He, L.; Khan, A.; Wei, S.; Akhtar, K.; Ali, I.; Ullah, S.; Munsif, F.; Zhao, Q.; Jiang, L. Organic manure coupled with inorganic fertilizer: An approach for the sustainable production of rice by improving soil properties and nitrogen use efficiency. *Agronomy* **2019**, *9*, 651. [CrossRef]
61. Nacoon, S.; Jogloy, S.; Riddech, N.; Mongkolthanaruk, W.; Kuyper, T.W.; Boonlue, S. Interaction between Phosphate Solubilizing Bacteria and Arbuscular Mycorrhizal Fungi on Growth Promotion and Tuber Inulin Content of *Helianthus tuberosus* L. *Sci. Rep.* **2020**, *10*, 4916. [CrossRef]
62. Boring, T.J.; Thelen, K.D.; Board, J.E.; De Bruin, J.L.; Lee, C.D.; Naeve, S.L.; Ross, W.J.; Kent, W.A.; Ries, L.L. Phosphorus and potassium fertilizer application strategies in corn–soybean rotations. *Agronomy* **2018**, *8*, 195. [CrossRef]
63. Chehab, H.; Tekaya, M.; Ouhibi, M.; Gouiaa, M.; Zakhama, H.; Mahjoub, Z.; Laamari, S.; Sfina, H.; Chihaoui, B.; Boujnah, D.; et al. Effects of compost, olive mill wastewater and legume cover cropson soil characteristics, tree performance and oil quality of olive trees cv. Chemlali grown under organic farming system. *Sci. Hortic. (Amsterdam)* **2019**, *253*, 163–171. [CrossRef]
64. Chang, E.H.; Chung, R.S.; Tsai, Y.H. Effect of different application rates of organic fertilizer on soil enzyme activity and microbial population: Original article. *Soil Sci. Plant Nutr.* **2007**, *53*, 132–140. [CrossRef]
65. Mierzwa-Hersztek, M.; Gondek, K.; Baran, A. Effect of poultry litter biochar on soil enzymatic activity, ecotoxicity and plant growth. *Appl. Soil Ecol.* **2016**, *105*, 144–150. [CrossRef]
66. Jabborova, D.; Annapurna, K.; Paul, S.; Kumar, S.; Saad, H.A.; Desouky, S.; Ibrahim, M.F.M.; Elkelish, A. Beneficial features of biochar and arbuscular mycorrhiza for improving spinach plant growth, root morphological traits, physiological properties, and soil enzymatic activities. *J. Fungi* **2021**, *7*, 571. [CrossRef] [PubMed]
67. Wei, X.; Wu, S.C.; Nie, X.P.; Yediler, A.; Wong, M.H. The effects of residual tetracycline on soil enzymatic activities and plant growth. *J. Environ. Sci. Heal. Part B Pestic. Food Contam. Agric. Wastes* **2009**, *44*, 461–471. [CrossRef] [PubMed]
68. Khan, H.A.; Naqvi, S.R.; Mehran, M.T.; Khoja, A.H.; Khan Niazi, M.B.; Juchelková, D.; Atabani, A. A performance evaluation study of nano-biochar as a potential slow-release nano-fertilizer from wheat straw residue for sustainable agriculture. *Chemosphere* **2021**, *285*, 131382. [CrossRef]
69. Nardi, S.; Schiavon, M.; Francioso, O. Chemical structure and biological activity of humic substances define their role as plant growth promoters. *Molecules* **2021**, *26*, 2256. [CrossRef]
70. Yang, B.; Kun, D.; Han, L.; Heo, K.; Biging, G. The effects of tree characteristics on rainfall interception in urban areas. *Landsc. Ecol. Eng.* **2019**, *15*, 289–296. [CrossRef]
71. Mishra, B.; Bose, S.K.; Sangwan, N.S. Comparative investigation of therapeutic plant Withania somnifera for yield, productivity, withanolide content, and expression of pathway genes during contrasting seasons. *Ind. Crops Prod.* **2020**, *154*, 112508. [CrossRef]
72. He, Y.; Xie, K.; Xu, P.; Huang, X.; Gu, W.; Zhang, F.; Tang, S. Evolution of microbial community diversity and enzymatic activity during composting. *Res. Microbiol.* **2013**, *164*, 189–198. [CrossRef]

agriculture

Article

Sustainable Utilization Strategy of Organic Waste via Fabrication of Bioelastomer with Antibacterial and Antioxidant Activities Using Mandarin Peel Extracts

Kang Hyun Lee [1,†], Youngsang Chun [2,†], Ja Hyun Lee [3], Jong Uk Lee [4], Taek Lee [5,*] and Hah Young Yoo [1,*]

[1] Department of Biotechnology, Sangmyung University, 20, Hongjimun 2-Gil, Jongno-Gu, Seoul 03016, Republic of Korea
[2] Department of Bio-Convergence Engineering, Dongyang Mirae University, 445-8, Gyeongin-ro, Guro-gu, Seoul 08221, Republic of Korea
[3] Department of Convergence Bio-Chemical Engineering, Soonchunhyang University, 22, Soonchunhyang-ro, Asan-si 31538, Republic of Korea
[4] Department of Chemical Engineering, Sunchon National University, 255, Jungang-ro, Suncheon-si 57922, Republic of Korea
[5] Department of Chemical Engineering, Kwangwoon University, 20 Kwangwoon-Ro, Nowon-gu, Seoul 01897, Republic of Korea
* Correspondence: tlee@kw.ac.kr (T.L.); y2h2000@smu.ac.kr (H.Y.Y.)
† These authors contributed equally to this work.

Abstract: Mandarin peels (MPs), a food-processing residue, have several restrictions on their disposal and can cause serious environmental pollution. In this study, MP was used to fabricate a functional bioelastomer with antioxidant and antibacterial activities. Bioactive compounds were recovered from MPs in liquid form and added to the bioelastomer during fabrication to maintain the mechanical strength of the bioelastomer. The radical scavenging activities of the fabricated bioelastomer (B–MPE 15%) were 3.3% for DPPH and 20.8% for ABTS, respectively. In addition, B–MPE 15% exhibited antibacterial activity against gram-positive (*Staphylococcus aureus*), gram-negative (*Escherichia coli*), and antibiotic-resistant bacteria (Methicillin-resistant *S. aureus* and Vancomycin resistant *Enterococcus*). The chemical properties of B–MPE 15% were not significantly different from those of the control group (bare PDMS). Tensile strength, elongation at break, and water vapor transmission rate of B–MPE 15% were found to be 5.1 N/mm^2, 649%, and 33.3 g/(m^2 day), respectively. Therefore, the addition of MP extracts did not significantly affect the physical properties. The fabricated bioelastomer with antibacterial and antioxidant activities is expected to be utilized in the food packaging, pharmaceutical, and medical industries. Our research is expected to represent a future-oriented strategy for realizing carbon neutrality by upcycling food waste.

Keywords: bioelastomer; mandarin peel; flavanone; antioxidant; antibacterial

Citation: Lee, K.H.; Chun, Y.; Lee, J.H.; Lee, J.U.; Lee, T.; Yoo, H.Y. Sustainable Utilization Strategy of Organic Waste via Fabrication of Bioelastomer with Antibacterial and Antioxidant Activities Using Mandarin Peel Extracts. *Agriculture* **2023**, *13*, 161. https://doi.org/10.3390/agriculture13010161

Academic Editor: Maria Roulia

Received: 2 November 2022
Revised: 28 December 2022
Accepted: 5 January 2023
Published: 9 January 2023

1. Introduction

Mandarin is one of the most popular fruits because of its sweet taste and ease of consumption, as well as its antioxidant, anticancer, antibacterial, and anti-adipogenic properties [1,2]. An estimated 630,000 tons of mandarins were produced in Korea [3], and it has been reported that approximately 50,000 tons of mandarin-processing residues are generated annually [4]. However, only 30% of the mandarin residues are used as medicinal herbs. While the remaining 70% used to be disposed of into the ocean [5], Korea has strictly forbidden this disposal method for food wastes since 2013 in accordance with the 1996 Protocol of the London Convention [6], as it leads to disruption of the marine food chain and loss of marine biodiversity [7]. Food waste is typically disposed of through landfilling, incineration, and composting [8]. However, in Korea, direct landfilling of food waste has been prohibited, by law, since 2005 to alleviate the shortage of landfills, protect groundwater and soil, and promote the conversion of food waste into value-added materials [9]. Meanwhile, incineration of food waste with a high moisture content can generate dioxins during combustion with other low-humidity wastes [10]. During composting, nitrate- and phosphorus-contacting leachates and greenhouse gases (GHGs) get generated, causing eutrophication and global warming, respectively [11]. As a result, various developed countries other than Korea, including the USA, EU, Japan, and China, are also suffering from social and environmental issues related to the disposal of food waste [12,13]. Therefore, it is necessary to propose a sustainable food-waste management plan that not only prevents environmental pollution but also converts food waste into value-added materials.

Traditionally, value-added materials, such as fuels and chemicals, have been produced in petroleum refineries [14]. However, these refineries are considered the main contributors to GHG emissions because they release carbon that was buried in the ground into the atmosphere. The CO_2 emissions from this sector alone were estimated to be 1079 million tons in 2015 [15]. Many countries around the world are seriously concerned about environmental pollution caused by these GHG emissions, so they participated in the UN Framework Convention on Climate Change (UNFCCC) and adopted the Kyoto Protocol (1997), the Lima Call for Climate Action (2014), the Paris Agreement (2015), etc. [16]. Consequently, biorefineries, which utilize biomass as raw materials instead of petroleum-based materials, have been attracting attention as a strategy to achieve net-zero CO_2 emissions [17]. In biorefineries, food waste is considered an ideal feedstock because it satisfies economic feasibility, owing to its low transport and storage costs, year-round availability, and ease of handling [18]. Mandarin peels (MPs), which account for approximately 7–11% of mandarins generated during juice processing, are mostly discarded because they are considered to have no economic value [19,20]. However, these residues contain cellulose, hemicellulose, pectin, essential oil, and flavonoid, which have the potential to be converted into value-added materials [21]. Although various studies have been carried out regarding the production of biofuels, such as ethanol [22] and methane [23], energy-based products have limitations that currently prevent them from completely replacing low-cost fossil fuels [24]. Therefore, in consideration of economic feasibility, it is necessary to produce bio-based products with high market values [25], such as flavonoids [26] and essential oil [27].

The predominant flavonoids in MP are hesperidin and narirutin [28], which have antioxidant [29], antibacterial [30], antidiabetic [31], and anti-inflammatory properties [32]. In addition, bioactive compounds derived from natural sources are in increasing demand, as alternatives to synthetic compounds, because of their safety and non-toxic effect on the human body [33]. Extraction techniques, including maceration, Soxhlet, microwave-assisted, ultrasound-assisted, and enzyme-assisted extraction, have been used to extract bioactive compounds from MP [34]. Among the various extraction methods, microwave-assisted extraction (MAE) has the distinct advantages of short extraction times, high extraction yields, and low solvent usage [35]. In addition, this technique is suitable for industrial-scale application [36]. The recovered bioactive substances are mixed with various polymers to fabricate bioelastomers that are used as functional materials in the food, pharmaceutical, and medical industries [37]. Dordevic et al. [38] produced edible chitosan films for food packaging using extracts of blueberries, red grapes, and parsley. Meanwhile, non-biodegradable synthetic polymers, such as polyethylene glycol (PEG), polyvinyl alcohol (PVA), polypropylene (PP), polyethylene (PE), and polydimethylsiloxane (PDMS), are also widely used in bioelastomer fabrication because they have stronger physicochemical properties than natural polymers [39].

Following these research trends, we previously fabricated a bioelastomer with antioxidant and antibacterial activities [40]. In our previous study [40], PDMS, which has higher flexibility, thermal stability, and biocompatibility, as well as lower toxicity than other synthetic polymers [41], was used as the polymer. In addition, by-products of aronia juice processing were used as the raw materials for natural bioactive compounds. However, the direct use of by-products in powder form dramatically reduced the mechanical strength of the fabricated bioelastomer. These barriers accentuate the disadvantage of PDMS, which is its relatively low mechanical strength due to its major structure composed of Si–O bonds, unlike the C–C bonds constituting PVA, PP, and PE [42]. Therefore, it is necessary to design a strategy for fabricating bioelastomers with antioxidant and antibacterial activities while maintaining their mechanical strength.

In this study, a strategy for the sustainable utilization of food processing residue was designed by fabricating a functional bioelastomer using bioactive compounds extracted from MPs. First, a mixing ratio of extraction solvents suitable for recovering the flavonoids, hesperidin and narirutin, from MP was selected. Microwave-assisted extraction was utilized to recover hesperidin and narirutin from MP with high efficiency in a short time. Furthermore, the effects of microwave power and irradiation time on flavonoid extraction were investigated. The recovered bioactive compounds were used to fabricate functional bioelastomers with antioxidant and antibacterial activities. Finally, the biological properties, chemical structure, and mechanical strength of the fabricated bioelastomer were compared with those of the control group. This study is the first attempt to produce a functional material by mixing flavonoid extracts obtained from MP, a waste resource, with PDMS to design a sustainable biorefinery.

2. Materials and Methods

2.1. Materials

Mandarin peels (MPs) were purchased from Cheongmyeongyagcho (Chungju-si, Chungcheongbuk-do, Korea). The MPs were ground with a blender and sieved to a size of 90 μm. Polydimethylsiloxane (PDMS; Elastosil E43) was obtained from Wacker (Munich, Germany). Hesperidin, narirutin, dimethyl sulfoxide (DMSO), 1,1-diphenyl-2-picrylhydrazyl (DPPH), 2,2′-azino-bis(3-ethylbenzothiazoline-6-sulphonic acid) (ABTS), and potassium persulfate were purchased from Sigma-Aldrich (St. Louis, MO, USA). Methanol (MeOH), phosphoric acid, acetonitrile, and heptane were purchased from Samchun Chemical (Seoul, Republic of Korea). All reagents used in this study were of analytical grades.

2.2. Preparation of Mandarin Peel Extracts

To select the most efficient extraction solvent, 1 g of MP was immersed in 10 mL of a MeOH:DMSO solution mixed in different proportions (1:9, 3:7, 5:5, 7:3, and 9:1, *v/v*). MP, then, was extracted in an incubator at 40 °C, with a shaking speed of 150 rpm, for 24 h. To design the MAE process for maximum flavanone recovery from MP, MP was immersed in the selected extraction solvent (1:10, *w/w*) and extracted at various microwave powers (70, 210, 350, 490, and 630 W) and irradiation times (5, 10, 15, 20, and 30 s). Each extract was centrifuged at 13,000 rpm for 10 min to separate the supernatant and, then, analyzed and used for bioelastomer fabrication.

2.3. Fabrication of Bioelastomer

PDMS was poured into a square Petri dish (24.5 × 24.5 cm), and 20 mL of heptane and mandarin peel extracts (MPEs) were recovered under the determined MAE conditions were added and mixed uniformly. The bioelastomer manufacturing conditions are listed in Table 1. The Petri dish was transferred to a vacuum oven and dried at 40 °C until the moisture was completely removed.

Table 1. Detailed manufacturing conditions of bioelastomers.

Sample	PDMS (g)	MPE (g)
B-MPE 0% (control, PDMS)	50	0
B-MPE 1% (*w/w*)	49.5	0.5
B-MPE 3%	48.5	1.5
B-MPE 5%	47.5	2.5
B-MPE 7%	46.5	3.5
B-MPE 10%	45	5
B-MPE 15%	42.5	7.5
B-MPE 20%	40	10

2.4. *Antioxidant Activity of Bioelastomer*

2.4.1. DPPH Radical Scavenging Activity

DPPH radical scavenging activity was determined using the DPPH assay with a slight modification [43]. The DPPH stock solution was prepared by dissolving the DPPH reagent in MeOH to a concentration of 0.5 mM. The DPPH working solution was prepared by diluting the prepared DPPH stock solution with methanol until the absorbance at 517 nm reached 1.2. Each bioelastomer (size: 1 × 1 cm) was immersed in 1 mL of the DPPH working solution and reacted at 25 °C for 30 min. The bioelastomer was then removed, and the absorbance of the supernatant was measured at 517 nm using a spectrophotometer (DU 730, Beckman Coulter, Brea, CA, USA). The blank was 1 mL of methanol, and the control was 1 mL of DPPH working solution without bioelastomer added. All experiments were performed in triplicate to obtain the standard deviations. Radical scavenging activity was calculated using the following Equation (1):

$$\text{Radical scavenging activity (\%)} = (1 - (OD_{sample}/OD_{control})) \times 100 \quad (1)$$

2.4.2. ABTS Radical Scavenging Activity

ABTS radical scavenging activity was measured using the ABTS assay with a slight modification [44]. An ABTS stock solution was prepared by reacting 7 mM ABTS solution and 2.45 mM potassium persulfate in a 1:1 ratio (*v/v*). ABTS working solution was prepared by diluting the prepared ABTS stock solution with methanol until the absorbance at 734 nm reached 1.0. Each bioelastomer (size: 1 × 1 cm) was immersed in 1 mL of the ABTS working solution and reacted at 25 °C for 30 min. The bioelastomer was then removed, and the absorbance of the supernatant was measured at 734 nm using a spectrophotometer; the blank was 1 mL of methanol, and the control was 1 mL of ABTS working solution without bioelastomer added. All experiments were performed in triplicate to obtain the standard deviations. The radical scavenging activity was calculated using Equation (1) above.

2.5. *Antibacterial Activity of Bioelastomer*

The antibacterial activity of the bioelastomer was determined following the method in our previous study [41]. *Staphylococcus aureus* was used as the gram-positive bacteria, *Escherichia coli* as the gram-negative bacteria, and methicillin-resistant *Staphylococcus aureus* (MRSA) and vancomycin-resistant *Enterococcus* (VRE) as the antibiotic-resistant bacteria. All the bacteria were cultured in a shaking incubator at 150 rpm, for 24 h, at 37 °C in 50 mL of nutrient broth. The bacteria were diluted to 10^6 CFU/mL and inoculated on nutrient agar plates. Bioelastomers (size: 1 × 1 cm) were placed in the center of the nutrient agar medium, inoculated with each bacterium, and then, incubated at 37 °C for 24 h. The antibacterial zone was determined using the Image J software (v1.52i, National Institutes of Health, Bethesda, MN, USA). All experiments were performed in triplicate to obtain the standard deviations.

2.6. *Characterization of Bioelastomer*

The chemical structures of the bioelastomers were investigated using Fourier-transform infrared spectroscopy (FT-IR; JASCO FTIR-4600, Jasco, Japan). Tensile strength and elongation at break were determined, according to ASTM D412, using a universal testing machine (Instron 3367; Norwood, MA, USA). Water vapor transmission rate (WVTR) was determined, according to ASTM F1249, using a Permatran-W 3/33 MA (Mocon, Minneapolis, MN, USA). The morphology of both bioelastomers was characterized using scanning electron microscopy (SEM, SNE-3000M, SEC Inc., Suwon, Republic of Korea) at a scanning voltage of 5 kV.

2.7. *Analytical Methods*

The hesperidin and narirutin contents in MPE were determined using high-performance liquid chromatography (HPLC). The analytical conditions were as follows: INNO column

C18 (5 μm, 4.6 mm × 250 mm, Young Jin Biochrom, Seongnam-si, Republic of Korea); diode array detector (DAD); wavelength, 250 nm; temperature, 25 °C; injection volume, 5 μL; flow rate, 0.8 mL/min. The gradient elution conditions were as follows: acetonitrile for solvent A and 0.03% (*v/v*) phosphoric acid in DW for solvent B; 0 min, 10% A; 0–15 min, 20% A; 15–28 min, 40% A; 28–36 min, 75% A; 36–38 min, 10% A; 38–50 min, 10% A. Standard curves for quantification were prepared using hesperidin and narirutin as the standard reagents.

3. Results and Discussion

3.1. Selection of Extraction Solvent for Flavanone Recovery from Mandarin Peels

To maximize flavanone recovery from MPs, appropriate mixing ratios of the extraction solvents were investigated. A MeOH:DMSO mixture was used to extract total flavonoids from MP [45]. Similarly, Magwaza et al. [46] demonstrated that a mixed MeOH:DMSO solution (1:1, *v/v*) effectively and rapidly extracts phenolic compounds, such as flavanone glycosides and phenolic acid, from mandarin rinds. Figure 1 shows the effects of the mixing ratio of MeOH:DMSO on the flavanone (hesperidin and narirutin) recovery from MPs. The extraction solvent was mixed with solutions of MeOH and DMSO that had MeOH:DMSO ratios of 1:9, 3:7, 5:5, 7:3, and 9:1 (*v/v*). The resulting flavanone recovery with each solution was 26.6, 31.9, 43.7, 28.0, and 18.4 mg/g-biomass for hesperidin and 5.2, 6.3, 8.7, 6.1, and 5.3 mg/g-biomass for narirutin, respectively. These results agree with those of a previous study [46], which showed that the MeOH:DMSO solution mixed in the same ratio is the most effective extraction solvent for recovering flavanone from MPs. Therefore, a 5:5 ratio (*v/v*) of the MeOH:DMSO mixture was selected as the extraction solvent for recovering flavanone from MPs (Figure A1).

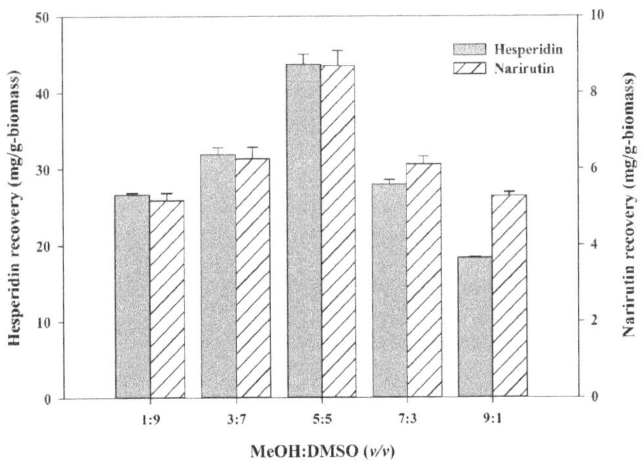

Figure 1. Effects of the mixing ratio of MeOH:DMSO solution on flavanone (hesperidin and narirutin) recovery from MPs.

3.2. Determination of Microwave-Assisted Extraction Conditions

The MAE process was designed to recover high yield of flavanone from MP in a short time. However, high microwave power and extended irradiation time can cause the thermal degradation of these phenolic compounds [47]; extraction conditions should be determined based on the maximum energy that can be input without causing thermal degradation to reduce the consequent loss of bioactive compounds. Therefore, the effects of microwave power and irradiation time on flavanone extraction from MPs were investigated (Figure 2). Figure 2a shows the results of hesperidin recovery from MPs. At a microwave power of 70 W, hesperidin recovery was not significantly affected by irradiation time, and it only slightly increased from 49.0 mg/g-biomass at 5 s to 53.2 mg/g-biomass at 30 s. At a microwave power of 210 W, hesperidin recovery steadily increased from 49.5 mg/g-biomass at 5 s to 66.8 mg/g-biomass at 30 s, but the maximum recovery was not achieved. At microwave powers of 350, 490, and 630 W, hesperidin recovery steadily increased for 20 s, after which it decreased with increasing irradiation time. The reduction significantly increased as the microwave power increased, which was presumed to be because of thermal degradation caused by excessive energy input. This phenomenon was confirmed by Ahmad and Langrish [48], who extracted phenolic acids from MPs. Finally, the maximum hesperidin recovery was found to be 71.6 mg/g-biomass at 490 W and 20 s. Recovery of narirutin from MPs showed a similar tendency to that of hesperidin (Figure 2b). At microwave powers of 350, 490, and 630 W, narirutin recovery steadily increased for 20 s, after which it decreased with increasing irradiation time. The maximum narirutin recovery was found to be 16.3 mg/g-biomass at 490 W and 20 s. At 490 W and 20 s, the energy input was approximately 9600 J (W × s); as it was estimated that energies higher than 9600 J caused thermal degradation of flavanones, 490 W and 20 s were chosen as the extraction conditions for flavanone recovery from MPs.

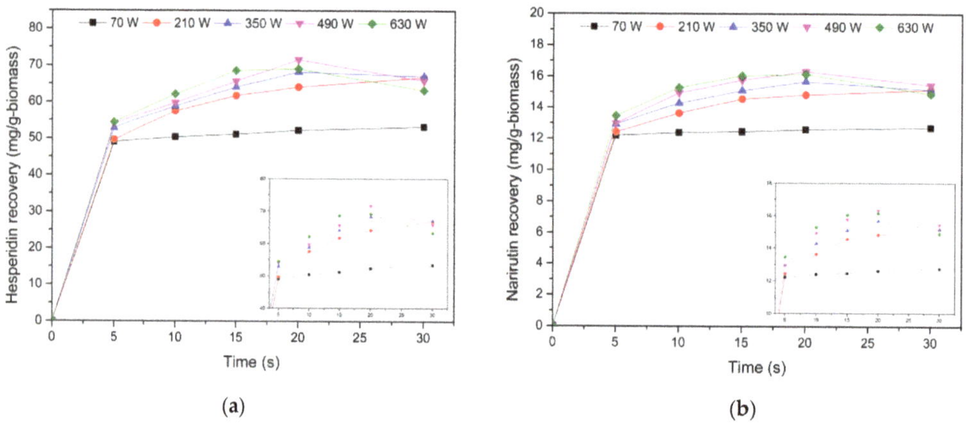

Figure 2. Effect of microwave power and irradiation time on the hesperidin (**a**) and narirutin (**b**) recovery from MPs.

3.3. *Antioxidant Activity of Bioelastomer*

Measuring the antioxidant activity of bioelastomers is important for preventing the negative effects of free radicals in biological and food packaging applications [49]. Table 2 shows the radical scavenging activity of the fabricated bioelastomer. The radical scavenging activity of the bioelastomer increased in proportion to the content of MP-derived flavanone extract (MPE). The DPPH radical scavenging activity of the bioelastomer increased sharply by 1.8-fold when 15% MPE was added, but it was still a low 3.3 ± 0.2%. The ABTS radical scavenging activity exceeded 20% at 15% MPE and reached 26.7 ± 1.2% at 20% MPE. The DPPH and ABTS radical scavenging activities of the hesperidin standard (100 ppm) were found to be 3.5 ± 0.4% and 17.8 ± 0.6%, respectively. These results imply that the ABTS assay is more sensitive than the DPPH assay for evaluating the antioxidant activity of the bioelastomers and hesperidin, a major component of MPE. Floegel et al. [50] reported that the DPPH assay is appropriate for hydrophobic systems, and the ABTS assay is suitable for hydrophilic, lipophilic, and highly pigmented systems. Flavonoid glycosides, such as hesperidin, are hydrophilic because the presence of sugars increases their polarity [51]. In addition, MPE recovered using a MeOH:DMSO mixture is hydrophilic and is presumed to contain high amounts of pigments such as flavonoids, carotenoids, and chlorophylls [52].

Table 2. Radical scavenging activity of the fabricated bioelastomer.

Sample	Radical Scavenging Activity (%)	
	DPPH	ABTS
B–MPE 0% (control, PDMS)	0	0
B–MPE 1%	1.0 ± 0.0	13.1 ± 0.3
B–MPE 3%	1.1 ± 0.1	16.8 ± 0.4
B–MPE 5%	1.2 ± 0.1	17.3 ± 0.6
B–MPE 7%	1.5 ± 0.1	18.8 ± 0.3
B–MPE 10%	1.8 ± 0.1	19.8 ± 0.6
B–MPE 15%	3.3 ± 0.2	20.8 ± 0.8
B–MPE 20%	4.8 ± 0.3	26.7 ± 1.2
Hesperidin 100 ppm	3.5 ± 0.4	17.8 ± 0.6

3.4. Antibacterial Activity of Bioelastomer

S. aureus can cause food poisoning and toxic shock syndrome, while *E. coli* can cause septicemia and cholecystitis [53]. In addition, infections of antibiotic-resistant bacteria, such as methicillin-resistant *S. aureus* (MRSA) and vancomycin-resistant *Enterococcus* (VRE), threaten human safety to the extent that the World Health Organization (WHO) has selected it as one of the top 10 threats to global public health [54]. Therefore, bioelastomers that exhibit antibacterial activity are expected to be highly useful in the food packaging, pharmaceutical, and medical fields.

B–MPE 0–10% did not exhibit antibacterial activity against any bacteria (data not shown). However, B–MPE 15% showed antibacterial activity against gram-positive (*S. aureus*), gram-negative (*E. coli*), and antibiotic-resistant bacteria (MRSA and VRE) (Figure 3). The antibacterial zone of B–MPE 15% was determined to be 20.4 ± 1.5 cm^2 for *S. aureus*, 16.0 ± 1.2 cm^2 for *E. coli*, 9.4 ± 0.4 cm^2 for MRSA, and 14.8 ± 1.0 cm^2 for VRE. The antibacterial zone of 50 ppm ampicillin, an antibiotic used as a positive control, was found to be 19.3 ± 0.6 cm^2 for *S. aureus* and 12.5 ± 0.5 cm^2 for *E. coli*. The antibacterial activity of B–MPE 15% was presumed to be due to the flavonoids present in MPE: flavonoids can cause bacterial cell death by inhibiting the metabolism and synthesis of DNA and RNA [55]. The antibacterial effect of B–MPE 15% was more sensitive against gram-positive bacteria (*S. aureus*) than against gram-negative bacteria (*E. coli*). Alexandre et al. [56] reported that gram-positive bacteria are more sensitive to interactions with phenolic compounds because they lack an outer membrane, causing the compounds to diffuse into them more quickly than in gram-negative bacteria. Similarly, Choi et al. [57] demonstrated that hesperidin, a major flavonoid in MPE, has lower values of minimum inhibitory concentration (MIC) and minimum bactericidal concentration (MBC) for *S. aureus* than for *E. coli*.

Therefore, the bioelastomer with antioxidant and antibacterial activity was determined to be B–MPE 15%, and the contents of hesperidin and narirutin added to the bioelastomer were estimated to be 53.7 mg/50 g-B–MPE 15% and 12.2 mg/50 g-B–MPE 15%, respectively.

3.5. IR Analysis of Bioelastomer

The chemical structures of B-MPE 0% (control, PDMS) and B-MPE 15% were measured using FTIR. In the FT-IR spectra (Figure 4), the peak was at 797 cm^{-1}, corresponding to the symmetric stretching of Si–O–Si, the peaks at 1020 cm^{-1} and 1100 cm^{-1} corresponded to the asymmetric stretching of the Si–O–Si of the PDMS backbone, the peak at 1257 cm^{-1} corresponded to the asymmetric stretching of CH_3, and 2926 cm^{-1} corresponded to the symmetric bending of the CH_3 of the PDMS side chain. The same peaks appeared for both PDMS layers, indicating that the PDMS monomers with MPE were completely cured. Furthermore, this proves that there are only physical interactions between the filler and the matrix without the formation of covalent bonds [58]. These results agreed with those of Shivangi et al. [59], who found that the addition of bioactive extracts did not significantly affect the surface chemical properties of the fabricated biofilm.

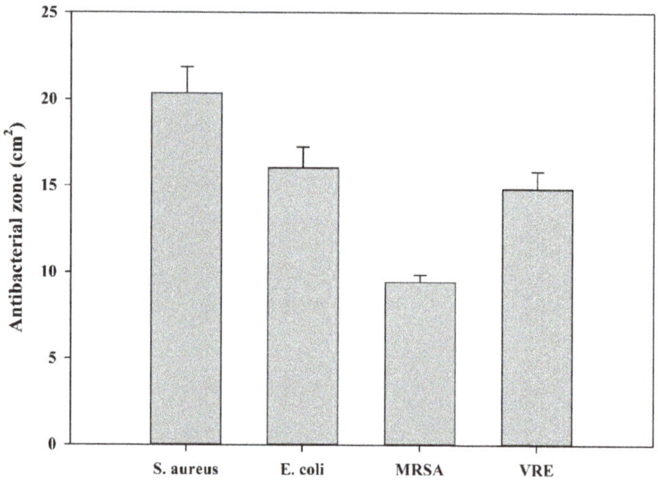

Figure 3. Antibacterial activity of Bioelastomer–MPE 15% against gram-positive (*S. aureus*), Gram-negative (*E. coli*), and antibiotic-resistant bacteria (MRSA and VRE).

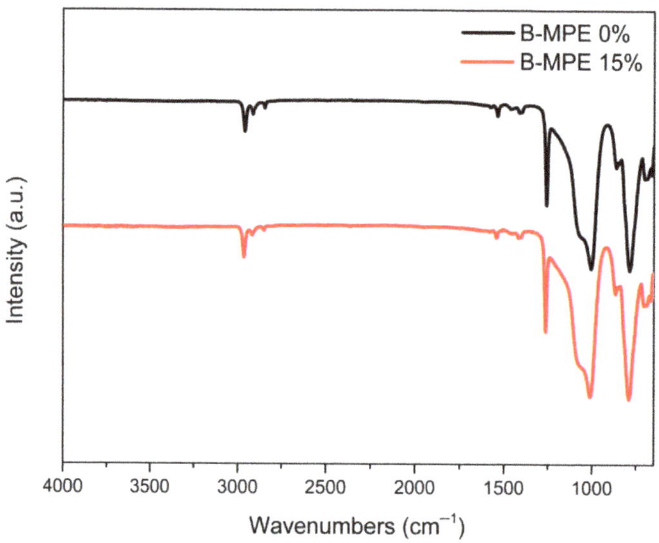

Figure 4. Fourier transform infrared (FT-IR) spectra of the Bioelastomer–MPE 0% (control, PMDS, black) and the Bioelastomer–MPE 15% (red).

3.6. SEM Image of Bioelastomer

Scanning electron microscopy (SEM) was used to observe the surface of the fabricated bioelastomer (Figure 5). The B–MPE 0% film was transparent with a smooth surface (Figure 5a). In contrast, the B–MPE 15% film exhibited a yellowish porous surface (Figure 5b). The difference in the morphology of the samples was attributed to the presence of insoluble flavonoid matter that was left behind after the evaporation of the extraction solvent [60]. In addition, the addition of the extracts changed the optical properties of the bioelastomer, resulting in differences in color and transparency.

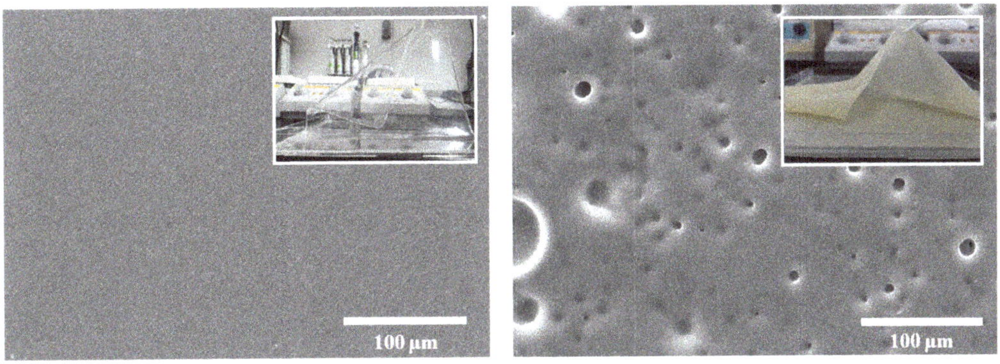

(a) (b)

Figure 5. SEM images of B–MPE 0% (**a**) and B–MPE 15% (**b**) (inset: film-type product).

3.7. Mechanical Strength of Bioelastomer

To evaluate the applicability of the selected B–MPE 15%, the mechanical properties of the fabricated bioelastomer, namely its tensility and flexibility, were investigated by measuring its tensile strength and elongation at break, respectively [61]; the results are shown in Table 3. There was no significant difference between the tensile strengths of B–MPE 0% (5.2 N/mm^2) and B–MPE 15% (5.1 N/mm^2). This indicated that the MPEs were uniformly mixed with the well-cured PDMS layer without reducing the mechanical strength of the fabricated bioelastomer. Meanwhile, the elongation at break of B–MPE 15% (694%) was significantly higher than that of B–MPE 0% (551%); these results are consistent with those reported by da Rosa et al. [62], who showed that phenolic compounds, derived from plant extracts, can increase elongation at break.

Table 3. Physical properties of Bioelastomer–MPE 0% (control, PMDS) and Bioelastomer–MPE 15%.

Sample	Tensile Strength (N/mm^2)	Elongation at Break (%)	Water vapor Transmission Rate (g/(m^2 day))
B–MPE 0%	5.2	551	26.6
B–MPE 15%	5.1	649	33.3

3.8. Water Vapor Transmission Rate of Bioelastomer

The water vapor transmission rate (WVTR) refers to the amount of water vapor that can permeate per unit area of a material per unit time. In the food packaging industry, lower values are considered advantageous [63]. Bourakadi et al. [64] reported that barrier properties, including the WVTR, are significantly affected by the chemical properties of the additives. From the WVTRs of B–MPE 0% and 15%, which were found to be 26.6 g/(m^2 day) and 33.3 g/(m^2 day), respectively (Table 3), it can be seen that the addition of MPE slightly increased the WVTR of the bioelastomers. This was probably due to the presence of polar compounds in the MPE: in general, extracts containing polar compounds, such as flavonoids and phenolic acids, improve the hydrophilicity of film materials, leading to an increase in their water vapor permeability [65].

4. Conclusions

Here, we proposed a biorefinery strategy based on the extraction of useful substances prior to the saccharification process of MPs, as well as the utilization of the extracted compounds. In this study, we recovered hesperidin and narirutin from MP and used them to fabricate bioelastomers. These bioelastomers that were fabricated using extracts exhibited significantly improved mechanical strength compared to bioelastomers that directly utilized biomass in powder form. In addition, the fabricated bioelastomer exhibited significant antioxidant and antibacterial activities and, thus, shows great potential for use in the food packaging, pharmaceutical, and medical industries. Our biorefinery strategy is expected to provide future direction for the realization of a sustainable society and carbon neutrality.

Author Contributions: Conceptualization, K.H.L. and Y.C.; methodology, J.H.L.; software, J.H.L.; validation, J.U.L. and T.L.; formal analysis, K.H.L. and Y.C.; investigation, J.U.L.; data curation, K.H.L. and Y.C.; writing—original draft preparation, K.H.L. and Y.C.; writing—review and editing, T.L. and H.Y.Y.; visualization, J.H.L.; supervision, H.Y.Y.; project administration, T.L. and H.Y.Y.; funding acquisition, T.L. and H.Y.Y. All authors have read and agreed to the published version of the manuscript.

Funding: This work was supported by the National Research Foundation of Korea (NRF) grant funded by the Ministry of Science and ICT (MSIT) (NRF-2020R1C1C1005060 and RS-2022-00165814).

Institutional Review Board Statement: Not applicable.

Data Availability Statement: The data presented in this study are available on request from the corresponding author.

Conflicts of Interest: The authors declare no conflict of interest.

Appendix A

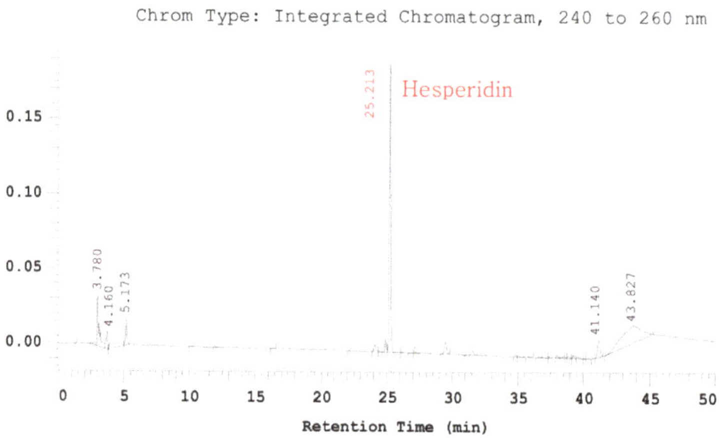

(**a**)

Figure A1. *Cont.*

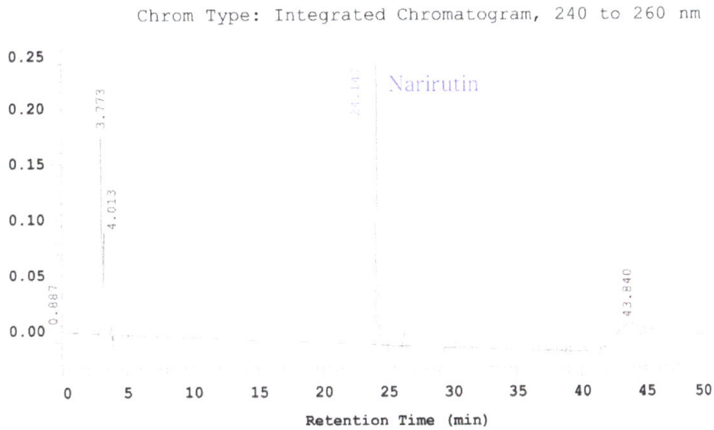

(**b**)

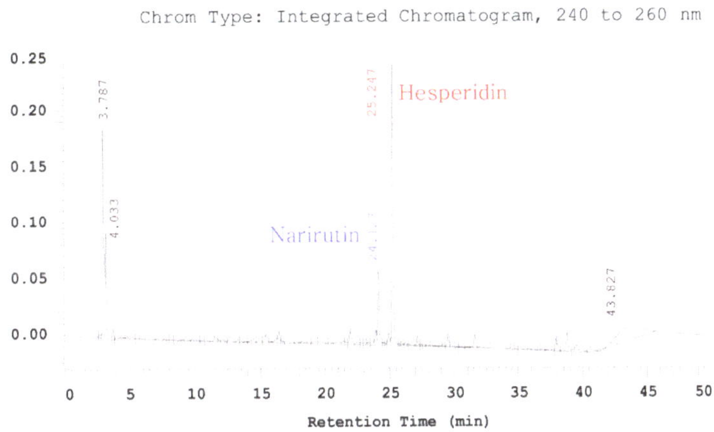

(**c**)

Figure A1. HPLC chromatogram for hesperidin standard (**a**); narirutin standard (**b**); mandarin peel extracts (**c**).

References

1. Jang, S.-K.; Jung, C.-D.; Seong, H.; Myung, S.; Kim, H. An integrated biorefinery process for mandarin peel waste elimination. *J. Clean. Prod.* **2022**, *371*, 133594. [CrossRef]
2. Falcinelli, B.; Famiani, F.; Paoletti, A.; D'Egidio, S.; Stagnari, F.; Galieni, A.; Benincasa, P. Phenolic Compounds and Antioxidant Activity of Sprouts from Seeds of *Citrus* Species. *Agriculture* **2020**, *10*, 33. [CrossRef]
3. Jeju special self-governing province. Citrus Annual Distribution Processing Analysis. Available online: http://citrus.or.kr/board/data/chang/2019.pdf (accessed on 21 March 2021).
4. Lee, H.H.; Heo, D.Y.; Han, H.R.; Park, H.R.; Ruangcharus, C.; Kim, S.U.; Seo, D.C.; Oh, T.-K.; Hong, C.O. Evaluation of the effects of mandarin (*Citrus reticulate*) by-products containing citric acid on immobilization of cadmium in arable soils. *Appl. Biol. Chem.* **2019**, *62*, 45. [CrossRef]
5. Choi, I.S.; Kim, J.-H.; Wi, S.G.; Kim, K.H.; Bae, H.-J. Bioethanol production from mandarin (*Citrus unshiu*) peel waste using popping pretreatment. *Appl. Energy* **2013**, *102*, 204–210. [CrossRef]
6. Shin, D.Y.; Cho, H.U.; Utomo, J.C.; Choi, Y.N.; Xua, X.; Park, J.M. Biodiesel production from *Scenedesmus bijuga* grown in anaerobically digested food wastewater effluent. *Bioresour. Technol.* **2015**, *184*, 215–221. [CrossRef]
7. Kadlimatti, H.M.; Raj Mohan, B.; Saidutta, M.B. Bio-oil from microwave assisted pyrolysis of food waste-optimization using response surface methodology. *Biomass Bioenergy* **2019**, *123*, 25–33. [CrossRef]
8. Jang, Y.W.; Lee, K.H.; Yoo, H.Y. Improved sugar recovery from orange peel by statistical optimization of thermo-alkaline pretreatment. *Processes* **2021**, *9*, 409. [CrossRef]

9. Kim, M.-H.; Kim, J.-W. Comparison through a LCA evaluation analysis of food waste disposal options from the perspective of global warming and resource recovery. *Sci. Total Environ.* **2010**, *408*, 3998–4006. [CrossRef]
10. Saqib, N.U.; Sharma, H.B.; Baroutian, S.; Dubey, B.; Sarmah, A.K. Valorisation of food waste via hydrothermal carbonisation and techno-economic feasibility assessment. *Sci. Total. Environ.* **2019**, *690*, 261–276. [CrossRef]
11. Wang, Y.; Tang, Y.; Li, M.; Yuan, Z. Aeration rate improves the compost quality of food waste and promotes the decomposition of toxic materials in leachate by changing the bacterial community. *Bioresour. Technol.* **2021**, *340*, 125716. [CrossRef]
12. Ryen, E.G.; Babbitt, C.W. The role of US policy in advancing circular economy solutions for wasted food. *J. Clean. Prod.* **2022**, *369*, 133200. [CrossRef]
13. Joshi, P.; Visvanathan, C. Sustainable management practices of food waste in Asia: Technological and policy drivers. *J. Environ. Manag.* **2019**, *247*, 538–550. [CrossRef] [PubMed]
14. Velvizhi, G.; Balakumar, K.; Shetti, N.P.; Ahmad, E.; Kishore Pant, K.; Aminabhavi, T.M. Integrated biorefinery processes for conversion of lignocellulosic biomass to value added materials: Paving a path towards circular economy. *Bioresour. Technol.* **2022**, *343*, 126151. [CrossRef] [PubMed]
15. Jia, F.R.; Jing, W.T.; Liu, G.X.; Yue, Q.; Wang, H.M.; Shi, L. Paraffin-based crude oil refining process unit-level energy consumption and CO_2 emissions in China. *J. Clean. Prod.* **2020**, *255*, 120347. [CrossRef]
16. Zang, Z. Are China's climate commitments in a post-Paris agreement sufficiently ambitious? *Wiley Interdiscip. Rev. Clim. Change* **2017**, *8*, e443. [CrossRef]
17. Lee, K.H.; Lee, S.K.; Lee, J.; Kim, S.; Kim, S.W.; Park, C.; Yoo, H.Y. Energy-efficient glucose recovery from chestnut shell by optimization of NaOH pretreatment at room temperature and application to bioethanol production. *Environ. Res.* **2022**, *208*, 112710. [CrossRef]
18. Yoo, H.Y.; Kim, S.W. The next-generation biomass for biorefining. *Bioresources* **2021**, *16*, 2188–2191. [CrossRef]
19. Mahato, N.; Sinha, M.; Sharma, K.; Koteswararao, R.; Cho, M.H. Modern extraction and purification techniques for obtaining high purity food-grade bioactive compounds and value-added co-products from citrus wastes. *Foods* **2019**, *8*, 523. [CrossRef]
20. Hiasa, S.; Iwamoto, S.; Endo, T.; Edashige, Y. Isolation of cellulose nanofibrils from mandarin (*Citrus unshiu*) peel waste. *Ind. Crops Prod.* **2014**, *62*, 280–285. [CrossRef]
21. Satari, B.; Karimi, K. Citrus processing wastes: Environmental impacts, recent advances, and future perspectives in total valorization. *Resour. Conserv. Recycl.* **2018**, *129*, 153–167. [CrossRef]
22. John, I.; Pola, J.; Thanabalan, M.; Appusamy, A. Bioethanol production from Musambi peel by acid catalyzed steam pretreatment and enzymatic saccharification: Optimization of delignification using Taguchi design. *Waste Biomass Valorization* **2020**, *11*, 2631–2643. [CrossRef]
23. Patsalou, M.; Samanides, C.G.; Protopapa, E.; Stavrinou, S.; Vyrides, I.; Koutinas, M. A Citrus peel waste biorefinery for ethanol and methane production. *Molecules* **2019**, *24*, 2451. [CrossRef] [PubMed]
24. Manaf, I.S.A.; Embong, N.H.; Khazaai, S.N.M.; Rahim, M.H.A.; Yusoff, M.M.; Lee, K.T.; Maniam, G.P. A review for key challenges of the development of biodiesel industry. *Energy Convers. Manag.* **2019**, *185*, 508–517. [CrossRef]
25. Koutra, E.; Economou, C.N.; Tsafrakidou, P.; Kornaros, M. Bio-based products from microalgae cultivated in digestates. *Trends Biotechnol.* **2018**, *36*, 819–833. [CrossRef]
26. Zhang, H.J.; Cui, J.F.; Tian, G.F.; DiMarco-Crook, C.; Gao, W.; Zhao, C.Y.; Li, G.Y.; Lian, Y.H.; Xiao, H.; Zheng, J.K. Efficiency of four different dietary preparation methods in extracting functional compounds from dried tangerine peel. *Food Chem.* **2019**, *289*, 340–350. [CrossRef]
27. Song, X.; Liu, T.; Wang, L.; Liu, L.; Li, X.; Wu, X. Antibacterial effects and mechanism of mandarin (*Citrus reticulata* L.) essential oil against *Staphylococcus aureus*. *Molecules* **2020**, *25*, 4956. [CrossRef]
28. Khan, M.K.; Zill, E.H.; Dangles, O. A comprehensive review on flavanones, the major citrus polyphenols. *J. Food Compos. Anal.* **2014**, *33*, 85–104. [CrossRef]
29. Smeriglio, A.; Cornara, L.; Denaro, M.; Barreca, D.; Burlando, B.; Xiao, J.; Trombetta, D. Antioxidant and cytoprotective activities of an ancient Mediterranean citrus (*Citrus lumia* Risso) albedo extract: Microscopic observations and polyphenol characterization. *Food Chem.* **2019**, *279*, 347–355. [CrossRef]
30. Ernawita; Wahyuono, R.A.; Hesse, J.; Hipler, U.-C.; Elsner, P.; Böhm, V. In vitro lipophilic antioxidant capacity, antidiabetic and antibacterial activity of citrus fruits extracts from Aceh, Indonesia. *Antioxidants* **2017**, *6*, 11. [CrossRef]
31. Gupta, A.; Jacobson, G.A.; Burgess, J.R.; Jelinek, H.F.; Nichols, D.S.; Narkowicz, C.K.; Al-Aubaidy, H.A. Citrus bioflavonoids dipeptidyl peptidase-4 inhibition compared with gliptin antidiabetic medications. *Biochem. Biophys. Res. Commun.* **2018**, *503*, 21–25. [CrossRef]
32. Chen, X.M.; Tait, A.R.; Kitts, D.D. Flavonoid composition of orange peel and its association with antioxidant and anti-inflammatory activities. *Food Chem.* **2017**, *218*, 15–21. [CrossRef] [PubMed]
33. Bensid, A.; El Abed, N.; Houicher, A.; Regenstein, J.M.; Özogul, F. Antioxidant and antimicrobial preservatives: Properties, mechanism of action and applications in food—A review. *Crit. Rev. Food Sci. Nutr.* **2020**, *32*, 1–17. [CrossRef] [PubMed]
34. Singh, B.; Singh, J.P.; Kaur, A.; Singh, N. Phenolic composition, antioxidant potential and health benefits of citrus peel. *Food Res. Int.* **2020**, *132*, 109114. [CrossRef] [PubMed]
35. Bustamante, J.; Van Stempvoort, S.; García-Gallarreta, M.; Houghton, J.A.; Briers, H.K.; Budarin, V.L.; Matharu, A.S.; Clark, J.H. Microwave assisted hydro-distillation of essential oils from wet citrus peel waste. *J. Clean. Prod.* **2016**, *137*, 598–605. [CrossRef]

36. Drinić, Z.; Vladić, J.; Koren, A.; Zeremski, T.; Stojanov, N.; Kiprovski, B.; Vidović, S. Microwave-assisted extraction of cannabinoids and antioxidants from *Cannabis sativa* aerial parts and process modeling. *J. Chem. Technol. Biotechnol.* **2020**, *95*, 831–839. [CrossRef]
37. de Oliveira Filho, J.G.; Braga, A.R.C.; de Oliveira, B.R.; Gomes, F.P.; Moreira, V.L.; Pereira, V.A.C.; Egea, M.B. The potential of anthocyanins in smart, active, and bioactive eco-friendly polymer-based films: A review. *Food Res. Int.* **2021**, *142*, 110202. [CrossRef]
38. Dordevic, S.; Dordevic, D.; Sedlacek, P.; Kalina, M.; Tesikova, K.; Antonic, B.; Tremlova, B.; Treml, J.; Nejezchlebova, M.; Vapenka, L.; et al. Incorporation of Natural Blueberry, Red Grapes and Parsley Extract By-Products into the Production of Chitosan Edible Films. *Polymers* **2021**, *13*, 3388. [CrossRef]
39. Magaña, I.; López, R.; Enríquez-Medrano, F.J.; Kumar, S.; Aguilar-Sanchez, A.; Handa, R.; de León, R.D.; Valencia, L. Bioelastomers: Current state of development. *J. Mater. Chem. A* **2022**, *10*, 5019–5043. [CrossRef]
40. Lee, K.H.; Chun, Y.; Jang, Y.W.; Lee, S.K.; Kim, H.R.; Lee, J.H.; Kim, S.W.; Park, C.; Yoo, H.Y. Fabrication of functional bioelastomer for food packaging from Aronia (*Aronia melanocarpa*) juice processing by-products. *Foods* **2020**, *9*, 1565. [CrossRef]
41. Liu, J.; Cheng, Y.; Xu, K.; An, L.; Su, Y.; Li, X.; Zhang, Z. Effect of nano-silica filler on microstructure and mechanical properties of polydimethylsiloxane-based nanocomposites prepared by "inhibition-grafting" method. *Compos. Sci. Technol.* **2018**, *167*, 355–363. [CrossRef]
42. Huang, N.; Zang, J.; Zhang, G.; Guan, L.; Li, S.; Zhao, L.; Tang, L. Efficient interfacial interaction for improving mechanical properties of polydimethylsiloxane nanocomposites filled with low content of graphene oxide nanoribbons. *RSC Adv.* **2017**, *7*, 22045–22053. [CrossRef]
43. Periakaruppan, R.; Li, J.; Mei, H.; Yu, Y.; Hu, S.; Chen, X.; Li, X.; Guo, G. Agro–waste mediated biopolymer for production of biogenic nano iron oxide with superparamagnetic power and antioxidant strength. *J. Clean. Prod.* **2021**, *311*, 127512. [CrossRef]
44. Lee, K.H.; Jang, Y.W.; Kim, H.; Ki, J.-S.; Yoo, H.Y. Optimization of lutein recovery from *Tetraselmis suecica* by response surface methodology. *Biomolecules* **2021**, *11*, 182. [CrossRef] [PubMed]
45. Inoue, T.; Tsubaki, S.; Ogawa, K.; Onishi, K.; Azuma, J.I. Isolation of hesperidin from peels of thinned *Citrus unshiu* fruits by microwave-assisted extraction. *Food Chem.* **2010**, *123*, 542–547. [CrossRef]
46. Magwaza, L.S.; Oara, U.L.; Cronje, P.J.R.; Landahl, S.; Ortiz, J.O.; Terry, L.A. Rapid methods for extracting and quantifying phenolic compounds in citrus rinds. *Food Sci. Nutr.* **2016**, *4*, 4–10. [CrossRef]
47. Ballard, T.S.; Mallikarjunan, P.; Zhou, K.; O'Keefe, S. Microwave-assisted extraction of phenolic antioxidant compounds from peanut skins. *Food Chem.* **2010**, *120*, 1185–1192. [CrossRef]
48. Ahmad, J.; Langrish, T.A.G. Optimisation of total phenolic acids extraction from mandarin peels using microwave energy: The importance of the Maillard reaction. *J. Food Eng.* **2012**, *109*, 162–174. [CrossRef]
49. Chen, C.-W.; Xie, J.; Yang, F.-X.; Zhang, H.-L.; Xu, Z.-W.; Liu, J.-L.; Chen, Y.-J. Development of moisture-absorbing and antioxidant active packaging film based on poly (vinyl alcohol) incorporated with green tea extract and its effect on the quality of dried eel. *J. Food Process. Preserv.* **2018**, *42*, e13374. [CrossRef]
50. Floegel, A.; Kim, D.-O.; Chung, S.-J.; Koo, S.I.; Chun, O.K. Comparison of ABTS/DPPH assays to measure antioxidant capacity in popular antioxidant-rich US foods. *J. Food Compos. Anal.* **2011**, *24*, 1043–1048. [CrossRef]
51. Karaçelik, A.A.; Küçük, M.; Iskefiyeli, Z.; Aydemir, S.; De Smet, S.; Miserez, B.; Sandra, P. Antioxidant components of *Viburnum opulus* L. determined by on-line HPLC–UV–ABTS radical scavenging and LC–UV–ESI-MS methods. *Food Chem.* **2015**, *175*, 106–114. [CrossRef]
52. Rodrigo, M.J.; Alquézar, B.; Alós, E.; Lado, J.; Zacarías, L. Biochemical bases and molecular regulation of pigmentation in the peel of *Citrus* fruit. *Sci. Hortic.* **2013**, *163*, 46–62. [CrossRef]
53. Zhang, Y.; Liu, X.; Wang, Y.; Jiang, P.; Quek, S. Antibacterial activity and mechanism of cinnamon essential oil against *Escherichia coli* and *Staphylococcus aureus*. *Food Control* **2016**, *59*, 282–289. [CrossRef]
54. Herraiz-Carboné, M.; Cotillas, S.; Lacasa, E.; Sainz de Baranda, C.; Riquelme, E.; Cañizares, P.; Rodrigo, M.; Sáez, C. Are We Correctly Targeting the Research on Disinfection of Antibiotic-Resistant Bacteria (ARB)? *J. Clean. Prod.* **2021**, *320*, 128865. [CrossRef]
55. Shehata, M.G.; Awad, T.S.; Asker, D.H.; Abd El-Aziz, N.M.; Youssef, M.M. Antioxidant and antimicrobial activities and UPLC-ESI-MS/MS polyphenolic profile of sweet orange peel extracts. *Curr. Res. Food Sci.* **2021**, *4*, 326–335. [CrossRef] [PubMed]
56. Alexandre, E.M.; Silva, S.; Santos, S.A.; Silvestre, A.J.; Duarte, M.F.; Saraiva, J.A.; Pintado, M. Antimicrobial activity of pomegranate peel extracts performed by high pressure and enzymatic assisted extraction. *Food Res. Int.* **2019**, *115*, 167–176. [CrossRef]
57. Choi, S.-S.; Lee, S.-H.; Lee, K.-A. A comparative study of hesperetin, hesperidin and hesperidin glucoside: Antioxidant, anti-inflammatory, and antibacterial activities in vitro. *Antioxidants* **2022**, *11*, 1618. [CrossRef]
58. Riaz, A.; Lagnika, C.; Luo, H.; Dai, Z.; Nie, M.; Hashim, M.M.; Liu, C.; Song, J.; Li, D. Chitosan-based biodegradable active food packaging film containing Chinese chive (*Allium tuberosum*) root extract for food application. *Int. J. Biol. Macromol.* **2020**, *150*, 595–604. [CrossRef]
59. Shivangi, S.; Dorairaj, D.; Negi, P.S.; Shetty, N.P. Development and characterisation of a pectin-based edible film that contains mulberry leaf extract and its bio-active components. *Food Hydrocoll.* **2021**, *121*, 107046. [CrossRef]
60. Abshirini, M.; Saha, M.C.; Altan, M.C.; Liu, Y. Synthesis and characterization of hierarchical porous structure of polydimethylsiloxane (PDMS) sheets via two-step phase separation method. *Mater. Des.* **2021**, *212*, 110194. [CrossRef]

61. Abdelwahab, M.A.; Jacob, S.; Misra, M.; Mohanty, A.K. Super-tough sustainable biobased composites from polylactide bioplastic and lignin for bio-elastomer application. *Polymer* **2021**, *212*, 123153. [CrossRef]
62. da Rosa, G.S.; Vanga, S.K.; Gariepy, Y.; Raghavan, V. Development of Biodegradable Films with Improved Antioxidant Properties Based on the Addition of Carrageenan Containing Olive Leaf Extract for Food Packaging Applications. *J. Polym. Environ.* **2020**, *28*, 123–130. [CrossRef]
63. Song, Z.; Xiao, H.; Zhao, Y. Hydrophobic-modified nano-cellulose fiber/PLA biodegradable composites for lowering water vapor transmission rate (WVTR) of paper. *Carbohydr. Polym.* **2014**, *111*, 442–448. [CrossRef] [PubMed]
64. El Bourakadi, K.; el Kacem Qaiss, A.; Bouhfid, R. Bio-films based on alginate/modified clay through spray drying: Mechanical, rheological, morphological, and transport properties for potential use as active food packaging. *Int. J. Biol. Macromol.* **2022**, *210*, 663–668. [CrossRef] [PubMed]
65. Cerqueira, M.A.; Souza, B.; Martins, J.T.; Teixeira, J.A.; Vicente, A.A. Seed extracts of *Gleditsia triacanthos*: Functional properties evaluation and incorporation into galactomannan films. *Food Res. Int.* **2010**, *43*, 2031–2038. [CrossRef]

agriculture

Article

Rabbit Manure Compost for Seedling Nursery Blocks: Suitability and Optimization of the Manufacturing Production Process

Rangling Li, Hongying Wang, Enze Duan, Jiayu Fan and Liangju Wang *

College of Engineering, China Agricultural University, Beijing 100083, China
* Correspondence: wangl@cau.edu.cn

Abstract: Using rabbit manure to prepare growing media is an effective method to solve environmental pollution and realize resource utilization. The solution to rabbit manure management is the composting process which could produce compost suitable for seedling nursery blocks, which could improve transplanting efficiency and seedlings' survival rate. Seedling nursery blocks were obtained by mixing rabbit manure compost, vermiculite, rice straw, and peat. The effect of cold pressing parameters, including moisture content (25–45%), binder content (1–5%), molding compression ratio (2.5–4.5:1), and strain maintenance time (0–120 s), were investigated on blocks quality (i.e., ventilatory porosity, relaxation density, compressive resistance, and specific energy consumption) through a general rotation combined experiment. These results showed there were significant interaction effects between molding compression ratio and moisture content, moisture and binder content, binder content and strain maintenance time, and molding compression ratio and binder content on block quality. The optimal parameters for manufacturing blocks were that the molding compression ratio, moisture content, binder content, and strain maintenance time were 4:1, 33.5%, 3.1%, and 60 s, and the relaxation density, ventilation porosity, and specific energy consumption were 363.31 kg/m^3, 18.72%, and 0.44 J/g, which could achieve emergence performance.

Keywords: waste management; sustainability; growing media; compression molding; peat substitute

Citation: Li, R.; Wang, H.; Duan, E.; Fan, J.; Wang, L. Rabbit Manure Compost for Seedling Nursery Blocks: Suitability and Optimization of the Manufacturing Production Process. *Agriculture* **2022**, *12*, 2156. https://doi.org/10.3390/agriculture12122156

Academic Editor: Maria Roulia

Received: 31 October 2022
Accepted: 14 December 2022
Published: 14 December 2022

1. Introduction

China has a long history of rabbit breeding and is the largest rabbit meat producer, consumer, and exporter in the world [1]. Rabbit breeding has become one of the pillar industries for targeted poverty alleviation in some areas due to its low input cost and considerable economic benefits [2,3]. With the continuous improvement of the large-scale and intensive breeding industry, the amount of rabbit manure is increasing. China's annual output of rabbit manure is about 18.1 million tons. Rabbit manure could represent an environmental issue due to the presence of antibiotics and high levels of nitrogen; the manure would release ammonia and nitrogen oxides into the atmosphere and cause other issues during the accumulation. Thereby, its valorization treatment has become a top priority because it is an organic waste and may be a cost for farmers who have not had to spread the manure (in landless intensive systems) [4–7].

Rabbit manure commonly has a good granule structure and low water content. This means rabbit manure is easier to collect than some livestock and poultry manure, which may be conducive to its resource utilization. Besides, it was shown in existing studies that rabbit manure contains nutrient content (N > 1.6%, P > 6.5%, and K > 1.2% of the dry matter) and minimal risk of potentially toxic elements [8]. Therefore, rabbit manure can be used as a high-quality organic fertilizer for flowers, fruits, and vegetables [9–11]. It has been proven that the growing media prepared by agricultural and forestry wastes, such as livestock and poultry manure and crop straw, have sound seedling effects. This can also partially or entirely replace non-renewable peat and rock wool to meet the increasing demand for growing media for facility agriculture and conform to the concept of green agricultural development [12,13]. In previous studies, the breeding feasibility of rabbit manure compound growing media for salt-tolerant calendula and salt-intolerant cucumber had been preliminarily verified [14]. Preparing rabbit manure for growing media may realize its high-value utilization nearby.

The seedling nursery block can replace the seedling tray, which has the advantage of improving the survival rate of seedlings and facilitating mechanized transplanting [15]. Therefore, pressing the granular growing media into the seedling nursery block has become the direction of developing high-grade growing media [16,17]. Rabbit manure contains more hemicellulose than other manure types. In contrast, as a viscoelastic composition, hemicellulose helps form the "solid bridge," reduces the energy consumption of the molding, and may promote the water absorption and ventilatory porosity of the manufactured seedling nursery blocks. That is, rabbit manure has good growing media-forming characteristics [18–20].

In the existing reports on the formation of seedling nursery blocks, the growing media materials, such as cow manure, vermicompost, straw, and peat, are pressed mainly by hydraulic pressure or pneumatic pressure through high pressure (4.5–247 kN) or elevated temperature (80–120 °C) [21–24]. However, their promotion and application are relatively limited due to increased equipment investment and large operating energy consumption. Rabbit manure with good forming characteristics may be pressed and formed at low pressure and normal temperature to achieve low energy consumption and high efficiency. However, it has not yet been systematically reported.

According to transportation and storage requirements, some parameters (e.g., relaxation density, dimensional stability, and compressive properties) may be used in the research of the seedling nursery block-forming process [25]. According to the requirements for seedling root growth, the block should have good ventilation porosity after water absorption [8]. Therefore, ventilation porosity should also be used to evaluate the block-forming effect. This has been relatively ignored in previous molding process studies or replaced only with an expansion ratio [25]. These indicators are affected by some internal factors (including types and additions of binders, moisture content of raw materials, length and additions of straws, and types of regulators) and external factors (e.g., molding pressure/load, molding temperature, compression ratio, and compression speed) [21–23]. There are also interactions among distinct factors. Therefore, exploring the influence of these parameters helps guide the research on the organic waste cold-pressing process. Seeking the optimal compression parameter combination to meet multiple objectives (e.g., transportation and seedling cultivation) will be conducive to the exploration and development of the seedling nursery block industry.

In this content, granular growing media were prepared with rabbit manure, rice straw, vermiculite, and peat. The forming characteristics under low pressure and normal temperature were explored, including the influence and interaction of moisture content, binder content, molding compression ratio, and strain maintenance time on the ventilatory porosity, relaxation density, compressive resistance, and specific energy consumption of the blocks. Then, regression equations of parameters and indexes were constructed, and multi-objective optimization was conducted to obtain a suitable combination of parameters for forming rabbit manure compost seedling nursery blocks. Finally, the seedling experiment was conducted to verify its feasibility. Theoretical support for the high-value utilization of rabbit manure and the optimization and development of the molding process of the seedling nursery block is provided by this study.

2. Materials and Methods

2.1. Growing Media Materials

The growing media materials mainly included decomposed rabbit manure compost, vermiculite, rice straw, and peat. The moisture and lignocellulose contents of growing media materials are shown in Table 1. Among them, rabbit manure was collected from a large-scale rabbit breeding herd in Henan Province, China, and the other materials were obtained from a local market. The materials were naturally air-dried, crushed, and passed through a 4.75 mm sieve. Since the ratio of length to width of rice straw is large and the ratio of width to height is small, it is challenging to obtain uniform rice straw directly. Therefore, a sieving instrument was used for screening. The selected sizes were 3.35 mm, 2.36 mm, 1.18 mm, 600 μm, 425 μm, and 150 μm. Finally, 50 μm–1.18 mm rice straw was selected for subsequent experiments. Additionally, a food-grade binder (sodium carboxymethyl cellulose) was used to improve the formation effect of the nursery block.

Table 1. Moisture and lignocellulose contents of main growing media materials.

	Moisture Content (%)	Cellulose (% Dry Matter)	Hemicellulose (% Dry Matter)	Lignin (% Dry Matter)
Rabbit manure compost	20.3	25.27	29.95	21.39
Rice straw	6.56	41.13	24.80	9.72
Vermiculite	2.4	/	/	/
Peat	37.5	36.87	22.17	18.26

2.2. Experimental Methods

2.2.1. Instruments and Procedures

The test system for cold compression molding is shown in Figure 1. It was mainly composed of a universal material testing machine (Instron-3367, Instron, Norwood, MA, USA), a punch-pin, a cavity mold, a die cushion, a stripper, and a control and display system. The pin diameter was 50 mm, and the protrusions were provided to form seeding troughs for substrate blocks while reducing surface runoff of moisture during seedling raising. The diameter of the cavity was 50 mm, the height was 90 mm, and the inner wall was polished to reduce the energy consumption of the molding. The system can control the compression procedures (by adjusting parameters such as compression force, displacement, time, speed, etc.) and simultaneously record the compression force and displacement of the growing media surface in real-time.

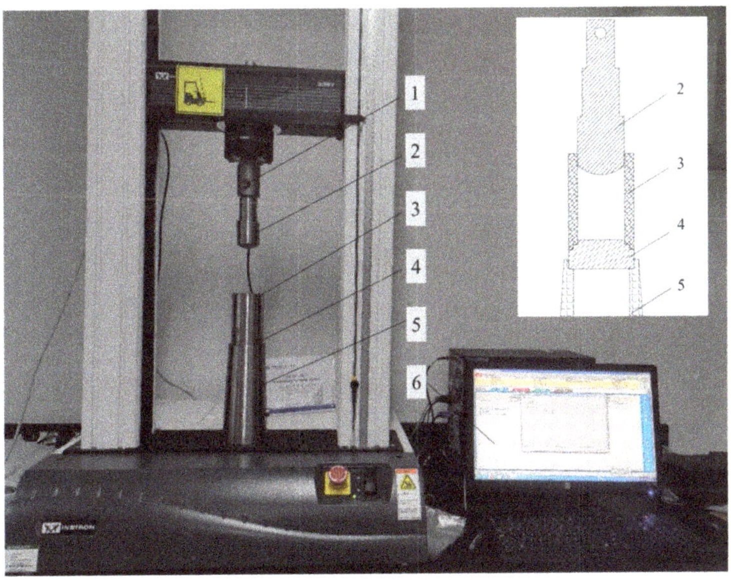

Figure 1. Test system for cold compression molding. 1. Positioning screws; 2. Punch-pin; 3. Cavity mold; 4. Die cushion; 5. Stripper; 6. Control and display system.

2.2.2. Experiment Design

In a previous experiment, the optimal proportion of rabbit manure compound growing media was determined as the rabbit manure compost:vermiculite:perlite:peat = 40:25:25:10 (v/v/v/v), according to the physiochemical properties of the growing media and plant growth performance [14]. However, the perlite may be unfavorable for molding as it becomes powder during compression [26]. Crop straw is often used as biomass for molding [27,28]. The influence of commonly used growing media materials (e.g., perlite, rice straw, and wheat straw) on the rabbit manure compost seedling nursery blocks forming effect was explored. Then, rice straw was determined according to mixing difficulty, water absorption performance, and forming energy consumption. Finally, the proportion of rabbit manure compost seedling nursery blocks was determined: rabbit manure compost:vermiculite:rice straw:peat = 40:25:25:10 (v/v/v/v). The pH of the blocks was 7.77, and the contents of C, N, P, K, Ca, and Mg were 35.75%, 1.36%, 0.78%, 2.07%, 1.67%, and 1.87% of the dry matter, respectively.

On this basis, the moisture content (factor A), binder mass fraction (factor B), molding compression ratio (factor C), and strain maintenance time (factor D) were selected as the testing factors. A general rotation combined experimental design (29 groups) was used to study the influence of these factors, interactions, and quadratic terms on the indicators through variance analysis to optimize process parameters. The actual values and codes of the tested factors are shown in Table 2.

Table 2. Actual and coded levels of cold compression molding test.

Levels	Factors			
	A Moisture Content (%)	B Binder Mass Fraction (%)	C Molding Compression Ratio	D Strain Maintenance Time (s)
2	45	5	4.5:1	120
1	40	4	4:1	90
0	35	3	3.5:1	60
−1	30	2	3:1	30
−2	25	1	2.5:1	0

The technological process for forming rabbit manure compost seedling nursery blocks is shown in Figure 2. First, rabbit manure compost, vermiculite, rice straw, and peat were mixed in a determined ratio. Second, the different contents of the binder (1%, 2%, 3%, 4%, or 5%) and deionized water (adjusting the moisture content to 25%, 30%, 35%, 40%, or 45%) were added to obtain granular growing media according to the experimental design. The mixture was then sealed and placed at 4 °C for 48 h to homogenize. Then, the granular growing media was loaded into the self-made forming mold. After the growing media was formed and filled the mold, the universal material testing machine was started to drive the punch for compression. The compression speed was 35 mm/min, and the compression displacement was converted by the compression ratio ((2.5, 3, 3.5, 4 or 4.5):1) of the experimental design. After reaching the set displacement, the strain was maintained for some time (0, 30, 60, 90, or 120 s). Then the die cushion was removed and demolded to obtain the rabbit manure compost seedling nursery blocks. The temperature during the test was room temperature, about 20 °C. Finally, seedling nursery blocks were dried at 60 °C for 4 h to obtain the finished products.

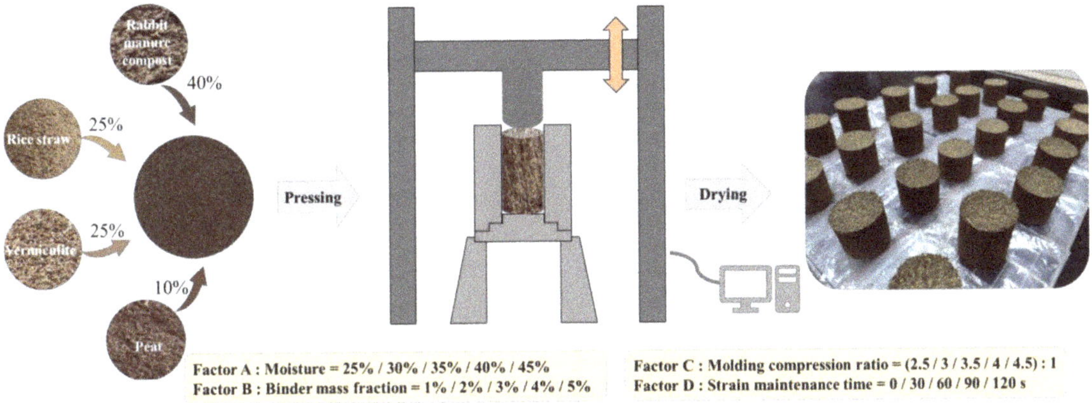

Figure 2. Technological process for forming rabbit manure compost seedling nursery blocks.

2.2.3. Evaluation Indices

(1) Relaxation density

The relaxation density (RD) reflects the stability of the seedling nursery blocks after 48 h of relaxation, which can reflect the support effect and anti-destructive strength of the seedling nursery block on seedlings [27]. This is an important evaluation index of the seedling nursery blocks molding effect, which was calculated using Equation (1) [28]:

$$RD = 4M/(\pi D^2 L) \tag{1}$$

where RD (kg/m^3) is the relaxation density of the rabbit manure compost seedling nursery blocks; M (kg) is the weight of blocks; D (m) and L (m) are the diameter and height of the blocks after relaxation.

(2) Ventilatory porosity

The aeration performance of the growing media, which directly affects the water, air, fertilizer, and other environments of the root system, is characterized by ventilatory porosity (VP). Therefore, its permeability should also be considered while ensuring the mechanical strength of rabbit manure compost seedling nursery blocks. In this study, the ring knife method [29] was used to measure ventilatory porosity and was expressed as follows:

$$VP = 100\% \times (M_1 - M_2)/V \tag{2}$$

where VP (%) is the ventilatory porosity of the rabbit manure compost seedling nursery blocks; V (cm^3) is the volume of the blocks; M_1 (g) and M_2 (g) are the weight of blocks when saturated water absorption and water no longer exudes.

(3) Specific energy consumption

Specific energy consumption (SEC) is the energy consumption per unit mass in block compression molding. This could reflect the difficulty of block molding and is related to the production cost [27]. Based on the real-time collected load-displacement data during the rabbit manure compost seedling nursery block forming process, the specific energy consumption is calculated according to Equation (3):

$$SEC = 10^{-3} \times \left(\int f \cdot dx\right)/M \tag{3}$$

where SEC (J/g) is the specific energy consumption of the rabbit manure compost seedling nursery blocks; f (N) is load; x (mm) is molding displacement; M (g) is the weight of growing media.

(4) Compressive resistance

The compressive resistance is to place the rabbit manure compost seedling nursery block sideways on the test bench of the universal material testing machine (Instron-3367, Instron, Norwood, MA, USA) and control the upper plate on the beam to move downward at a uniform speed. After the block is destroyed, the system automatically unloads and moves upward in the opposite direction. The peak value of the force received during the deformation of the block is the compressive resistance of the sample [30], and the machine automatically records this value.

2.2.4. Emergence Performance of Rabbit Manure Compost Seedling Nursery Blocks

The optimized rabbit manure compost seedling nursery blocks were used for seedling-raising experiments in order to test the seedling emergence performance. The test seeds were Beijing new No. 3 cabbage seeds (Jingyan Yinong). Four Chinese cabbage seeds were placed in the seeding hole of each seedling nursery block and covered with vermiculite. Low-level immersion irrigation combined with micro spraying was used to make the block fully absorb water without adding fertilizer. After 10 days of seedling raising, the seedling emergence rate of 76 seeds was counted.

2.2.5. Chemical Analysis

The pH value of the growing media was measured at a ratio of fresh sample to distilled water of 1:10 using a pH meter (Sartorius PB-10, Sartorius AG, Göttingen, Germany). Lignocellulose content (including cellulose, hemicellulose, and lignin) was determined according to the methods described in [31]. The content of P, K, Ca, and Mg was measured using an inductively coupled plasma spectrometer (ICPOES730, Agilent Technologies Inc., Palo Alto, CA, USA). The content of C and N was determined using an elemental analyzer (Vario EL cube, Elementar, Hanau, Germany).

2.2.6. Data Analysis

The quaternion quadratic general rotation combination test was designed, and the data (n = 3) were analyzed using Design Expert 10. The parameters were optimized using MATLAB R2021b. All figures were drawn using Origin 2021.

3. Results and Discussion

3.1. Results and Analysis of Combined Tests

The results of relaxation density, ventilatory porosity, specific energy consumption, and compressive resistance of rabbit manure compost seedling nursery blocks obtained as per the testing design and requirements are listed in Table 3. The ranges of relaxation density, ventilatory porosity, specific energy consumption, and compressive resistance were 220.93–350.04 kg/m^3, 9.18–20.98%, 0.10–0.71 J/g, and 27.65–207.53 N, respectively. The maximum forming pressure of rabbit manure compost seedling nursery blocks was 2.97 kN. Cao et al. [21] found that the seedling nursery block density of cattle manure was 769–1125 kg/m^3, and the forming pressure was 5–25 kN. Liu et al. [24] showed that the nutritious peat block density after pressing was 1801–1917.6 kg/m^3, and the forming pressure was 162.4–247.3 kN. The density and forming pressure of seedling nursery blocks in the above two studies were larger than our results, as the density was about 2–4 times larger, and the forming pressure was about 7–82 times larger. These differences were attributed to differences in material types, moisture content, and forming conditions [27]. In the above literature, the cattle manure seedling nursery block was made by mixing with straw or corncob, with a moisture content of 10–30%. The materials of the nutritious peat block included peat, vermiculite, perlite, sand, added molding curing agent, super absorbent resin, and micronutrients, with a moisture content of 0.5–2%. The materials used in this experiment were rabbit manure, vermiculite, rice straw, and peat, which were easily compressed (consistent with the expectations in the Introduction). They had a relatively high moisture content ranging from 25% to 45%. Moreover, the cattle manure seedling nursery block adopted hot pressing (80–160 °C), but cold pressing was used in this experiment.

Table 3. Test scheme and results.

Test Number	Factors and Levels				Evaluation Indicators			
	A Moisture Content (%)	B Binder Mass Fraction (%)	C Molding Compression Ratio	D Strain Maintenance Time (s)	Relaxation Density (kg·m^{-3})	Ventilatory Porosity (%)	Specific Energy Consumption (J·g^{-1})	Compressive Resistance (N)
1	−1	−1	−1	−1	261.87	11.83	0.32	27.65
2	−1	−1	1	−1	295.19	12.44	0.71	58.60
3	1	−1	−1	−1	249.80	11.10	0.20	35.34
4	1	−1	1	−1	280.04	11.70	0.49	57.68
5	−1	1	−1	−1	284.18	12.23	0.26	40.47
6	−1	1	1	−1	334.22	12.50	0.62	179.33
7	1	1	−1	−1	260.64	10.37	0.19	42.77
8	1	1	1	−1	282.60	10.39	0.48	77.22
9	−1	−1	−1	1	261.87	10.63	0.28	28.72
10	−1	−1	1	1	304.93	11.67	0.69	71.44
11	1	−1	−1	1	253.94	11.30	0.20	32.97
12	1	−1	1	1	275.59	9.18	0.49	71.94
13	−1	1	−1	1	287.76	13.33	0.26	76.90
14	−1	1	1	1	350.04	13.23	0.60	207.53
15	1	1	−1	1	262.08	10.47	0.19	44.59
16	1	1	1	1	282.40	11.48	0.45	76.19
17	0	0	−2	0	220.93	10.66	0.10	27.16
18	0	0	2	0	276.41	10.29	0.66	62.82
19	−2	0	0	0	308.90	16.27	0.60	62.47
20	2	0	0	0	272.06	12.47	0.35	46.51
21	0	−2	0	0	274.32	12.14	0.56	51.12
22	0	2	0	0	310.00	14.27	0.55	154.09
23	0	0	0	−2	246.29	10.71	0.34	30.17
24	0	0	0	2	271.87	11.32	0.36	65.24
25	0	0	0	0	270.01	20.98	0.34	62.04
26	0	0	0	0	268.12	18.09	0.34	59.84
27	0	0	0	0	265.68	18.76	0.33	66.02
28	0	0	0	0	258.31	18.50	0.36	63.84
29	0	0	0	0	262.90	18.96	0.31	60.00

3.2. *Analysis of Variance and Regression Equations*

Analysis of variance is a standard method for evaluating model significance and accuracy [32]. The above test results were analyzed by variance analysis, and regression models between factors and evaluation indicators were constructed. The model fit for relaxation density, ventilatory porosity, specific energy consumption, and compressive resistance are shown in Table 4. The R^2 of relaxation density, ventilatory porosity, and specific energy consumption were all $\geq$0.92, indicating that more than 92% of variable variations could be explained by the models. The differences between Adjusted R^2 and Predicted R^2 were all <0.2, and the Adeq Precision were all >4, indicating that the five regression models fit well and could be used to predict relaxation density, ventilatory porosity, and specific energy consumption [33]. In comparison, the model of compressive resistance could explain only 85% of the variation and lacked fitting. Therefore, the influence of factors on compressive resistance was analyzed, but no regression model was established.

Table 4. Model fit for relaxation density, ventilatory porosity, specific energy consumption, and compressive resistance.

Response Variable	R^2	Adjusted R^2	Predicted R^2	Adeq Precision
Relaxation density	0.92	0.89	0.81	21.91
Ventilatory porosity	0.95	0.93	0.89	18.99
Specific energy consumption	0.98	0.97	0.95	43.72
Compressive resistance	0.85	0.79	0.68	15.04

The insignificant items ($\alpha = 0.1$) of the regression equation were excluded, and the simplified regression equations of relaxation density, ventilatory porosity, and specific energy consumption are shown in Equations (4)–(6):

$$RD = 262.7 - 12.78 \times A + 9.67 \times B + 16.41 \times C - 6.5 \times AB - 5.91 \times AC + 8.43 \times A^2 + 8.85 \times B^2 \quad (4)$$

$$VP = 19.06 - 0.81 \times A + 0.35 \times B + 0.46 \times BD - 1.3 \times A^2 - 1.59 \times B^2 - 2.27 \times C^2 - 2.14 \times D^2 \quad (5)$$

$$SEC = 0.34 - 0.064 \times A - 0.015 \times B + 0.16 \times C - 0.022 \times AC + 0.029 \times A^2 + 0.049 \times B^2 \quad (6)$$

where *RD*, *VP*, and *SEC* are the relaxation density (kg/m^3), ventilatory porosity (%), and specific energy consumption (J/g) of the rabbit manure compost seedling nursery blocks; *A* and *B* are the moisture and binder content (%) of rabbit manure growing media; *C* and *D* are the molding compression ratio and strain maintenance time (s) of rabbit manure compost seedling nursery blocks. The *A*, *B*, *C*, and *D* are coded values.

3.3. Effects of Different Factors on Evaluation Indicators

3.3.1. On Relaxation Density

In Equation (4), it was shown that the moisture content, binder mass fraction, and molding compression ratio all significantly affected the relaxation density. The influence of distinct factors on the relaxation density of rabbit manure compost seedling nursery blocks was ranked as molding compression ratio > moisture content > binder mass fraction. Molding compression ratio and binder mass fraction were significantly positively correlated with relaxation density, while moisture content was significantly negatively correlated with relaxation density. In terms of interaction, the interaction between the moisture content and binder mass fraction and the moisture content and molding compression ratio significantly affected relaxation density.

The effect of moisture content and binder mass fraction on relaxation density is shown in Figure 3a. When the binder mass fraction was constant, and the moisture content was less than 35%, the relaxation density of the seedling nursery blocks decreased with the increment of the moisture content. When the moisture content was more than 35%, the relaxation density changed slightly with the increment of the moisture content. Similarly, when the moisture content was constant, the relaxation density of the seedling nursery blocks increased with increments in the binder mass fraction. The lower the moisture content and the higher the binder mass fraction, the greater the relaxation density of the rabbit manure compost seedling nursery blocks.

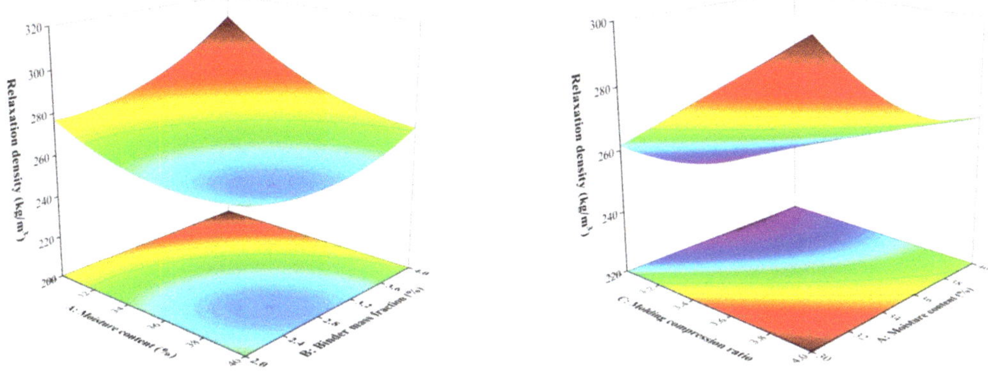

(**a**) (**b**)

Figure 3. Interaction effect on relaxation density. (**a**) Effect of moisture content and binder mass fraction on relaxation density with the molding compression ratio of 3.5:1 and strain maintenance time of 60 s; (**b**) Effect of molding compression ratio and moisture content on relaxation density with the binder mass fraction of 3% and strain maintenance time of 60 s.

This was because an appropriate amount of water could dissolve the binder, and the water evaporated during the drying process, which was conducive to the adhesion of the matrix materials [15], and also enhanced the role of the "solid bridge" of rabbit manure and rice straw. However, too much water diluted the binder and weakened the adhesion on the particle surface [34]. Within the test level range, relaxation density varied from 220.93 kg/m^3 to 350.04 kg/m^3, which was light in weight and met the requirements for plant germination and growth [35].

The effect of moisture content and molding compression ratio on relaxation density is shown in Figure 3b. When the molding compression ratio was constant, the relaxation density of rabbit manure compost seedling nursery blocks decreased with an increment in moisture content. This was due to part of the moisture attached to the particle surface, which increased the degree of relaxation of the matrix after molding [36]. When the moisture content of the matrix was constant, the relaxation density increased with an increase in the molding compression ratio. Especially when the moisture content was less than 35%, and the compression ratio was greater than 3.5:1, the relaxation density reached its maximum value. This phenomenon was consistent with the study reported by Cao et al. [21], where the bulk density of compressed straw-cattle manure block changed from 0.958 g/cm^3 to 1.092 g/cm^3 when the moisture content increased from 10% to 30%. In that study, the bulk density of the compressed straw-cattle manure block changed from 0.884 g/cm^3 to 1.258 g/cm^3 when the molding pressure increased from 5 kN to 10 kN.

3.3.2. On Ventilatory Porosity

In Equation (5), it was shown that the ventilatory porosity of rabbit manure compost seedling nursery blocks was significantly affected by the moisture content and binder mass fraction, and the moisture content was more important. The binder mass fraction was significantly positively correlated with ventilatory porosity, while moisture content was significantly negatively correlated. In terms of interaction, the interaction between binder mass fraction and strain maintenance time significantly affected the ventilatory porosity.

The effect of binder mass fraction and strain maintenance time on ventilatory porosity is shown in Figure 4. When the binder content was constant, the ventilatory porosity of rabbit manure compost seedling nursery blocks first increased and then decreased with an increment in the strain maintenance time. Similarly, when the strain maintenance time was constant, the air porosity first increased and then decreased with the increment of binder content. When the binder content was between 2.8% and 3.3% and the strain maintenance time was between 54 and 66 s, the ventilatory porosity reached its maximum value.

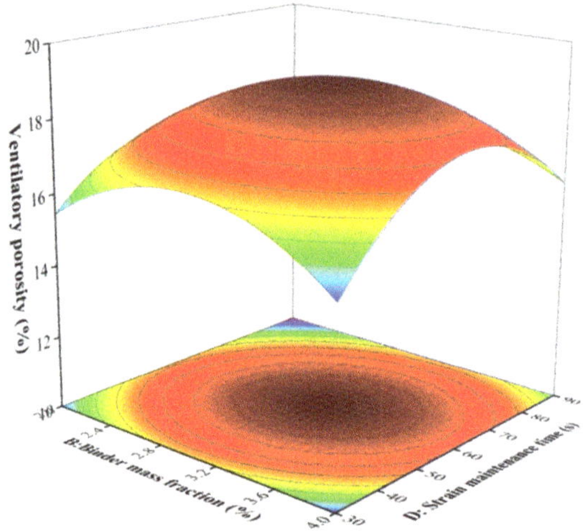

Figure 4. Interaction effect of binder mass fraction and strain maintenance time on ventilatory porosity with the moisture content of 35% and molding compression ratio of 3.5:1.

This was due to the two stages of seedling nursery block molding. The first was the compression molding stage with variable displacement, where a large amount of air between particles was discharged. The second was the strain maintenance stage, in which the rebounding of pressed growing media was relieved through the relaxation of residual stress [27]. When the strain maintenance time increased, the stress of the seedling nursery blocks gradually relaxed, the block maintained its original shape, and the expansion rate was relatively large after sufficient water absorption. When the strain maintenance time was longer than a specific range, the plastic deformation of the block was high, and swelling was slight after sufficient water absorption [28]. Since the binder had the function of water retention, which also affected the water absorption expansion of the block, its effect on ventilatory porosity was similar to the strain maintenance time.

3.3.3. On Specific Energy Consumption

In Equation (6), it was shown that the moisture content, binder mass fraction, and molding compression ratio all significantly affected the specific energy consumption, and the effects were ranked as molding compression ratio > moisture content > binder mass fraction. The molding compression ratio was significantly positively correlated with specific energy consumption, while moisture content and binder mass fraction were significantly negatively correlated. Besides, the interaction between molding compression ratio and moisture content significantly affected the specific energy consumption.

With the increase in the molding compression ratio, the work of the friction and extrusion between the matrix particles, between the matrix and the die, increased significantly, increasing the total energy consumption of the block compression molding. The moisture in the matrix material played a specific lubricating role. At the same time, the cushioning properties of the matrix can be improved by absorbing water through the matrix, thereby reducing the energy consumption of molding [37]. Yang et al. [38] reported that increasing water content could improve the uniformity of the adhesive, which was conducive to the low-pressure densification of biomass.

The effect of moisture content and molding compression ratio on specific energy consumption is shown in Figure 5. When the molding compression ratio was constant, the specific energy consumption of block compression molding decreased with increased moisture content. This phenomenon was more apparent when the moisture content was lower. When the water content was constant, the specific energy consumption increased with the molding compression ratio, which might be due to the incompressibility of water. It was found that the effect of the interaction between molding compression ratio and moisture content on energy consumption was consistent with the relaxation density, which was also shown by the research of Chen et al. [27]. Therefore, when the moisture content was more than 30%, the specific energy consumption of rabbit manure compost seedling nursery block formation was less than 0.6 J/g. Within this experiment, the maximum molding load of the rabbit manure compost seedling nursery blocks was 2974 N and did not require processes such as heating and cooling. It has relatively low energy consumption compared with the current high-pressure or high-temperature pressing [21–24].

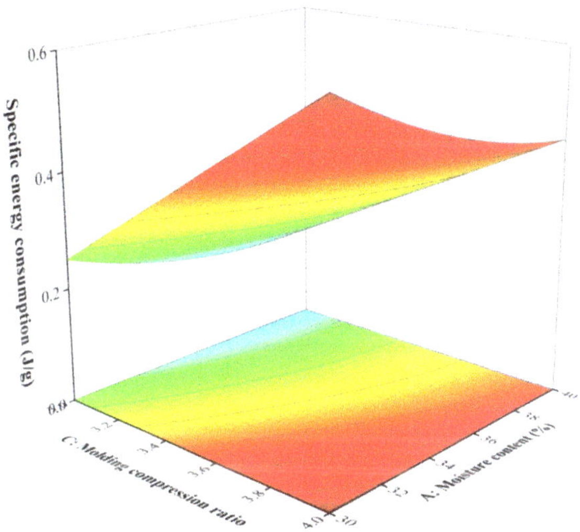

Figure 5. Interaction effect of moisture content and molding compression ratio on specific energy consumption with the binder mass fraction of 3% and strain maintenance time of 60 s.

3.3.4. On Compressive Resistance

It was shown by the variance analysis of compressive resistance that the compressive resistance was significantly affected by the moisture content, binder mass fraction, and molding compression ratio. The influence of distinct factors on the compressive resistance of rabbit manure compost seedling nursery blocks was ranked as molding compression ratio > binder mass fraction > moisture content. The molding compression ratio and binder mass fraction were significantly positively correlated with compressive resistance, while moisture content was significantly negatively correlated. Furthermore, these three factors had interactive effects on compressive resistance.

The effect of moisture content and binder mass fraction on compressive resistance is shown in Figure 6a. When the moisture content was constant, the compressive resistance of rabbit manure compost seedling nursery blocks increased with an increase in the binder content, especially when the moisture content was low. When the content of the binder was constant, the compressive resistance of rabbit manure compost seedling nursery blocks decreased with an increase in moisture content, especially when the binder content was high. Great compressive resistance of the rabbit manure compost seedling nursery blocks was obtained at lower matrix moisture content and higher binder content due to the greater relaxation density at this time (Figure 3a).

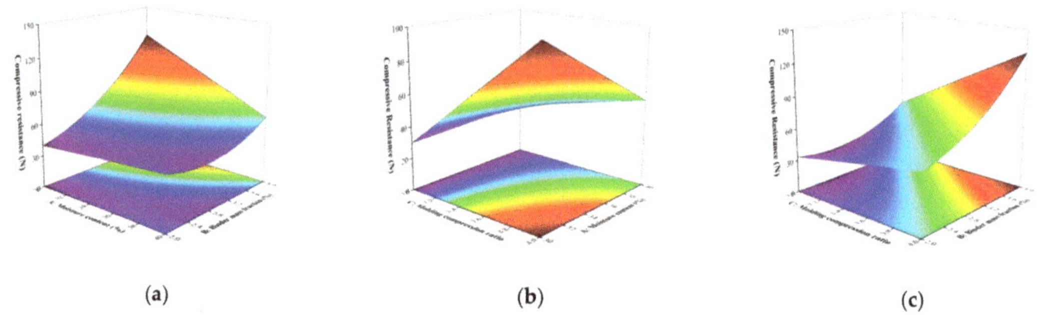

Figure 6. Interaction effect on compressive resistance. (**a**) Effect of moisture content and binder mass fraction on compressive resistance with the molding compression ratio of 3.5:1 and strain maintenance time of 60 s; (**b**) Effect of moisture content and molding compression ratio on compressive resistance with the binder mass fraction of 3% and strain maintenance time of 60 s; (**c**) Effect of binder mass fraction and molding compression ratio on compressive resistance with the moisture content of 35% and strain maintenance time of 60 s.

The effect of moisture content and molding compression ratio on compressive resistance is shown in Figure 6b. When the moisture content was constant, the compressive resistance of rabbit manure compost seedling nursery blocks increased gradually with an increase in the compression ratio, especially when the moisture content was low. When the molding compression ratio was constant, the compressive resistance decreased gradually with an increase in moisture content. The highest compressive resistance of rabbit manure compost seedling nursery blocks was obtained under molding conditions of a high compression ratio and low moisture content. This is consistent with the variation law of the relaxation density (Figure 3a) and specific energy consumption (Figure 5) of rabbit manure compost seedling nursery blocks.

The effect of binder mass fraction and molding compression ratio on compressive resistance is shown in Figure 6c. When the binder content was constant, the compressive resistance increased with an increase in the compression ratio, especially when the binder content was higher. When the compression ratio was constant, the compressive resistance increased with the increase in the binder content, especially when the molding compression ratio was large. The compressive resistance increased with binder content and compression ratio due to increased molding pressure and bonding force to improve the molding effect.

3.4. Comprehensive Optimization and Verification Tests

In order to obtain rabbit manure compost seedling nursery blocks with good molding effect and good use effect, it is required that the seedling nursery blocks have sufficient stability (e.g., the bulk density of the seedling matrix should be 200–600 g/cm [35]), adequate aeration (e.g., the requirement of ventilatory porosity is 15–30% [35]), and low energy consumption in the molding process. Although the compressive resistance of the block was an important quality indicator, the influence of the molding parameters on compressive resistance was consistent with the relaxation density, and the accuracy of the regression model of compressive resistance was low. Therefore, the following five objective functions (F_1, F_2, F_3, F_4, and F_5) were constructed, and the regression equations (Equations (4)–(6)) of relaxation density, ventilation porosity, and specific energy consumption were brought into the objective functions to obtain the optimal parameter combination.

$$\begin{cases} F_1 = RD_{\max} \\ F_2 = VP_{\max} \\ F_3 = SEC_{\min} \\ F_4 = \left(\frac{RD-200}{600-200} + \frac{VP-15}{30-15} - \frac{SEC-0.2}{0.6-0.2}\right)_{max} \\ F_5 = \left(2 \times \frac{RD-200}{600-200} + 2 \times \frac{VP-15}{30-15} - \frac{SEC-0.2}{0.6-0.2}\right)_{max} \\ 1 \le A \le 2 \\ -2 \le B \le 2 \\ -0.5 \le C \le 2 \\ -1 \le D \le 1 \\ 200 \le RD \le 600 \\ 15 \le VP \le 30 \\ 0.2 \le SEC \le 0.6 \end{cases}$$

Among them, F_1, F_2, F_3, F_4, and F_5 represented the maximum relaxation density, maximum ventilatory porosity, and minimum specific energy consumption, the comprehensive maximum value when the indicator weights are consistent. The comprehensive maximum value when the density and porosity weights were doubled on the premise that other indicators met the requirements. *RD*, *VP*, and *SEC* are the relaxation density (kg/m^3), ventilatory porosity (%), and specific energy consumption (J/g) of the rabbit manure compost seedling nursery blocks; *A* and *B* are the moisture and binder content (%) of rabbit manure growing media; C and *D* are the molding compression ratio and strain maintenance time (s) of rabbit manure compost seedling nursery blocks. The *A*, *B*, *C*, and *D* are coded values. The optimization solution was conducted using MATLAB, and the five optimized parameter combinations were used for the test. The test results and theoretical prediction values are shown in Table 5.

Table 5. Prediction and test results of rabbit manure compost seedling nursery blocks after parameter optimization.

	Moisture Content (%)	Binder Mass Fraction (%)	Molding Compression Ratio	Strain Maintenance Time (s)	Relaxation Density (kg·m^{-3})		Ventilatory Porosity (%)		Specific Energy Consumption (J·g^{-1})	
					Test Value	Theoretical Value	Test Value	Theoretical Value	Test Value	Theoretical Value
1	4.1:1	34.5	3.8	60	364.80	298.38	18.85	15.12	0.54	0.56
2	4.0:1	39.0	2.8	48	338.47	269.01	18.13	14.87	0.39	0.45
3	4.0:1	39.5	3.0	45	348.67	269.12	18.19	14.47	0.41	0.45
4	3.5:1	39.0	2.8	54	321.61	257.33	16.37	17.38	0.26	0.31
5	4.0:1	39.0	2.9	57	351.30	269.19	16.60	15.24	0.36	0.45

It can be seen from Table 5 that the theoretical values of RD and VP were generally higher than the test values, and the theoretical values of SEC were usually lower than the test values. Therefore, the constant term of the model was modified according to Equation (7) [36]. The final modified models are shown in Equations (8)–(10).

$$c = c_0 + \sum_{1}^{n} (I_t - I_e)/n \quad (7)$$

where c is the corrected value of the model constant term; c_0 is the original constant term of the model; I_t is the test value of indices; I_e is the theoretical value of indices; n is the number of tests.

$$RD = 335.06 - 12.78 \times A + 9.67 \times B + 16.41 \times C - 6.5 \times AB - 5.91 \times AC + 8.43 \times A^2 + 8.85 \times B^2 \quad (8)$$

$$VP = 21.27 - 0.81 \times A + 0.35 \times B + 0.46 \times BD - 1.3 \times A^2 - 1.59 \times B^2 - 2.27 \times C^2 - 2.14 \times D^2 \quad (9)$$

$$SEC = 0.29 - 0.064 \times A - 0.015 \times B + 0.16 \times C - 0.022 \times AC + 0.029 \times A^2 + 0.049 \times B^2 \quad (10)$$

where *RD*, *VP*, and *SEC* are the relaxation density (kg/m^3), ventilatory porosity (%), and specific energy consumption (J/g) of the rabbit manure compost seedling nursery blocks; *A* and *B* are the moisture and binder content (%) of rabbit manure growing media; *C* and *D* are the molding compression ratio and strain maintenance time (s) of rabbit manure compost seedling nursery blocks. The *A*, *B*, *C*, and *D* are coded values.

When pressing loose growing media into seedling nursery blocks while seeking higher mechanical strength of blocks, their ventilatory porosity should be improved as much as possible to improve the root environment of seedlings. After the models were modified, the optimization conditions of indicators were set: the ventilation porosity was the largest, the compression ratio was 4, the water content was 33.5%, the binder mass fraction was 3.1%, and the strain maintenance time was 60 s. The test was conducted using the optimized parameter values. The verification results are shown in Table 6.

Table 6. Prediction and test results of rabbit manure compost seedling nursery blocks optimized with ventilatory porosity as the target.

	Relaxation Density (kg·m^{-3})	**Ventilatory Porosity (%)**	**Specific Energy Consumption (J·g^{-1})**
Test value	363.31	18.72	0.44
Theoretical value	354.55	19.09	0.45
Error	2.47%	−1.91%	−3.23%

The relative error predicted by the three models was less than 5%. This indicated that the modified model could realize the prediction of relaxation density, ventilatory porosity, and specific energy consumption of rabbit manure seedling matrix blocks and provided a reference and basis for the production of blocks that met the requirements of seedling raising and transplanting operations.

3.5. Seedling Emergence Performance of Rabbit Manure Compost Seedling Nursery Blocks

The seedling experiments were conducted with rabbit manure compost seedling nursery blocks prepared after the above optimization parameters. It was shown by the results that the emergence rate of Chinese cabbage seeds was 97.4%, which met the requirements of the vegetable seedling matrix (i.e., 90%, which is usually the minimum requirement for commercially applied substrates) [35]. During the nursery period, the rabbit manure compost seedling nursery blocks repeatedly absorbed water more than 40 times without loose or deformed situations (Figure 7). Generally, a mature commercial seedling-raising substrate can maintain a stable structure under the seedling-raising cycle of >20 days [39,40], with watering >30 times. The seedling emergence rate and the overall state of the blocks preliminarily proved the feasibility of the rabbit manure compost seedling nursery blocks. Plant growth, not only germination, may also need to be further evaluated in future studies for comprehensive evaluation. Then, how to meet the needs of different seedlings and be more conducive to market promotion and commercial application of the matrix block needs further research.

Figure 7. Nursery test of rabbit manure compost seedling nursery blocks.

4. Conclusions

In this study, the decomposed rabbit manure, vermiculite, rice straw, and peat were mixed and molded to manufacture rabbit manure compost seedling nursery blocks for replacing plug trays and improving the mechanization of seedling transplanting. The influences of cold pressing process parameters (e.g., moisture content, binder content, molding compression ratio, and strain maintenance time) on the evaluation indices (e.g., ventilatory porosity, relaxation density, compressive resistance, and specific energy consumption) of the blocks were analyzed. The regression equations were constructed among molding compression ratio, moisture content, binder content, strain maintenance time and relaxation density, ventilation porosity, and specific energy consumption. The relative errors between the experimental and predicted values of each model were within 5%. The optimal combination of parameters for forming rabbit manure compost seedling nursery blocks was that the molding compression ratio, moisture content, mass fraction of the binder, and strain maintenance time were 4:1, 33.5%, 3.1%, and 60 s. The relaxation density, ventilation porosity, and specific energy consumption were 363.31 kg/m^3, 18.72%, and 0.44 J/g, which had an adequate seedling performance.

Author Contributions: Conceptualization, R.L. and E.D.; methodology, L.W.; software, J.F.; writing—original draft preparation, R.L.; writing—review and editing, L.W.; supervision, L.W. and H.W.; funding acquisition, H.W. All authors have read and agreed to the published version of the manuscript.

Funding: This study was supported by the China Agriculture Research System of MOF and MARA (CARS-43-D-3).

Institutional Review Board Statement: Not applicable.

Informed Consent Statement: Not applicable.

Data Availability Statement: The data that supported the findings of this study are available from the corresponding author upon reasonable request.

Conflicts of Interest: The authors declare no conflict of interest.

References

1. Zhang, M.Y.; Wu, L.P. Consumer demand for rabbit meat in urban China: 2011–2018. In Proceedings of the 12th World Rabbit Congress, Nantes, France, 3–5 November 2021.
2. Zhang, S.L.; Zhu, J.F. Cost and benefit analysis of meat rabbit breeding based on 2018 national survey data of 13 provinces [of China]. In Proceedings of the 12th World Rabbit Congress, Nantes, France, 3–5 November 2021.
3. Wu, L.P.; Qin, Y.H. Production overview of my country's rabbit industry in 2020 and development prospects in 2021. *Chin. J. Anim. Sci.* **2021**, *57*, 258–262.

4. Wen, B.; Guo, X.Y.; Fu, X.C.; Liu, N.; Bai, L.; Liu, H.Z.; Wang, P.; Xiong, Z.B.; Zhang, K.; Xiao, B.X. Application of three kinds of microbial agent in high-temperature composting of rabbit manure. *J. Grassl. Forage Sci.* **2016**, *2*, 43–51. (In Chinese with English Abstract)
5. Duan, E.Z.; Ma, J.X.; Chang, C.C.; Chen, J.Y.; Wang, H.Y. Physicochemical properties and utilization of fermented organic fertilizer of rabbit manure. *Chin. J. Rabbit Farming* **2018**, *6*, 36–39. (In Chinese with English Abstract)
6. National Bureau of Statistics. *China Rural Statistical Yearbook 2021*; China Statistics Press: Beijing, China, 2021.
7. Bao, W.Q.; Liu, J.J.; An, J.; Xie, G.H. Discussion on value-taking of relative parameters for assessment of livestock and poultry excrement resource in China. *Trans. CSAE* **2018**, *34*, 314–322. (In Chinese with English Abstract)
8. Li, R.L.; Hao, H.Y.; Sun, H.; Wang, L.J.; Wang, H.Y. Composted rabbit manure as organic matrix for manufacturing horticultural growing media: Composting process and seedling effects. *Sustainability* **2022**, *14*, 5146. [CrossRef]
9. Ikrarwati; Syamsi, N.A.; Sastro, Y.; Rusbana, T.B.; Sudolar, N.R.; Romadhonah, Y. Quality of Growth Media and Yields of *Allium ascolanicum* L. On Ultisol Soil Combined with Rabbit Manure. *IOP Conf. Ser. Earth Environ. Sci.* **2021**, *715*, 012039. [CrossRef]
10. Adekiya, A.O.; Ejue, W.S.; Olayanju, A.; Dunsin, O.; Aboyeji, C.M.; Aremu, C.; Adegbite, K.; Akinpelu, O. Different organic manure sources and NPK fertilizer on soil chemical properties, growth, yield and quality of okra. *Sci. Rep.* **2020**, *10*, 16083. [CrossRef]
11. Islas-Valdez, S.; Lucho-Constantino, C.A.; Beltrán-Hernández, R.I.; Gómez-Mercado, R.; Vázquez-Rodríguez, G.A.; Herrera, J.M.; Jiménez-González, A. Effectiveness of rabbit manure biofertilizer in barley crop yield. *Environ. Sci. Pollut. Res.* **2017**, *24*, 25731–25740. [CrossRef]
12. Pêgo, R.G.; Antunes, L.F.d.S.; Silva, A.R.C. Vigor of zinnia seedlings produced in alternative substrate in trays with different cell size. *Ornam. Hortic.* **2019**, *25*, 417–424. [CrossRef]
13. Kazemi, F.; Rabbani, M.; Jozay, M. Investigating the plant and air-quality performances of an internal green wall system under hydroponic conditions. *J. Environ. Manag.* **2020**, *275*, 111230. [CrossRef]
14. Li, R.L.; Hao, H.Y.; Yang, C.C.; Wang, L.J.; Wang, H.Y. Rabbit manure compound growing media: Proportioning optimization according to physiochemical characteristic and seedling effects. *Front. Plant Sci.* **2022**, *13*, 1008089. [CrossRef] [PubMed]
15. Wang, G.Q.; Qu, P.; Huang, H.Y.; Wu, G.F.; Yan, H.J. Evaluation of Coconut Coir Dust/Modified Urea–Formaldehyde Resins as a Growing Medium for Pepper Seedlings. *HortTechnology* **2020**, *30*, 322–330. [CrossRef]
16. Cui, Z.C.; Guan, C.S.; Xu, T.; Yang, Y.T.; Xu, B.X.; Chen, Y.S. Development status of technology and equipment of substrate block seedling cultivation and transplanting. *J. Chin. Agric. Mech.* **2021**, *43*, 29–34. (In Chinese with English Abstract)
17. Yang, L.Y.; Cao, H.L.; Yuan, Q.X.; Luoa, S.; Liu, Z.G. Component Optimization of Dairy Manure Vermicompost, Straw, and Peat in Seedling Compressed Substrates Using Simplex-Centroid Design. *J. Air Waste Manag. Assoc.* **2018**, *68*, 215–226. [CrossRef]
18. Kong, L.; Xiong, Y.; Liu, T.; Tu, Y.; Tian, S.; Sun, L.; Chen, T. Effect of fiber natures on the formation of "solid bridge" for preparing wood sawdust derived biomass pellet fuel. *Fuel Process. Technol.* **2016**, *144*, 79–84. [CrossRef]
19. Li, H.; Jiang, L.B.; Li, C.Z.; Liang, J.; Yuan, X.Z.; Xiao, Z.H.; Xiao, Z.H.; Wang, H. Co-pelletization of sewage sludge and biomass: The energy input and properties of pellets. *Fuel Process. Technol.* **2015**, *132*, 55–61. [CrossRef]
20. Hu, J.J. Straw Pellet Fuel Cold Molding by Compression: Experimental Study and Numerical Simulation. Ph.D. Thesis, Dalian University of Technology, Dalian, China, 2008. (In Chinese with English Abstract)
21. Cao, H.L.; Yang, L.Y.; Yuan, Q.X.; Huang, C.Q. Experimental research of seedling substrate compressed of cattle manures. *Trans. CSAM* **2015**, *46*, 197–202. (In Chinese with English Abstract)
22. Luo, S. Molding Process and Molding Machine Design and Experiment of Vermicompost Substrate. Ph.D. Thesis, Huazhong Agricultural University, Wuhan, China, 2019. (In Chinese with English Abstract)
23. Xin, M.J.; Chen, T.Y.; Zhang, Q.; Jiao, J.K.; Bai, X.W.; Song, Y.Q.; Zhao, R.; Wei, C. Parameters optimization for molding of vegetable seedling substrate nursery block with rice straw. *J. Agric. Eng.* **2017**, *33*, 219–225. (In Chinese with English Abstract)
24. Liu, Y.; Fan, T.; Qu, Z.X.; Wang, D.Z. Experimental study on nutritious peat block press molding. *For. Mach. Woodwork. Equip.* **2014**, *42*, 10–12. (In Chinese with English Abstract)
25. Liu, D.J.; Zhou, Y.J.; Liu, K.; Cheng, Q.; Bai, X.W.; Gong, Y.J. Experimental study on molding technology for making seedling block based on maize stovers. *Trans. CSAE* **2020**, *36*, 241–248. (In Chinese with English Abstract)
26. Elghezal, L.; Jamei, M.; Georgopoulos, I.O. DEM Simulations of Stiff and Soft Materials with Crushable Particles: An Application of Expanded Perlite as a Soft Granular Material. *Granul. Matter* **2013**, *15*, 685–704. [CrossRef]
27. Chen, T.Y.; Jia, H.L.; Zhang, S.W.; Sun, X.M.; Song, Y.Q.; Yuan, H.F. Optimization of cold pressing process parameters of chopped corn straws for fuel. *Energies* **2020**, *13*, 652. [CrossRef]
28. Ma, L.; Sha, L.; Liu, X.X.; Zhang, S.T. Study of molding and drying characteristics of compressed municipal sludge-corn stalk fuel pellets. *Energies* **2021**, *14*, 3116. [CrossRef]
29. Zhang, L.; Sun, X.Y.; Tian, Y.; Gong, X.Q. Composted green waste as a substitute for peat in growth media: Effects on growth and nutrition of *Calathea insignis*. *PLoS ONE* **2013**, *8*, e78121. [CrossRef] [PubMed]
30. Kaliyan, N.; Morey, R.V. Factors affecting strength and durability of densified biomass products. *Biomass Bioenergy* **2009**, *33*, 337–359. [CrossRef]
31. Van Soest, P.J.; Robertson, J.B.; Lewis, B.A. Methods for Dietary Fiber Neutral Detergent Fiber, and Nonstarch Polysaccharides in Relation to Animal Nutrition. *J. Dairy Sci.* **1991**, *74*, 3583–3597. [CrossRef]

32. Patel, G.B.; Shah, K.R.; Shindhal, T.; Rakholiya, P.; Varjani, S. Process parameter studies by central composite design of response surface methodology for lipase activity of newly obtained actinomycete. *Environ. Technol. Innov.* **2021**, *23*, 101724. [CrossRef]
33. Fang, M.; Yu, Z.H.; Zhang, W.J.; Cao, J.; Liu, W.H. Friction coefficient calibration of corn stalk particle mixtures using Plackett-Burman design and response surface methodology. *Powder Technol.* **2022**, *396*, 731–742. [CrossRef]
34. Xu, C.; Yuan, Q.X.; Qin, C.N.; Xie, G.R.; He, T.; Song, N. Effect of wood vinegar on physicochemical properties and seeding capability of cow manure aerobic composting. *Trans. CSAM* **2020**, *51*, 353–360. (In Chinese with English Abstract)
35. *NY/T 2118-2012;* Plug Seedling Substrate of Vegetables. China Agricultural Press: Beijing, China, 2012.
36. Li, W.Z.; Jiang, Y.; Rao, S.; Yin, X.L.; Wang, M.F.; Jiang, E.C. Optimization parameters of mixture pelleting of eucalyptus sawdust and corn stover. *Acta Energ. Sol. Sin.* **2020**, *41*, 332–338. (In Chinese with English Abstract)
37. Zhang, C.W. Forming Technology and Properties of Biomass Composites with Fully Open Cell Structure. Ph.D. Thesis, Shandong University, Jinan, China, 2020. (In Chinese with English Abstract)
38. Yank, A.; Ngadi, M.; Kok, R. Physical properties of rice husk and bran briquettes under low pressure densification for rural applications. *Biomass Bioenergy* **2016**, *84*, 22–30. [CrossRef]
39. Chang, R.; Guo, Q.; Pandey, P.; Li, Y.; Chen, Q.; Sun, Y. Pretreatment by Composting Increased the Utilization Proportion of Pig Manure Biogas Digestate and Improved the Seedling Substrate Quality. *Waste Manag.* **2021**, *129*, 47–53. [CrossRef] [PubMed]
40. Di Lonardo, S.; Cacini, S.; Becucci, L.; Lenzi, A.; Orsenigo, S.; Zubani, L.; Rossi, G.; Zaccheo, P.; Massa, D. Testing New Peat-Free Substrate Mixtures for the Cultivation of Perennial Herbaceous Species: A Case Study on *Leucanthemum vulgare* Lam. *Sci. Hortic.* **2021**, *289*, 110472. [CrossRef]

agriculture

Article

Deteriorating Harmful Effects of Drought in Cucumber by Spraying Glycinebetaine

El-Saied E. Metwaly [1], Hatim M. Al-Yasi [2], Esmat F. Ali [2,*], Hamada A. Farouk [3] and Saad Farouk [4,*]

1 Vegetable and Floriculture Department, Faculty of Agriculture, Mansoura University, Mansoura 35516, Egypt
2 Department of Biology, College of Science, Taif University, P.O. Box 11099, Taif 21944, Saudi Arabia
3 Faculty of Agriculture, Department of Plant Pathology, University of Assiut, Assiut 71526, Egypt
4 Agricultural Botany Department, Faculty of Agriculture, Mansoura University, Mansoura 35516, Egypt
* Correspondence: a.esmat@tu.edu.sa (E.F.A.); gadalla@mans.edu.eg (S.F.); Tel.: +20-1021721645 (S.F.)

Abstract: In order to alleviate the shortage of irrigation water in dry regions, refining water use efficiency (WUE) is a key issue in sustainable productivity. Furthermore, glycinebetaine (GlyBet) is a vital osmoprotectant produced in crops for improving drought tolerance; however, little is known about its role in improving plant WUE under field conditions in non-accumulating plants such as cucumber. In order to elucidate the effectiveness of GlyBet concentrations (0, 2000, 4000, and 6000 mg/L) in mitigating the deleterious effects of drought (e.g., well-watered (1250 m^3/fed), moderate drought (950 m^3/fed), and severe drought (650 m^3/fed)), field experiments were conducted at Elmia village, Dakahlia, Egypt in the 2020 and 2021 seasons on vegetative growth, some physiological attributes, as well as yield and quality. Drought considerably decreased vegetative growth, yield and its components, leaf relative water content, and photosynthetic pigment concentrations compared with well-watered plants while increasing electrolyte leakage. The most harmful causes were severe drought. However, exogenous spraying with GlyBet substantially boosted the mentioned attributes, but reduced electrolyte leakage within well-watering. Commonly 6000 mg/L contributed to the maximum growth and productivity, preserving cucumber plant water status above other concentrations or untreated plants. Under extreme drought, the application of 6000 mg/L GlyBet had a beneficial effect on moderating the damage of water deficit on cucumber plant growth and productivity. Overall, using GlyBet as a cost-effective and eco-friendly biostimulant six times (10, 20, 30, 40, 50, and 60 days from sowing) has the potential to mitigate drought damage while also increasing yield; however, more research is needed to determine the optimal rate and timing of application.

Keywords: cucumber; glycinebetaine; water stress; yield

Citation: Metwaly, E.-S.E.; Al-Yasi, H.M.; Ali, E.F.; Farouk, H.A.; Farouk, S. Deteriorating Harmful Effects of Drought in Cucumber by Spraying Glycinebetaine. *Agriculture* **2022**, *12*, 2166. https://doi.org/10.3390/agriculture12122166

Academic Editor: Maria Roulia

Received: 8 October 2022
Accepted: 6 December 2022
Published: 16 December 2022

1. Introduction

After tomato, cabbage, and onion, cucumber (*Cucumis sativus* L., Cucurbitaceae) is the fourth most popularly grown vegetable worldwide [1]. It is native to India and presumably came from the foothills of the Himalayan Mountains [2,3]. Cucumber is now widely grown around the world in both temperate and tropical climates; cucumbers require relatively high light intensity (20–30 mol/m^2/day), temperature (18–29 °C), and relative humidity (60–80%) [4]. According to FAO statistics (www.fao.org/faostat/en/ '3 February 2021'), the globally produced quantity was 87,805,086 tons of cucumber in total, with Asia producing 84.9% of that production. Cucumber has numerous uses in food, medicine, and cosmetics due to their abundance of water, nutrients, and phytochemicals [5,6]. Superior hydration and phytochemicals found in cucumbers have a variety of health benefits, including weight loss and treatment of eczema, constipation, hypertension, atherosclerosis, and cancer [6,7]. According to recent studies, cucumbers contain kaempferol, a key anti-diabetic substance [8]. In addition, cucumber is frequently used for skin treatments and natural attractiveness [6,9].

Due to abrupt climate changes and anthropogenic activities, more severe episodes of moisture unavailability, uneven rainfall distribution, and increases in average and maximum temperatures are predicted for the future, which will affect farming systems and agriculture productivity [10,11]. Drought stress (DS) has cost the global economy between USD 30 and USD 44 billion over the previous ten years [12]. In dry and semi-arid environments, DS represents the main ecological disorder that has a negative impact on plant growth and restricts sustainable production [13–15]. Owing to their direct effect on photoassimilate allocation, as well as an alteration in a sequence of biochemical processes and molecular reactions, DS undoubtedly reduced crop productivity by up to 70% [13–17]. It adjusts several metabolic processes, such as photosynthesis, water absorption, and reactive oxygen species (ROS) accumulation, leading to growth and yield deterioration [13,18,19].

Agricultural extension in Egypt requires a huge amount of water, which is now scant to fulfill the anticipated requisite, as 85% of accessible water is consumed in agriculture and utmost of the on-farm irrigation systems have little proficiency, along with poor irrigation administration. The cost-effective novel technique for enhancing water use efficiency (WUE) and boosting plant productivity has received increased attention due to the growing competition for scarce water resources. Through the use of water-saving and drought-tolerant varieties, effective agronomic practices and management, WUE can be increased [20,21]. Moreover, it is believed that the most important regulator for controlling plants' water usage within DS is chemical signaling, which involves phytohormones or osmoprotectants [22]. Plants under stress factors rapidly accumulate osmoprotectants to maintain water status and sustain typical functioning [23]. Osmoprotectants may be adjusting cellular osmotic pressure, detoxifying ROS, and preserving and stabilizing membrane integrity [24]. These findings show that the use of osmoprotectants may be a realistic strategy for increasing crop WUE under water scarcity.

Glycine betaine (N, N″, N″-trimethyl-glycine, GlyBet) is one of the most effective osmoprotectants, and it shields cellular components via maintaining an osmotic equilibrium and stabilizing the quaternary structures of complex proteins [18,25,26]. It is low-cost, reachable, eco-friendly, and provides both economic and environmental efficiency. GlyBet application increases the growth, and tolerance of a wider range of crops under stressful disorders [13,18,25,26] by regulating a number of physiological and biochemical processes [18,27], maintaining turgor pressure [27], enhancing net CO_2 assimilation rate [18], shielding the effectual proteins and enzymes, and lipids of the chloroplasts and sustaining electron stream over thylakoid membranes [28], as well as regulation of photosynthetic machinery and ion homeostasis [11]. Moreover, GlyBet may act as anti-transpirant, which permits the plant to enter extra water for a long period and facilitates photosynthesis [29]. Shemi et al. [18] mentioned that foliar-applied GlyBet increases several morphological features, yield, and accumulation of several metabolites of drought-affected maize plants. On canola, Dawood and Sadak [30] found that the application of GlyBet increased drought tolerance by enhancing shoot and root systems growth, photosynthetic pigment concentration, improving IAA, proline, soluble sugars, yield, and its components and quality (seed yield/plant, carbohydrate %, phenolic %, flavonoid level).

Different plant species have been shown to be able to tolerate drought better when GlyBet is applied exogenously. The available data infrequently differ on the plant species and water deficit severity. This study's objective was to assess how exogenously applied GlyBet affected cucumber grown under DS in terms of WUE and drought resistance. Under various irrigation regimes, the impact of GlyBet on cucumber yield and its components, photosynthetic pigments, ion percentage, relative water content, and membrane permeability was additionally tested. The findings of this study offer a new innovative water-saving technique in arid and semi-arid regions. It was assumed that GlyBet application can be used as an appropriate method for improving the growth and production of drought-affected cucumber plants.

2. Materials and Methods

2.1. Experimental Site and Soil Depiction

At a private farm in Elmia village, Dekernes district, Dakahlia governorate, Egypt (latitude 31°5′18″ N, longitude 31°35′49″ E, 8 m above sea level), two field trials were conducted in 2020 and 2021. The experiment region is classified as a semi-arid area with average precipitation and temperature of 5.39–7.52 mm and 21.3–24.7 °C, respectively, for both seasons. The physical and chemical characteristics of the experimental soil's profile, which displays clay loamy texture down to a depth of 0–60 cm, were provided in a Table 1.

Table 1. Physicochemical attributes of the experimental soil during 2020 and 2021 seasons.

Seasons	Silt %	Clay %	Sand %	Soil Texture	F.C %	W.P %	AW %	pH	E.C (dSm^{-1})	O.M %	$CaCO_3$ %	N ppm	P ppm	K ppm
2020	40.5	37.2	22.3	Clay loamy	35.7	18.9	16.8	8.22	1.51	1.8	3.39	51.9	5.7	288
2021	41.1	36.9	22.0	Clay loamy	35.2	18.4	16.8	8.13	1.78	2.o	3.45	54.1	6.2	294

F.C.—field capacity; W.P.—welting point; AW—available water; pH—potential of hydrogen; E.C—electrical conductivity; OM—organic matter; $CaCO_3$—calcium carbonate; N—nitrogen; P—phosphorus; K—potassium.

2.2. Experimental Layout

A split-plot based on a randomized complete block design and a drip irrigation system was used for the present investigation. The key plots were devoted to different three irrigation regimes: well watering (WW; 1250 m^3/fed), moderate drought (MD; 950 m^3/fed), and severe drought (SD; 650 m^3/fed), and the irrigation intervals were recognized consistent with the growth stage and local recommendation. Alternatively, the sub-plots were allocated to spraying treatments (water as control, 2000 mg/L GlyBet; 4000 mg/L GlyBet; and 6000 mg/L GlyBet). GlyBet as a pure water-soluble chemical was obtained from Sigma-Aldrich Co., LLC Company Profile | Saint Louis, MO, USA. The 12 treatments were replicated three times, making a total of 36 plots (each plot was 24 m^2 with 1.5 m wide bordered regions).

2.3. Crop Husbandry

The experimental field was deeply tilled, and the soil smoothed before seeding. Prior to sowing, 10 m^3/fed (4200 m^2) of farmyard manure had been added and properly mixed with the top 0–30 cm of soil. As fertigation at 2-day intervals starting one week after planting, 70, 45, and 65 kg fed^{-1} of ammonium nitrate (33.5% N), phosphoric acid (50% P_2O_5), and potassium sulfate (50% K_2O), were delivered correspondingly according to the recommendations of Ministry of Agriculture and Land Reclamation, Egypt. On the first and third of August in both seasons, respectively, cucumber seeds (*C. sativus* L., cv. *JABBAR, F1*, were secured from Fine Seeds Company, Giza, Egypt) were manually sown at two sides of the dripper.

Irrigation was started once sowing for identical seedling emergence 14 days before irrigation regime treatment. The GlyBet concentrations with 0.01% (*v*/*v*) Tween-20 (a witting agent) were sprayed six times at 10, 20, 30, 40, 50, and 60 days from sowing. The spraying (15 L/plot) was performed to a run-off in the early morning via a back-sprayer. For each treatment, irrigation was used to replenish the water loss accumulated every 2 days.

2.4. Data Recording and Measurements

At 35 days from sowing, five randomly chosen plants from each plot were selected to evaluate vegetative growth trials as well as some physiological trials within the shoot.

2.4.1. Vegetative Growth Characteristics

Vine length (cm), number of branches and leaves per plant, leaf area (cm^2), leaf dry matter percentage, foliage fresh weight (g/plant).

2.4.2. Photosynthetic Pigment Concentration

Photosynthetic pigments were extracted for 48 h with ice-cold methanol at lab temperature, and subsequently the optical density of extraction was read, and then its concentration was calculated (mg g^{-1} FW) following the equation of Lichtenthaler [31].

2.4.3. Leaf Chemical Composition

The N, P, and K percent were analyzed according to AOAC [32].

2.4.4. Leaf Relative Water Content (LRWC)

The Farouk et al. [33] protocol was used to assess LRWC. Leaves from the middle of each plant were independently collected and the fresh weight was assessed. Following this, leaves were kept for 24 h in closed Petri dishes containing distilled water, and their turgid weight was recorded. To determine the dry weight, the completely turgid leaves were dried in an oven at 80 °C until a consistent weight was reached. Finally, the following equation was used to calculate LRWC (%):

$$\text{LRWC}\ (\%) = \frac{\text{fresh weight} - \text{Dry weight}}{\text{Turgid weight} - \text{Dry weight}} \times 100$$

2.4.5. Electrolyte Leakage (EL)

The EL was deliberate next to the scheme designated by Lutts et al. [34]. Leaves were cut into 1 cm segments and were carefully washed with deionized water (DW) thrice to eradicate any surface contamination. Then, leaf segments were incubated for 24 h at 25 °C on a rotary shaker by keeping the samples in closed and precleared vials containing 20 mL of DW. The electrical conductivity (EC1) of each solution was recorded using a conductivity meter. Then the samples were autoclaved for 20 min at 120 °C, and the electrical conductivity of each solution was measured (EC2) once incubated solutions were cooled down at 25 °C. Electrolyte leakage (%) was deliberated by the subsequent equation:

$$\text{EL}\ (\%) = \frac{\text{EC1}}{\text{EC2}} \times 100$$

2.4.6. Water Use Efficiency (WUE)

The WUE was considered following the equation of Howell [35]:

$$\text{WUE} = \frac{\text{Fruit yield (kg/fed)}}{\text{Crop water consumption } (\text{m}^3/\text{fed/season})}$$

2.4.7. Sex Expression, Fruit Yield, and Their Components

Five plants from each plot were selected for counting the number of male and female flowers for each treatment at two day intervals up to the end of the experiment. Sex ratio was considered as male flowers/female flowers. Additionally, these selected plants were used for estimating fruits' weight and numbers per plant, and total yield (ton/fed).

2.4.8. Fruit Chemical Quality

Fruit quality parameters, such as dry matter percentage, ascorbic acid concentration, and total soluble solids (TSS), were assessed [32]. Fruit dry matter % was estimated by taking a known weight of the fruit and drying at 105 °C; then, the fruit dry matter % was calculating via dividing the dry weight by the fresh weight and indicated as a percent. TSS (°Brix) of each fruit was assessed with a digital refractometer (Model HI96801, Hanna Instruments, Woonsocket, RI, USA). Meanwhile, ascorbic acid (vitamin C) concentration (mg/100 g fruit fresh weight) was estimated using 2,6-dichlorophenol indophenol reagent.

2.5. Statistical Analysis

Utilizing Costat software, a two-way ANOVA was carried out for statistical analysis (CoHortSoftware, 2006; Birmingham, UK). All data were examined using a split-plot methodology, with the replications treated as a random variable in the model and the interactions between GlyBet and irrigation water treatment treated as permanent effects. Data were tested for outstanding normality previous to analysis, and when an ANOVA showed that there were significant treatment effects, means were separated at $p \leq 0.05$ using the LSD pair-wise comparison test.

3. Results

3.1. Vegetative Growth Characters

The application GlyBet concentration had a significant impact on plant growth ($p \leq 0.05$) (Table 2). In both seasons, compared to well-watered cucumber plants, severe drought significantly reduced vine length (21.46 and 21.12%), foliage fresh weight (46.48 and 46.25%), branches number per plant (29.05 and 28.7%), leaves number per plant (26.2 and 25.9%), and leaves area per plants (27.9 and 27.6%) (Table 2).

The plant growth trials were greatly improved by exogenous GlyBet administration. In comparison to untreated plants, 6000 mg/L GlyBet recorded the maximum vine length (42.12 and 42.05%), foliage fresh weight (25.06 and 24.89%), branches per plant (36.05 and 35.95%), leaves per plant (20.16 and 20.22%) and leaves area per plants (12.11 and 14.55%) in both seasons (Table 2).

In comparison to untreated, well-watered plants, the spraying of GlyBet levels in particular 6000 mg/L under mild drought lessened the harmful effects of drought. The application of 6000 mg/L GlyBet resulted in the greatest vine length (38.78 and 38.91%), foliage fresh weight (49.33 and 49.23%), branches per plant (61.66 and 61.47%), leaves per plant (19.96 and 19.86%), and leaves area per plants (14.84 and 14.84%) in the 1st and 2nd years, in comparison to untreated, severely drought-affected plants (Table 2).

3.2. Photosynthetic Pigments

The data in Table 3 demonstrate that within drought stress, the concentration of photosynthetic pigments in cucumber leaves drastically decreased. Well-watered plants had the maximum concentration, which was followed in both seasons by moderate and then severe drought stress.

Additionally, Table 3 also proves that cucumber plants elicited with GlyBet displayed an encouraging impact on photosynthetic pigment accumulation related to control plants. The prime efficient was 6000 mg/L, which boosted Chl a (11.25 and 11.24%), Chl b (11.23 and 11.24%), and carotenoids (11.25 and 11.26%) compared with untreated plants.

The current study found that, in comparison to untreated plants at such drought levels, the application of GlyBet concentrations increased the concentrations of photosynthetic pigment. By boosting Chl a (11.79 and 11.77%), Chl b (11.75 and 11.82%), and carotenoid (11.83 and 11.80%) levels over untreated plants growing in extreme drought, foliar application of 6000 mg/L GlyBet reduces the negative impacts of drought.

3.3. Ion Content

Drought stress significantly reduced the percentages of N, P, and K, with the maximum reductions of N (28.01 and 25.99%), P (27.96 and 17.92%), and K (28.12 and 22.83%) recorded under severe drought linked to well-watered in both years, respectively (Table 4). Spraying GlyBet at all concentrations significantly raises the shoot's N, P, and K% compared to untreated plants (Table 4). In both growing seasons, spraying with 6000 mg/L GlyBet produced the maximum N, P, and K%. Data in Table 4 show spraying with GlyBet in special 6000 mg/L under moderate or severe drought nullified the depression impacts of drought above non-treated plants under such drought levels.

Table 2. Cucumber plant growth as affected by GlyBet concentration, irrigation regimes, and their interactions at 35 days from sowing in both seasons 2020 and 2021.

Treatments		Vine Length (cm)		Foliage FW g/Plant		Branches No/Plant		Leaves No /Plant		Leaves Area (cm^2)/Plant	
		S1	S2	S1	S2	S1	S2	S1	S2	S1	S2
Irrigation regimes (m^3/fed)											
WW		125.3 ± 42.6 ^{a}	108.4 ± 36.7 ^{a}	540 ± 83 ^{a}	467 ± 71 ^{a}	8.33 ± 1.60 ^{a}	7.20 ± 1.40 ^{a}	78.0 ± 15.9 ^{a}	67.5 ± 14.4 ^{a}	5062 ± 623 ^{a}	4379 ± 533 ^{a}
MD		108.9 ± 30.0 ^{b}	94.3 ± 25.0 ^{b}	433 ± 62 ^{b}	375 ± 51 ^{b}	6.91 ± 1.37 ^{b}	5.99 ± 1.16 ^{b}	65.3 ± 8.7 ^{b}	56.5 ± 7.7 ^{b}	4457 ± 530 ^{b}	3861 ± 449 ^{b}
SD		98.4 ± 25.9 ^{c}	85.5 ± 22.2 ^{c}	289 ± 92 ^{c}	251 ± 79 ^{c}	5.91 ± 2.18 ^{c}	5.13 ± 1.89 ^{c}	57.5 ± 9.3 ^{c}	50.0 ± 8.1 ^{c}	3646 ± 499 ^{c}	3167 ± 401 ^{c}
Glycinebetaine (mg/L)											
GlyBet 0		90.9 ± 17.8 ^{d}	78.7 ± 13.6 ^{d}	367 ± 231 ^{c}	318 ± 195 ^{c}	5.88 ± 2.61 ^{d}	5.09 ± 2.19 ^{d}	60.5 ± 17.2 ^{d}	52.4 ± 14.4 ^{c}	4093 ± 1175 ^{c}	3544 ± 968 ^{b}
GlyBet 2000		103.4 ± 18.5 ^{c}	89.6 ± 15.4 ^{c}	415 ± 225 ^{b}	360 ± 196 ^{b}	6.77 ± 2.05 ^{c}	5.87 ± 1.79 ^{c}	65.1 ± 9.8 ^{c}	56.4 ± 8.8 ^{b}	4283 ± 1229 ^{c}	3713 ± 1085 ^{b}
GlyBet 4000		120.2 ± 34.6 ^{b}	104.2 ± 30.6 ^{b}	441 ± 210 ^{a}	383 ± 185 ^{a}	7.55 ± 1.79 ^{b}	6.55 ± 1.64 ^{b}	69.5 ± 22.5 ^{b}	60.3 ± 20.5 ^{a}	4488 ± 1288 ^{b}	3891 ± 1156 ^{a}
GlyBet 6000		129.1 ± 32.5 ^{a}	111.8 ± 25.8 ^{a}	459 ± 215 ^{a}	397 ± 180 ^{a}	8.00 ± 2.11 ^{a}	6.92 ± 1.73 ^{a}	72.7 ± 23.9 ^{a}	63.0 ± 20.1 ^{a}	4689 ± 1346 ^{a}	4060 ± 1109 ^{a}
Interaction											
WW	GlyBet 0	100.0 ± 11.1 $^{d–g}$	86.0 ± 7.5 ^{de}	487 ± 42 ^{bc}	419 ± 29 ^{c}	7.33 ± 0.39 ^{de}	6.30 ± 0.32 ^{cd}	70.3 ± 3.7 ^{b}	60.4 ± 4.7 ^{bc}	4716 ± 330 ^{cd}	4056 ± 288 ^{bc}
	GlyBet 2000	113.0 ± 12.6 $^{c–e}$	97.7 ± 8.4 $^{b–d}$	532 ± 46 ^{ab}	460 ± 32 ^{b}	8.00 ± 0.42 ^{c}	6.92 ± 0.32 ^{b}	71.3 ± 3.7 ^{b}	61.7 ± 4.8 ^{b}	4939 ± 345 ^{bc}	4272 ± 303 ^{ab}
	GlyBet 4000	140.3 ± 15.6 ^{ab}	122.8 ± 10.6 ^{a}	559 ± 48 ^{a}	489 ± 34 ^{ab}	8.66 ± 0.45 ^{b}	7.58 ± 0.39 ^{a}	83.0 ± 4.3 ^{a}	72.6 ± 5.6 ^{a}	5178 ± 362 ^{ab}	4531 ± 322 ^{a}
	GlyBet 6000	148.0 ± 16.5 ^{a}	127.2 ± 11.1 ^{a}	582 ± 50 ^{a}	501 ± 35 ^{a}	9.33 ± 0.49 ^{a}	8.02 ± 0.41 ^{a}	87.6 ± 4.6 ^{a}	75.3 ± 5.8 ^{a}	5415 ± 379 ^{a}	4656 ± 331 ^{a}
MD	GlyBet 0	90.6 ± 10.1 ^{fg}	78.4 ± 6.7 ^{ef}	390 ± 34 ^{de}	337 ± 23 ^{d}	6.00 ± 0.31 ^{g}	5.19 ± 0.27 ^{f}	60.7 ± 3.2 ^{c}	52.4 ± 4.1 ^{de}	4166 ± 291 $^{e–g}$	3604 ± 256 $^{d–f}$
	GlyBet 2000	102.3 ± 11.3 $^{d–f}$	89.5 ± 7.8 $^{c–e}$	436 ± 38 ^{cd}	381 ± 26 ^{c}	6.66 ± 0.34 ^{f}	5.83 ± 0.30 ^{de}	62.3 ± 3.2 ^{c}	54.5 ± 4.2 $^{c–e}$	4354 ± 304 $^{d–f}$	3809 ± 270 $^{c–e}$
	GlyBet 4000	117.2 ± 13.1 ^{cd}	100.8 ± 8.8 ^{bc}	447 ± 38 ^{c}	384 ± 26 ^{c}	7.33 ± 0.39 ^{de}	6.30 ± 0.32 ^{cd}	68.3 ± 3.6 ^{b}	58.7 ± 4.5 $^{b–d}$	4555 ± 318 $^{c–e}$	3917 ± 278 $^{b–d}$
	GlyBet 6000	125.7 ± 14.0 ^{bc}	108.7 ± 9.4 ^{b}	458 ± 39 ^{c}	396 ± 27 ^{c}	7.66 ± 0.40 ^{cd}	6.63 ± 0.34 ^{bc}	70.0 ± 3.7 ^{b}	60.5 ± 4.7 ^{bc}	4754 ± 332 $^{b–d}$	4112 ± 292 ^{bc}
SD	GlyBet 0	82.0 ± 9.1 ^{g}	71.7 ± 6.2 ^{f}	225 ± 19 ^{h}	197 ± 13 ^{g}	4.33 ± 0.23 ^{h}	3.79 ± 0.19 ^{g}	50.6 ± 2.6 ^{d}	44.3 ± 3.4 ^{f}	3396 ± 237 ^{i}	2971 ± 211 ^{h}
	GlyBet 2000	95.0 ± 10.6 $^{e–g}$	81.7 ± 7.1 ^{ef}	277 ± 24 ^{gh}	238 ± 16 ^{f}	5.67 ± 0.29 ^{g}	4.87 ± 0.25 ^{f}	61.6 ± 3.2 ^{c}	53.0 ± 4.1 ^{de}	3557 ± 249 ^{hi}	3059 ± 217 ^{gh}
	GlyBet 4000	103.0 ± 11.5 $^{d–f}$	89.1 ± 7.8 $^{c–e}$	319 ± 27 ^{fg}	276 ± 19 ^{ef}	6.66 ± 0.34 ^{f}	5.76 ± 0.30 ^{e}	57.3 ± 3.0 ^{c}	49.5 ± 3.8 ^{ef}	3730 ± 261 $^{g–i}$	3226 ± 229 $^{f–h}$
	GlyBet 6000	113.8 ± 12.6 ^{cd}	99.6 ± 8.7 ^{bc}	336 ± 29 ^{ef}	294 ± 20 ^{e}	7.00 ± 0.37 ^{ef}	6.12 ± 0.31 ^{de}	60.7 ± 3.2 ^{c}	53.1 ± 4.1 ^{de}	3900 ± 273 $^{f–h}$	3412 ± 242 $^{e–g}$

WW—well-watered (1250 m^3/fed); MD—moderate drought (950 m^3/fed); SD—severe drought (650 m^3/fed); S1—first season; S2—second season; GlyBet—glycinebetaine, FW—fresh weight; fed—feddan. Means values ± standard error within each column for every trial with a similar lower-case letter are not significantly different following Tukey's HSD at $p \leq 0.05$.

Table 3. Photosynthetic pigment concentration of cucumber plant as affected by GlyBet concentration, irrigation regimes and their interactions at 35 days from sowing in both season 2020 and 2021.

Treatments		Chl. a (mg/100 FW)		Chl. b (mg/100 FW)		Carotenoids (mg/100g FW)	
		S1	S2	S1	S2	S1	S2
Irrigation regimes (m^3/fed).							
WW		76.09 ± 9.12 h	65.82 ± 7.28 h	33.04 ± 4.56 h	32.91 ± 3.51 h	21.36 ± 2.22 h	18.47 ± 2.03 h
MD		67.17 ± 6.55 b	58.19 ± 4.91 b	33.59 ± 3.28 b	29.09 ± 2.30 b	18.85 ± 1.46 b	16.33 ± 1.36 b
SD		53.12 ± 6.14 c	46.15 ± 4.76 c	26.56 ± 3.07 c	23.07 ± 2.28 c	14.91 ± 1.48 c	12.95 ± 1.33 c
Glycinebetaine (mg/L)							
GlyBet 0		61.40 ± 19.19 b	53.16 ± 15.64 b	30.70 ± 9.60 b	26.58 ± 7.78 c	17.23 ± 5.28 c	14.92 ± 4.38 b
GlyBet 2000		65.61 ± 21.41 h	56.89 ± 18.64 h	32.81 ± 10.70 h	28.45 ± 9.28 b	18.41 ± 5.90 b	15.97 ± 5.22 h
GlyBet 4000		66.52 ± 20.03 h	57.68 ± 17.65 h	33.25 ± 10.01 h	28.84 ± 8.78 ab	18.67 ± 5.51 ab	16.19 ± 4.95 h
GlyBet 6000		68.31 ± 22.14 h	59.14 ± 18.07 h	34.15 ± 11.07 h	29.57 ± 8.99 h	19.17 ± 6.10 h	16.60 ± 5.07 h
Interaction							
WW	GlyBet 0	70.79 ± 6.17 b–d	60.88 ± 3.79 bc	35.40 ± 3.08 b–d	30.44 ± 1.58 b	19.87 ± 1.05 b	17.09 ± 1.05 bc
	GlyBet 2000	76.31 ± 6.64 a–c	66.01 ± 4.12 ab	38.15 ± 3.33 a–c	33.01 ± 1.71 h	21.42 ± 1.13 h	18.53 ± 1.13 ab
	GlyBet 4000	76.91 ± 6.70 ab	67.29 ± 4.20 h	38.45 ± 3.35 ab	33.64 ± 1.75 h	21.59 ± 1.14 h	18.89 ± 1.16 h
	GlyBet 6000	80.36 ± 7.00 h	69.11 ± 4.31 h	40.18 ± 3.50 h	34.55 ± 1.79 h	22.55 ± 1.18 h	19.39 ± 1.19 h
MD	GlyBet 0	63.74 ± 5.56 de	55.13 ± 3.44 d	31.87 ± 2.77 de	27.56 ± 1.43 c	17.89 ± 0.95 c	15.47 ± 0.95 d
	GlyBet 2000	67.93 ± 5.91 cd	59.44 ± 3.71 cd	33.96 ± 2.96 cd	29.72 ± 1.54 bc	19.07 ± 1.00 bc	16.68 ± 1.02 cd
	GlyBet 4000	67.98 ± 5.91 cd	58.46 ± 3.65 cd	33.99 ± 2.96 cd	29.23 ± 1.51 bc	19.08 ± 1.01 bc	16.41 ± 1.00 cd
	GlyBet 6000	69.05 ± 6.01 b–d	59.72 ± 3.72 cd	34.53 ± 3.00 b–d	29.86 ± 1.54 b	19.38 ± 1.02 b	16.76 ± 1.02 cd
SD	GlyBet 0	49.68 ± 4.33 f	43.47 ± 2.71 e	24.84 ± 2.16 f	21.73 ± 1.13 e	13.94 ± 0.72 e	12.20 ± 0.741 e
	GlyBet 2000	52.61 ± 4.58 f	45.28 ± 2.83 e	26.30 ± 2.28 f	22.62 ± 1.17 de	14.76 ± 0.78 de	12.70 ± 0.77 e
	GlyBet 4000	54.66 ± 4.76 f	47.28 ± 2.95 e	27.33 ± 2.38 f	23.64 ± 1.23 de	15.34 ± 0.80 de	13.27 ± 0.81 e
	GlyBet 6000	55.54 ± 4.84 ef	48.59 ± 3.02 e	27.76 ± 2.42 ef	24.30 ± 1.26 d	15.59 ± 0.82 d	13.64 ± 0.84 e

WW—well-watered (1250 m^3/fed); MD—moderate drought (950 m^3/fed); SD—severe drought (650 m^3/fed); S1—first season; S2—second season; GlyBet—glycinebetaine; Chl. A—chlorophyll a; Chl. B—chlorophyll b; FW—fresh weight; fed—feddan. Means values ± standard error within each column for every trial with a similar lower-case letter are not significantly different following Tukey's HSD at $p \leq 0.05$.

3.4. Leaf Relative Water Content and Electrolyte Leakage

LRWC decreased under drought, and the lowest value (19.16 and 18.79%) was observed once severe drought was present compared to control (Table 4). Application of GlyBet had a favorable impact on the plant's LRWC %. In comparison to untreated control plants in both seasons, more than 12.15 and 12.17% were produced by the application of 6000 mg/L GlyBet (Table 4). The damages of DS on LRWC % were lightened by GlyBet spraying, resulting in an enhancement in LRWC % under moderate or severe drought, as compared with untreated plants grown under drought single (Table 4).

The data in the same table verified that cucumber plant EL % was dramatically enhanced by intensified drought, with the maximum EL % reported under severe drought, which increased by 11.98 and 12.43% in comparison to well-watered plants in the first and second seasons. Additionally, the use of GlyBet resulted in a non-significant reduction in EL %. The data additionally noted that, in general, the GlyBet spraying under irrigation levels increased EL % compared to untreated plants under such irrigation regimes when considering the interaction between irrigation treatment and GlyBet.

3.5. Water Use Efficiency

Data shown in Figure 1 demonstrated that, in comparison to well-watered plants, WUE increased dramatically as DS increased. When there was a severe drought, the WUE was at its highest. The maximum WUE was obtained by applying GlyBet, which enhanced it by 38.52% in the 1st season and by 39.18% in the 2nd season. Moreover, Figure 1 displays that GlyBet usage under moderate or severe drought enhanced WUE compared with untreated well-watered plants. The highest WUE in both seasons was acquired by application of 6000 mg/L GlyBet under moderate or severe drought stress, respectively.

Table 4. Ion percentage, relative water content and electrolyte leakage of cucumber plant as affected by GlyBet concentration, irrigation regimes and their interactions at 35 days from sowing in both 2020 and 2021 seasons.

Treatments		Nitrogen %		Phosphorus %		Potassium %		Relative Water Content %		Electrolyte Leakage %	
		S1	S2	S1	S2	S1	S2	S1	S2	S1	S2
Irrigation regimes (m^3/fed).											
WW		3.07 ± 0.35 ^{a}	2.77 ± 0.38 ^{a}	0.372 ± 0.046 ^{a}	0.318 ± 0.037 ^{a}	3.84 ± 0.44 ^{a}	3.46 ± 0.41 ^{a}	87.21 ± 9.30 ^{a}	75.43 ± 6.99 ^{a}	68.03 ± 7.77 ^{b}	58.85 ± 7.48 ^{b}
MD		2.70 ± 0.29 ^{b}	2.60 ± 0.28 ^{b}	0.327 ± 0.039 ^{b}	0.298 ± 0.023 ^{b}	3.38 ± 0.38 ^{b}	3.25 ± 0.27 ^{b}	79.09 ± 6.36 ^{b}	68.51 ± 3.92 ^{b}	75.00 ± 6.59 ^{a}	64.98 ± 6.71 ^{a}
SD		2.21 ± 0.24 ^{c}	2.05 ± 0.24 ^{c}	0.268 ± 0.033 ^{c}	0.261 ± 0.023 ^{c}	2.76 ± 0.32 ^{c}	2.67 ± 0.32c	70.50 ± 13.16 ^{c}	61.25 ± 11.13 ^{c}	76.18 ± 7.28 ^{a}	66.17 ± 6.29 ^{a}
Glycinebetaine (mg/L)											
GlyBet 0		2.48 ± 0.70 ^{d}	2.31 ± 0.62 ^{b}	0.301 ± 0.086 ^{b}	0.274 ± 0.044 ^{c}	3.10 ± 0.88 ^{d}	2.99 ± 0.68 ^{b}	74.48 ± 17.89 ^{c}	64.50 ± 14.18 ^{c}	74.88 ± 6.55 ^{a}	64.90 ± 6.51 ^{a}
GlyBet 2000		2.60 ± 0.73 ^{c}	2.51 ± 0.69 ^{a}	0.315 ± 0.090 ^{b}	0.297 ± 0.051 ^{ab}	3.25 ± 0.92 ^{c}	3.14 ± 0.84 ^{a}	77.04 ± 19.05 ^{bc}	66.79 ± 16.44 ^{c}	74.57 ± 15.17 ^{a}	64.63 ± 13.32 ^{a}
GlyBet 4000		2.72 ± 0.76 ^{b}	2.58 ± 0.76 ^{a}	0.330 ± 0.094 ^{a}	0.305 ± 0.060 ^{a}	3.40 ± 0.96 ^{b}	3.26 ± 0.83 ^{a}	80.68 ± 11.96 ^{ab}	69.95 ± 10.46 ^{b}	72.02 ± 3.25 ^{b}	62.42 ± 3.82 ^{ab}
GlyBet 6000		2.84 ± 0.80 ^{a}	2.47 ± 0.69 ^{ab}	0.345 ± 0.099 ^{a}	0.293 ± 0.052 ^{b}	3.56 ± 1.01 ^{a}	3.19 ± 0.57 ^{a}	83.53 ± 14.50 ^{a}	72.35 ± 10.96 ^{a}	70.81 ± 10.58 ^{b}	61.39 ± 10.23 ^{b}
Interaction											
WW	GlyBet 0	2.86 ± 0.12 ^{cd}	2.56 ± 0.26 ^{b}	0.3470 ± 0.024 $^{b–d}$	0.294 ± 0.015 ^{cd}	3.58 ± 0.19 ^{cd}	3.20 ± 0.20 ^{cd}	82.53 ± 6.75 $^{a–c}$	70.97 ± 3.70 ^{bc}	71.53 ± 3.11 ^{d}	61.51 ± 3.86 ^{cd}
	GlyBet 2000	2.99 ± 0.12 ^{bc}	2.79 ± 0.29 ^{ab}	0.3633 ± 0.025 ^{bc}	0.321 ± 0.016 ^{ab}	3.75 ± 0.20 ^{bc}	3.50 ± 0.21 ^{ab}	87.08 ± 7.12 ^{ab}	75.33 ± 3.92 ^{ab}	64.97 ± 2.83 ^{e}	56.20 ± 3.53 ^{de}
	GlyBet 4000	3.14 ± 0.14 ^{ab}	2.96 ± 0.30 ^{a}	0.3807 ± 0.026 ^{ab}	0.340 ± 0.017 ^{a}	3.93 ± 0.21 ^{ab}	3.70 ± 0.22 ^{a}	87.18 ± 7.13 ^{ab}	76.29 ± 3.97 ^{a}	71.55 ± 3.11 ^{d}	62.60 ± 3.93 ^{bc}
	GlyBet 6000	3.28 ± 0.14 ^{a}	2.76 ± 0.28 ^{ab}	0.3983 ± 0.028 ^{a}	0.317 ± 0.016 ^{b}	4.11 ± 0.21 ^{a}	3.45 ± 0.21 $^{a–c}$	92.04 ± 7.53 ^{a}	79.15 ± 4.12 ^{a}	64.08 ± 2.79 ^{e}	55.10 ± 3.46 ^{e}
MD	GlyBet 0	2.53 ± 0.11 ^{fg}	2.46 ± 0.25 ^{bc}	0.3063 ± 0.022 $^{e–g}$	0.283 ± 0.014 $^{c–e}$	3.16 ± 0.17 ^{fg}	3.08 ± 0.19 ^{de}	77.44 ± 6.33 ^{c}	66.99 ± 3.49 ^{cd}	78.35 ± 3.41 ^{ab}	67.77 ± 4.26 ^{ab}
	GlyBet 2000	2.64 ± 0.12 ^{ef}	2.65 ± 0.27 ^{ab}	0.3203 ± 0.023 $^{d–f}$	0.305 ± 0.016 ^{bc}	3.30 ± 0.18 ^{ef}	3.33 ± 0.20 $^{b–d}$	77.97 ± 6.38 ^{bc}	68.22 ± 3.54 ^{cd}	77.15 ± 3.36 $^{a–c}$	67.51 ± 4.23 ^{ab}
	GlyBet 4000	2.76 ± 0.11 ^{de}	2.65 ± 0.28 ^{ab}	0.3350 ± 0.023 $^{c–e}$	0.305 ± 0.015 ^{bc}	3.45 ± 0.18 ^{de}	3.32 ± 0.21 $^{b–d}$	79.68 ± 6.52 ^{bc}	68.52 ± 3.57 ^{cd}	71.31 ± 3.10 ^{d}	61.32 ± 3.85 ^{cd}
	GlyBet 6000	2.88 ± 0.13 ^{cd}	2.62 ± 0.27 ^{ab}	0.3497 ± 0.024 $^{b–d}$	0.302 ± 0.015 ^{bc}	3.61 ± 0.19 ^{cd}	3.28 ± 0.20 $^{b–d}$	81.29 ± 6.65 ^{bc}	70.31 ± 3.65 ^{bc}	73.19 ± 3.19 ^{cd}	63.31 ± 3.97 ^{bc}
SD	GlyBet 0	2.05 ± 0.08 ^{j}	1.93 ± 0.20 ^{d}	0.2500 ± 0.017 ^{i}	0.246 ± 0.013 ^{g}	2.58 ± 0.14 ^{i}	2.48 ± 0.16 ^{g}	63.47 ± 5.19 ^{e}	55.54 ± 2.89 ^{e}	74.77 ± 3.25 $^{b–d}$	65.42 ± 4.10 $^{a–c}$
	GlyBet 2000	2.15 ± 0.08 ^{ij}	2.08 ± 0.22 ^{cd}	0.2617 ± 0.018 ^{hi}	0.266 ± 0.014 $^{e–g}$	2.70 ± 0.14 ^{hi}	2.60 ± 0.16 ^{fg}	66.08 ± 5.40 ^{de}	56.83 ± 2.95 ^{e}	81.60 ± 3.55 ^{a}	70.17 ± 4.41 ^{a}
	GlyBet 4000	2.26 ± 0.09 ^{hi}	2.13 ± 0.22 ^{cd}	0.2743 ± 0.020 $^{g–i}$	0.272 ± 0.014 $^{d–f}$	2.83 ± 0.15 ^{hi}	2.76 ± 0.17 ^{fg}	75.18 ± 6.15 ^{cd}	65.03 ± 3.39 ^{d}	73.21 ± 3.19 ^{cd}	63.33 ± 3.98 ^{bc}
	GlyBet 6000	2.36 ± 0.09 ^{gh}	2.04 ± 0.21 ^{d}	0.2867 ± 0.020 $^{f–h}$	0.260 ± 0.013 ^{fg}	2.96 ± 0.15 ^{gh}	2.84 ± 0.17 ^{ef}	77.25 ± 6.32 ^{c}	67.60 ± 3.52 ^{cd}	75.15 ± 3.27 $^{b–d}$	65.76 ± 4.12 $^{a–c}$

WW—well-watered (1250 m^3/fed); MD—moderate drought (950 m^3/fed); SD—severe drought (650 m^3/fed); S1—first season; S2—second season; GlyBet—glycinebetaine; fed—feddan. Means values ± standard error within each column for every trial with a similar lower-case letter are not significantly different following Tukey's HSD at $p \leq 0.05$.

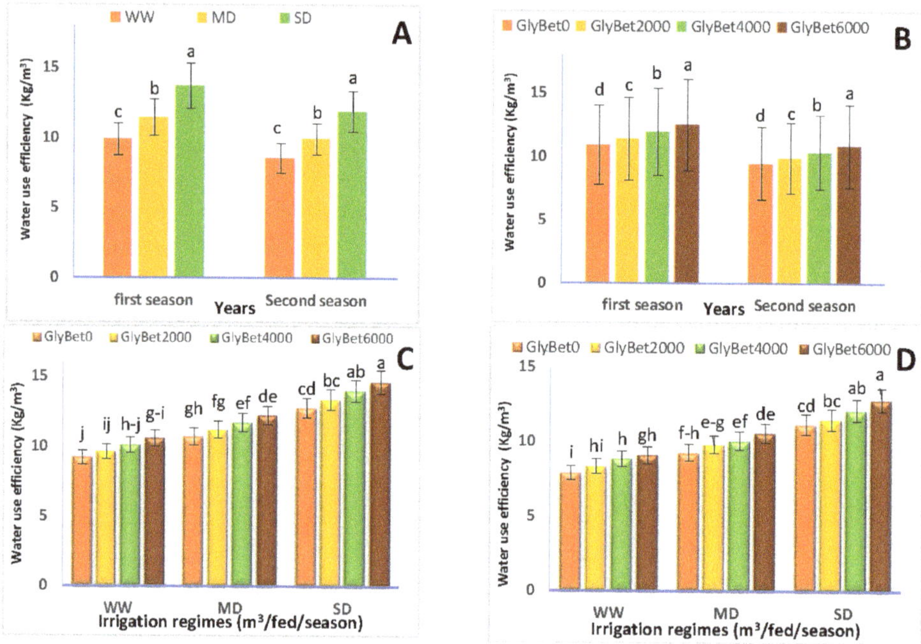

Figure 1. Water use efficiency of cucumber plant as affected by irrigation regimes (**A**), glycinebetaine concentration (**B**), and their interactions (**C**,**D**) in both seasons—2020 and 2021. WW—well-watered (1250 m^3/fed); MD—moderate drought (950 m^3/fed); SD—severe drought (650 m^3/fed); fed—feddan. Values followed by the same lower-case letter are not significantly different following Tukey's HSD at $p \leq 0.05$.

3.6. Sex Expression, Fruit Yield, and Their Components and Quality Parameters

The data in Table 5 clearly show that DS caused a dramatic reduction in yield and its component. In comparison to well-watered plants, there was a substantial decrease in fruit number per plant (55.87 and 55.31%), fruit weight/plant (28.01 and 27.70%), and total yield (27.99 and 27.68%) in the 1st and 2nd seasons, correspondingly. GlyBet spraying significantly improved all yield and its components over untreated plants. The utmost effective treatment was 6000 mg/L GlyBet, which boosted fruit number/plant (53.69 and 47.84%), fruit weight/plant (14.49 and 14.56%), and total yield (14.50 and 14.68%) in both seasons, correspondingly, compared with untreated plants (Table 5). GlyBet spraying moderates the drastic injuries of DS on cucumber crop yield. Since the supplementation of 6000 mg/L GlyBet during an extreme drought, all yield parameters in the first and second seasons have been decreased in relation to untreated drought-affected plants.

The data on the sex ratio indicates that it rose with drought and fell with the use of GlyBet application. The findings also showed that the use of GlyBet under moderate or severe drought lessens the negative effects of drought on sex ratio (Table 5).

Data in Table 6 show that drought levels increased fruit dry matter % and decreased both ascorbic acid and TSS of fruits relative to control plants. The maximum fruit dry matter was noted within severe drought. On the other hand, the greatest concentration of vitamin C and TSS was obtained under normal conditions in both seasons. Data indicated in Table 6 show that applying GlyBet considerably affected the previous parameters. The highest values were obtained while adding GlyBet at a rate of 6000 mg/L in both seasons, relative to other concentrations or untreated plants. Data presented in Table 6 indicate the interaction effects between irrigation regimes and GlyBet rates. Results show that application of GlyBet concentration under irrigation regimes significantly increased fruit dry matter %, vitamin

C, and TSS as compared with untreated plants under each irrigation regime. The highest percentages of fruit dry matter were evidently achieved with severe drought in combination with 6000 mg/L GlyBet. meanwhile, the greatest concentration of vitamin C and TSS was recorded under normal conditions and sprayed with 6000 mg/L GlyBet.

Table 5. Sex expression and yield components of cucumber plant as affected by GlyBet concentration, irrigation regimes, and their interactions in both seasons 2020 and 2021.

Treatments		Sex Ratio		Fruit Weight (g)/Plant		Fruit No/Plant		Total Yield (ton/fed)	
		S1	S2	S1	S2	S1	S2	S1	S2
Irrigation regimes (m^3/fed).									
WW		5.42 ± 4.48 ^{c}	5.50 ± 4.50 ^{c}	589 ± 69 ^{h}	509 ± 62 ^{h}	12.08 ± 0.94 ^{h}	10.25 ± 0.84 ^{h}	12.36 ± 1.47 ^{h}	10.69 ± 1.27 ^{h}
MD		10.85 ± 5.72 ^{b}	11.02 ± 5.75 ^{b}	518 ± 58 ^{b}	449 ± 50 ^{b}	7.08 ± 0.62 ^{b}	6.00 ± 0.25 ^{b}	10.89 ± 1.25 ^{b}	9.43 ± 1.02 ^{b}
SD		16.97 ± 4.51 ^{h}	17.29 ± 4.56 ^{h}	424 ± 49 ^{c}	368 ± 45 ^{c}	5.33 ± 0.60 ^{c}	4.58 ± 0.33 ^{c}	8.90 ± 1.06 ^{c}	7.73 ± 0.92 ^{c}
Glycinebetaine (mg/L)									
GlyBet 0		14.58 ± 9.60 ^{h}	14.85 ± 9.91 ^{h}	476 ± 135 ^{d}	412 ± 111 ^{d}	6.22 ± 0.66 ^{c}	5.33 ± 0.24 ^{c}	10.00 ± 2.85 ^{d}	8.65 ± 2.33 ^{d}
GlyBet 2000		10.94 ± 11.27 ^{b}	11.10 ± 11.32 ^{b}	498 ± 141 ^{c}	432 ± 125 ^{c}	8.11 ± 0.52 ^{b}	7.00 ± 0.26 ^{b}	10.46 ± 2.98 ^{c}	9.07 ± 2.62 ^{c}
GlyBet 4000		9.81 ± 9.44 ^{c}	9.95 ± 9.52 ^{c}	522 ± 148 ^{b}	452 ± 133 ^{b}	8.78 ± 0.48 ^{ab}	7.55 ± 0.31 ^{ab}	10.96 ± 3.12 ^{b}	9.50 ± 2.79 ^{b}
GlyBet 6000		8.99 ± 10.49 ^{d}	9.17 ± 10.79 ^{d}	545 ± 155 ^{h}	472 ± 128 ^{h}	9.56 ± 0.81 ^{h}	7.88 ± 0.33 ^{a}	11.45 ± 3.27 ^{h}	9.92 ± 2.67 ^{h}
Interaction									
WW	GlyBet 0	8.77 ± 0.66 ^{e}	8.860 ± 0.46 ^{e}	548 ± 29 ^{cd}	472 ± 29 ^{cd}	8.00 ± 0.88 ^{d}	6.660 ± 0.70 ^{de}	11.52 ± 0.70 ^{cd}	9.910 ± 0.51 $^{c–e}$
	GlyBet 2000	5.77 ± 0.44 ^{f}	5.860 ± 0.31 ^{f}	574 ± 30 ^{bc}	497 ± 31 ^{bc}	10.33 ± 0.75 ^{c}	8.670 ± 0.67 ^{c}	12.06 ± 0.73 ^{bc}	10.43 ± 0.54 ^{bc}
	GlyBet 4000	3.89 ± 0.29 ^{g}	3.990 ± 0.21 ^{g}	602 ± 32 ^{ab}	527 ± 33 ^{ab}	14.00 ± 0.83 ^{b}	12.00 ± 0.58 ^{b}	12.65 ± 0.76 ^{ab}	11.07 ± 0.57 ^{ab}
	GlyBet 6000	3.27 ± 0.25 ^{g}	3.300 ± 0.16 ^{g}	630 ± 33 ^{h}	541 ± 34 ^{h}	16.00 ± 1.02 ^{h}	13.67 ± 0.49 ^{h}	13.23 ± 0.80 ^{h}	11.37 ± 0.59 ^{h}
MD	GlyBet 0	15.25 ± 1.15 ^{c}	15.48 ± 0.80 ^{c}	485 ± 25 ^{fg}	419 ± 26 ^{ef}	5.660 ± 0.24 ^{fg}	5.000 ± 0.28 ^{fg}	10.18 ± 0.61 ^{ef}	8.800 ± 0.45 ^{fg}
	GlyBet 2000	8.820 ± 0.67 ^{e}	9.030 ± 0.46 ^{e}	506 ± 27 ^{ef}	443 ± 28 $^{d e}$	8.660 ± 0.33 ^{cd}	7.660 ± 0.61 ^{d}	10.63 ± 0.64 ^{de}	9.300 ± 0.48 ^{ef}
	GlyBet 4000	10.93 ± 0.82 ^{d}	11.04 ± 0.57 ^{d}	530 ± 28 ^{de}	455 ± 29 $^{c–e}$	6.330 ± 0.41 ^{f}	5.330 ± 0.50 ^{f}	11.13 ± 0.67 $^{c–e}$	9.570 ± 0.49 ^{de}
	GlyBet 6000	8.390 ± 0.63 ^{e}	8.520 ± 0.44 ^{e}	553 ± 28 ^{cd}	478 ± 30 ^{cd}	7.670 ± 0.66 ^{e}	6.000 ± 0.37 ^{e}	11.61 ± 0.70 ^{c}	10.04 ± 0.52 ^{cd}
SD	GlyBet 0	19.73 ± 1.49 ^{h}	20.23 ± 1.06 ^{h}	395 ± 21 ^{i}	346 ± 22 ^{h}	5.000 ± 0.15 ^{i}	4.330 ± 0.29 ^{h}	8.290 ± 0.50 ^{h}	7.260 ± 0.37 ^{i}
	GlyBet 2000	18.23 ± 1.37 ^{b}	18.41 ± 0.96 ^{b}	413 ± 22 ^{hi}	356 ± 22 ^{gh}	5.330 ± 0.25 ^{h}	4.660 ± 0.31 ^{g}	8.690 ± 0.52 ^{gh}	7.470 ± 0.38 ^{i}
	GlyBet 4000	14.60 ± 1.09 ^{c}	14.82 ± 0.76 ^{c}	433 ± 23 ^{hi}	375 ± 23 ^{gh}	6.000 ± 0.22 ^{g}	5.330 ± 0.62 ^{f}	9.110 ± 0.55 ^{gh}	7.880 ± 0.41 ^{hi}
	GlyBet 6000	15.32 ± 1.15 ^{c}	15.70 ± 0.82 ^{c}	454 ± 24 ^{gh}	397 ± 25 ^{fg}	5.000 ± 0.40 ^{i}	4.000 ± 0.33 ^{i}	9.530 ± 0.57 ^{fg}	8.330 ± 43 ^{gh}

WW—well-watered (1250 m^3/fed); MD—moderate drought (950 m^3/fed); SD—severe drought (650 m^3/fed); S1—first season; S2—second season; GlyBet—glycinebetaine; fed—feddan. Means values ± standard error within each column for every trial with a similar lower-case letter are not significantly different following Tukey's HSD at $p \leq 0.05$.

Table 6. Some fruit quality trials of cucumber as affected by GlyBet concentration, irrigation regimes, and their interactions in both seasons—2020 and 2021.

Treatments		Fruit Dry Matter (%)		Vit. C (mg/100 g Fresh Weight)		TSS (°Brix)	
		S1	S2	S1	S2	S1	S2
Irrigation regimes (m^3/fed).							
WW		3.07 ± 0.46 ^{b}	2.56 ± 0.40 ^{c}	29.31 ± 7.36 ^{h}	25.35 ± 6.32 ^{h}	3.03 ± 0.33 ^{h}	2.66 ± 0.31 ^{h}
MD		3.27 ± 0.55 ^{h}	2.83 ± 0.47 ^{b}	27.67 ± 6.38 ^{b}	23.96 ± 5.35 ^{b}	2.85 ± 0.24 ^{b}	2.47 ± 0.21 ^{b}
SD		3.38 ± 0.54 ^{h}	2.93 ± 0.47 ^{h}	24.51 ± 6.80 ^{c}	21.30 ± 5.90 ^{c}	2.49 ± 0.26 ^{c}	2.16 ± 0.21 ^{c}
Glycinebetaine (mg/L)							
GlyBet 0		2.91 ± 0.23 ^{d}	2.52 ± 0.23 ^{d}	22.76 ± 4.17 ^{d}	19.71 ± 3.24 ^{d}	2.62 ± 0.47 ^{b}	2.27 ± 0.36 ^{b}
GlyBet 2000		3.16 ± 0.38 ^{c}	2.74 ± 0.33 ^{c}	25.91 ± 4.80 ^{c}	22.46 ± 4.23 ^{c}	2.80 ± 0.54 ^{h}	2.43 ± 0.52 ^{h}
GlyBet 4000		3.35 ± 0.32 ^{b}	2.90 ± 0.25 ^{b}	28.66 ± 4.27 ^{b}	24.85 ± 3.77 ^{b}	2.84 ± 0.48 ^{h}	2.45 ± 0.45 ^{h}
GlyBet 6000		3.54 ± 0.36 ^{h}	3.06 ± 0.35 ^{h}	31.31 ± 4.97 ^{h}	27.12 ± 3.77 ^{h}	2.91 ± 0.55 ^{h}	2.53 ± 0.45 ^{h}
Interaction							
WW	GlyBet 0	2.81 ± 0.17 ^{g}	2.42 ± 0.15 ^{f}	24.57 ± 1.73 $^{e–g}$	21.13 ± 1.09 ^{ef}	2.83 ± 0.19 $^{b–e}$	2.43 ± 0.23 $^{b–e}$
	GlyBet 2000	2.94 ± 0.18 $^{e–g}$	2.54 ± 0.15 ^{ef}	28.08 ± 1.98 ^{cd}	24.29 ± 1.25 ^{cd}	3.03 ± 0.21 ^{ab}	2.63 ± 0.23 $^{a–c}$
	GlyBet 4000	3.19 ± 0.19 $^{d–h}$	2.79 ± 0.17 $^{c–e}$	30.69 ± 2.16 ^{bc}	26.85 ± 1.39 ^{b}	3.07 ± 0.21 ^{ab}	2.70 ± 0.20 ^{ab}
	GlyBet 6000	3.34 ± 0.20 $^{b–d}$	2.87 ± 0.18 $^{b–d}$	33.89 ± 2.39 ^{h}	29.14 ± 1.51 ^{h}	3.20 ± 0.22 ^{h}	2.76 ± 0.11 ^{h}
MD	GlyBet 0	2.90 ± 0.18 ^{fg}	2.51 ± 0.15 ^{f}	23.50 ± 1.66 ^{fg}	20.33 ± 1.06 ^{f}	2.70 ± 0.18 $^{c–f}$	2.36 ± 0.11 $^{c–f}$
	GlyBet 2000	3.22 ± 0.20 $^{c–e}$	2.82 ± 0.17 ^{cd}	26.65 ± 1.88 $^{d e}$	23.32 ± 1.21 ^{cd}	2.86 ± 0.20 $^{b–d}$	2.53 ± 0.23 $^{a–d}$
	GlyBet 4000	3.38 ± 0.21 $^{b–d}$	2.91 ± 0.18 ^{bc}	29.07 ± 2.05 $^{b–d}$	25.00 ± 1.30 ^{c}	2.90 ± 0.20 $^{b–d}$	2.46 ± 0.11 $^{b–e}$
	GlyBet 6000	3.59 ± 0.22 ^{ab}	3.10 ± 0.19 ^{ab}	31.46 ± 2.22 ^{ab}	27.21 ± 1.41 ^{b}	2.93 ± 0.20 $^{a–c}$	2.53 ± 0.23 $^{a–d}$
SD	GlyBet 0	3.01 ± 0.18 $^{e–g}$	2.64 ± 0.16 $^{d–f}$	20.21 ± 1.42 ^{h}	17.69 ± 0.91 ^{g}	2.33 ± 0.16 ^{g}	2.06 ± 0.11 ^{g}
	GlyBet 2000	3.31 ± 0.20 $^{b–d}$	2.85 ± 0.17 $^{b–d}$	23.00 ± 1.62 ^{gh}	19.78 ± 1.02 ^{f}	2.46 ± 0.17 ^{fg}	2.13 ± 0.20 ^{fg}
	GlyBet 4000	3.49 ± 0.21 $^{a–c}$	3.02 ± 0.18 $^{a–c}$	26.24 ± 1.85 $^{d–f}$	22.70 ± 1.17 $^{d e}$	2.56 ± 0.17 $^{e–g}$	2.20 ± 0.20 $^{e–g}$
	GlyBet 6000	3.68 ± 0.23 ^{h}	3.22 ± 0.20 ^{h}	28.60 ± 2.02 ^{cd}	25.02 ± 1.30 ^{c}	2.60 ± 0.18 $^{d–g}$	2.30 ± 0.11 $^{c–f}$

WW—well-watered (1250 m^3/fed); MD—moderate drought (950 m^3/fed); SD—severe drought (650 m^3/fed); S1—first season; S2—second season; GlyBet—glycinebetaine; Vit. C—vitamin C or ascorbic acid; TSS—total soluble solid; fed—feddan. Means values ± standard error within each column for every trial with a similar lower-case letter are not significantly different following Tukey's HSD at $p \leq 0.05$.

4. Discussion

Over 30–70% of crop yields are lost due to drought, one of the major obstacles that climate change has posed to crop production [10,11,15]. Therefore, it is essential to increase plants' resistance to drought in order to protect them from such yield losses and sustain productivity for food security [16,17]. Applying exogenous materials, especially those that are already compatible with plants, i.e., GlyBet, is one of the low-cost and eco-friendly, innovative water-saving techniques for improving plants' drought tolerance.

Drought commonly causes significant injury to plants, as the current study and earlier findings have shown [13–15]. In general, the occurrence of DS induces numerous physio–biochemical, morphological, and molecular alterations [13,14], such as the contraction of vascular tissues, decreased water absorption [36], and photoassimilate translocation [37]. Additionally, DS inhibited ion absorption, induced the accumulation of ROS [13,38]; and impaired ATP biosynthesis, which accelerated oxidative injury and subsequently reduced plant development [39]. Additionally, according to González-Villagra et al. [40], DS disrupts the production of endogenous phytohormones by increasing ABA concentration, lowering IAA and GAs, and rapidly decreasing zeatin concentration. The hormonal imbalance slowed down the growth of plant cells by reducing their turgor, elongation, and volume, leading to a decline in growth attributes [36]. In comparison to well-watered or unsprayed drought-affected plants, the current study has shown that GlyBet supplementation exhibits exceptional impacts on plant growth under normal or stress settings [13,18,25]. Tisarum et al. [41] also noted that by enhancing growth vigor, exogenous GlyBet treatment could mitigate the negative impacts of drought. Several potential strategies coupled with stress moderation by GlyBet have been accepted: (1) preserving water status, as demonstrated by an increase in LRWC in the recent study [18,25]; (2) an acceleration of growth promoters (IAA, GAs, salicylic acid, and cytokinin), and a reduction in ABA [42,43]; (3) increasing cell division and enlargement due to activation of water absorption and rising P concentration [44]. As for the interactions, there are a few studies confirming the current outcomes [13,18] that indicate application of GlyBet under drought mitigates the harmful effects of drought on plant growth.

Our findings demonstrated that the DS caused a significant decrease in photosynthetic pigments, which was reversed by the addition of GlyBet (Table 3). Drought-induced chlorophyll loss in several crops is a frequently seen occurrence [13,14,18]. The reduction in photosynthetic pigments within water deficit may be attributable to: (1) inhibition of the assimilation of the chlorophyll pigment complexes encoded by the cab gene family [28]; (2) destruction of chiral macro-aggregates of the light-harvesting pigment–protein complexes that offer defense to chloroplasts [45] as well as the formation of chlorophyllase [46]; and/or (3) the defeat of chloroplast membranes, exciting enlargement, modification of the lamellae vasculation and the presence of plastoglobules [47,48]. Previous studies have supported GlyBet's dramatic increase in photosynthetic pigment [13,18,25,41]. These rises could be attributable to well-organized ROS illumination mechanisms, as antioxidant enzymes and solutes would otherwise have destroyed the chlorophyll [13,18,25]. Additionally, carotenoids have the capacity to accumulate as a light receptor and photosystem shield against ROS [49]. As a result, the application of GlyBet [13,25] hastens the over-abundance of carotene in photosynthetic tissues. Finally, GlyBet protects the chloroplasts, with RUBISCO stabilizing membrane structure under drought [27]. Although the physiological effect of GlyBet is not clear, recent research showed that GlyBet acts a crucial function similar to cytokinin in enhancing the chlorophyll accumulation [50], increasing the number of chloroplasts [27].

Numerous crops have previously been shown to exhibit a decrease in ion percentage within DS [51,52]. At the soil-root interface, factors such as root form and growth rate, ion absorption kinetics, and soil nutrient supply dominate ion absorption [53]. This loss in N % is able to be connected to a decline in nitrate reductase activity that is interrelated with photosynthetic activity and decreased availability of carbon skeletons within DS [54]. The decline in K % under DS may be elucidated via the statement that a lack of water disturbs stomatal control, which reduces photosynthetic capability, and likewise the uptake of K for sustaining and regulating turgidity and stomatal control as recorded by Sarani et al. [55]. The role of GlyBet in increasing N, P, and K % is not totally implicit and there are plentiful comparable investigators who have recognized the present research. Estaji et al. [56] and Khoshkharam et al. [57] postulated that GlyBet supplementation increases ion concentrations in plant tissue. This encouraging impact might be ascribed to enhanced ion uptake, preserving membrane permeability (Table 4), and/or possibly providing a better-developed root system [27].

Drought cessations affect plant—water balance firstly, hence disturbing the plant's typical physio-biochemical occupations. LRWC % was first presented as a useful criterion for plant water status under water deficiency in the middle of the 1980s. The injuries of DS on LRWC % were alleviated by exogenous application of GlyBet, leading to an improvement in LRWC % under DS, as compared with untreated plants grown under drought only (Table 4). Previously, it was discovered that LRWC % decreased in this area during a drought [14,18,19,58]. Currently, the application of GlyBet mitigated the reducing trend of LRWC % of cucumber under DS, which was approved by Dustgeer et al. [25], Shemi et al. [18], and Yang et al. [26]. Genard et al. [59] stated that GlyBet not only preserves plant water in an arid site that may be owing to its solid hydrophilicity and solubility, but also plays a role in osmotic defense of plant tissues. Alasvandyari et al. [60] recorded that GlyBet can support plants' ability to maintain their leaves' water content by encouraging sodium elimination and K^+ accretion under drought conditions. Moreover, GlyBet spraying motivated the development of the root system and reinforces the capability of water absorption in addition to upregulation of aquaporin genes, so as to boost water preservation and enhanced WUE [26,27].

Drought causes an overabundance of ROS, which speeds up membrane lipid peroxidation and hence raises membrane permeability [14,18,19]. In this research, DS distinctly boosted EL % in cucumber leaves, which was in line with the findings of Nawaz and Wang [61] and Nazar et al. [62]. Such mutilation can be caused by oxidation and cross-linkage of protein thiols, and inhibition of key membrane proteins such as H^+-ATPase [63]. However, GlyBet application lightened the adverse effects of DS by decreasing EL % under stress or under well-watered conditions [18,25]. Current outcomes approved that GlyBet spraying could decreasing EL % by adjusting ROS homeostasis and lowering lipid peroxidation to defend cell micro-organelles from the negative injuries of drought [26]. This suggests that the use of GlyBet could maintain the stabilities of membranes in wheat plants under DS.

Typically, plant biomass significantly decreases in response to DS [48]. As a result, the total water losses from transpiration were cut in half, which significantly increased WUE [64]. The goal is to increase plant WUE in drought conditions, which can be accomplished in two ways: by improving the plant's ability to adapt and by increasing the crop's ability to produce biomass per unit of water. However, the impact of a drought on a plant's WUE often depends on the plant's cultivar and drought severity [65]. A foliar spray of GlyBet, on the other hand, increases cucumber WUE by influencing photosynthesis, enhancing root development, which leads to better water absorption, and increasing the plant's resistance to water scarcity. In some plants, such as wheat [66], increased WUE brought on by GlyBet treatment under well-watered or water-deficient conditions has already been documented.

When compared to well-watered plants, crop yield is consistently reduced by up to 70% under DS [13–15,18,41]. This decrease could be instigated by decreasing branch number and leaf size, which would decline biomass production, hinder the movement of photoassimilate to the developing fruits, and/or cause flower and fruit abortion [14]. Additionally, Song et al. [67] provided that DS caused pollen to swell, filament growth to reduce filament fertility, and grain production to decrease. According to Anjum et al. [68], drought stress decreased agricultural output by reducing photosynthetic pigments and Calvin cycle enzyme activity [18,69]. The drought-induced decline in the yield might have resulted from a diminished photosynthetic rate [14,18] and disturbed assimilate partitioning [70]. GlyBet's effects on stimulating metabolic processes and morphological modification and anatomical changes may be the cause of the increase in cucumber yield [18,27,41]. These findings agreed with the values provided by other authors [13,18,57]. The preservation of a greater net photosynthetic rate and an improvement in the source–sink relationship were both correlated with the yield enhancement by GlyBet [18,27]. GlyBet stimulates plant growth and yield owing to its osmoprotective influence on photosynthetic machinery and control of ion homeostasis [11] along with enhancing drought-affected plant CO_2 assimilation [18,57], and because of its role in biosynthesis and transport of hormones such as cytokinins that may have a role in the transport of photoassimilates [49]. Adak [71] found that TSS and vitamin C decreased by drought meanwhile increasing with GlyBet application. As for the interaction effects, several research [13,18,29,57] confirmed our results, which proved that application of GlyBet alleviated the drastic effect of drought on crop yield

5. Conclusions

Our results unequivocally show that applying GlyBet is a successful strategy for reducing drought injury and enhancing plant performance within water scarcity. Overall, spraying drought-affected cucumber plants six times at 10, 20, 30, 40, 50, 60 days after planting with 6000 mg/L GlyBet may be a potential method for reducing the effects of water deficit and therefore improving water use efficiency as well as crop yield and quality. This has a significant impact on both regional and national economic development, as well as water conservation in dry and semi-dry regions in the context of climate change adaptation efforts.

Author Contributions: Conceptualization, E.-S.E.M. and S.F.; methodology, E.-S.E.M., S.F., E.F.A., H.M.A.-Y. and H.A.F.; software, S.F. and H.A.F.; validation, E.-S.E.M. and S.F.; formal analysis, E.-S.E.M. and H.A.F.; investigation, E.-S.E.M. and S.F.; resources, E.-S.E.M., H.A.F. and S.F.; data curation, E.-S.E.M., H.A.F. and S.F.; writing—original draft preparation, E.-S.E.M., E.F.A., H.M.A.-Y. and H.A.F.; writing—review and editing, S.F.; visualization, E.-S.E.M. and S.F.; supervision, E.-S.E.M., S.F. and H.M.A.-Y.; project administration, E.-S.E.M., S.F., H.M.A.-Y. and E.F.A.; funding acquisition, H.M.A.-Y., E.-S.E.M. and S.F. All authors have read and agreed to the published version of the manuscript.

Funding: This research was funded by Taif University Researchers Supporting Project, grant number TURSP-2020/199.

Data Availability Statement: Not applicable.

Acknowledgments: The authors express their gratitude to Mansoura University, Egypt, for supporting our manuscript. The authors are also thankful to Taif University Researchers Supporting Project number (TURSP-2020/199), Taif University, Saudi Arabia, for the financial support and research facilities.

Conflicts of Interest: The authors declare no conflict of interest.

References

1. Jamir, M.; Sharma, A. A Sustainable production and marketing of cucumber crop in the hilly zone of Nagaland. *Technofame A J. Multidiscip. Adv. Res.* **2014**, *3*, 61–66.
2. Sebastian, P.; Schaefer, H.; Telford, I.R.H.; Renner, S.S. Cucumber (*Cucumis sativus*) and melon (*C. melo*) have numerous wild relatives in Asia and Australia, and the sister species of melon is from Australia. *Proc. Natl. Acad. Sci. USA* **2010**, *107*, 14269–14273. [CrossRef]
3. Lv, J.; Qi, J.J.; Shi, Q.X.; Shen, D.; Zhang, S.P.; Shao, G.J. Genetic diversity and population structure of cucumber (*Cucumis sativus* L.). *PLoS ONE* **2012**, *7*, e46919. [CrossRef] [PubMed]
4. Vora, J.D. Biochemical, Anti-microbial and organoleptic studies of cucumber (*Cucumis sativus*). *Int. J. Sci. Res.* **2014**, *3*, 662–664.
5. Muruganantham, N.; Solomon, S.; Senthamilselvi, M.M. Anti-cancer activity of *Cucumis sativus* (cucumber) flowers against human liver cancer. *Int. J. Clin. Pharmacol. Res.* **2016**, *8*, 39–41.
6. Saeed, H.; Waheed, A. A Review on cucumber (Cucumis sativus). *Int. J. Tech. Res. Sci.* **2017**, *2*, 402–405.

7. Oboh, G.; Ademiluyi, A.O.; Ogunsuyi, O.B.; Oyeleye, S.I.; Dada, A.F.; Boligon, A.A. Cabbage and cucumber extracts exhibited anticholinesterase, antimonoamine oxidase and antioxidant properties. *J. Food Biochem.* **2017**, *41*, e12358. [CrossRef]
8. Ibitoye, O.B.; Uwazie, J.N.; Ajiboye, T. Bioactivity-guided isolation of kaempferol as the antidiabetic principle from *Cucumis sativus* L. fruits. *J. Food Biochem.* **2018**, *42*, e12479. [CrossRef]
9. Fiume, M.M.; Bergfeld, W.F.; Belsito, D.V.; Hill, R.A.; Klaassen, C.D.; Liebler, D.C. Safety assessment of *Cucumis sativus* (cucumber)-derived ingredients as used in cosmetics. *Int. J. Toxicol.* **2014**, *33*, 47–64. [CrossRef]
10. Lesk, C.; Rowhani, P.; Ramankutty, N. Influence of extreme weather disasters on global crop production. *Nature* **2016**, *529*, 84–87. [CrossRef]
11. Raza, A.; Razzaq, A.; Mehmood, S.S.; Zou, X.; Zhang, X.; Lv, Y.; Xu, J. Impact of climate change on crops adaptation and strategies to tackle its outcome: A review. *Plants* **2019**, *8*, 34. [CrossRef]
12. Gupta, A.; Rico-Medina, A.; Caño-Delgado, A.I. The physiology of plant responses to drought. *Science* **2020**, *368*, 266–269. [CrossRef]
13. Shafiq, S.; Akram, N.A.; Ashraf, M.; García-Caparrós, P.; Ali, O.M.; Latef, A.A.H.A. Influence of glycine betaine (natural and synthetic) on growth, metabolism and yield production of drought_stressed maize (*Zea mays* L.) plants. *Plants* **2021**, *10*, 2540. [CrossRef]
14. Alam, A.; Ullaha, H.; Thuenproma, N.; Tisarumc, R.; Cha-umc, S.; Datta, A. Seed priming with salicylic acid enhances growth, physiological traits, fruit yield, and quality parameters of cantaloupe under water-deficit stress. *S. Afr. J. Bot.* **2022**, *150*, 1–12. [CrossRef]
15. Eid, M.A.M.; El-hady, M.A.A.; Abdelkader, M.A.; Abd-Elkrem, Y.M.; El-Gabry, Y.A.; El-temsah, M.E.; El-Areed, S.R.M.; Rady, M.M.; Alamer, K.H.; Alqubaie, A.I. Response in physiological traits and antioxidant capacity of two cotton cultivars under water limitations. *Agronomy* **2022**, *12*, 803. [CrossRef]
16. Farouk, S.; Al-Huqail, A.A. Sodium nitroprusside application regulates antioxidant capacity, improves phytopharmaceutical production and essential oil yield of marjoram herb under drought. *Ind. Crops Prod.* **2020**, *158*, 113034. [CrossRef]
17. Farouk, S.; Al-Ghamdi, A.A.M. Sodium nitroprusside application enhances drought tolerance in marjoram herb by promoting chlorophyll biosynthesis and enhancing osmotic adjustment capacity. *Arab. J. Geosci.* **2021**, *14*, 1–13. [CrossRef]
18. Shemi, R.; Wang, L.; Gheith, E.M.; Hussain, H.A.; Hussain, S.; Irfan, M.; Cholidah, L.; Zhang, K.; Zhang, S.; Wang, L. Effects of salicylic acid, zinc and glycine betaine on morphophysiological growth and yield of maize under drought stress. *Sci. Rep.* **2021**, *11*, 3195. [CrossRef] [PubMed]
19. Yadav, P.K.; Singh, A.K.; Tripathi, M.K.; Tiwari, S.; Rathore, J. Morpho_physiological characterization of maize (*Zea mays* L.) genotypes against drought. *Biol. Forum—Int. J.* **2022**, *14*, 573–581.
20. Jensen, C.R.; Ørum, J.E.; Pedersen, S.M.; Andersen, M.N.; Plauborg, F.; Liu, F.; Jacobsen, S.E.; Jensen, C.C.R. A short overview of measures for securing water resources for irrigated crop production. *J. Agron. Crop Sci.* **2014**, *200*, 333–343. [CrossRef]
21. Kang, S.; Hao, X.; Du, T.; Tong, L.; Su, X.; Lu, H.; Li, X.; Huo, Z.; Li, S.; Ding, R. Improving agricultural water productivity to ensure food security in China under changing environment: From research to practice. *Agric. Water Manag.* **2017**, *179*, 5–17. [CrossRef]
22. Saradadevi, R.; Palta, J.A.; Siddique, K.H.M.; Basu, P.S.; Jabran, K.; Saradadevi, R.; Palta, J.A.; Siddique, K.H.M. ABA-mediated stomatal response in regulating water use during the development of terminal drought in wheat. *Front. Plant Sci.* **2017**, *8*, 1251. [CrossRef]
23. Anjum, S.A.; Ashraf, U.; Tanveer, M.; Khan, I.; Hussain, S.; Shahzad, B.; Zohaib, A.; Abbas, F.; Saleem, M.F.; Ali, I. Drought induced changes in growth, osmolyte accumulation and antioxidant metabolism of three maize hybrids. *Front. Plant Sci.* **2017**, *8*, 69. [CrossRef]
24. Ashraf, M.; Foolad, M.R. Roles of glycinebetaine and proline in improving plant abiotic stress tolerance. *Environ. Exp. Bot.* **2007**, *59*, 206–216. [CrossRef]
25. Dustgeer, Z.; Seleiman, M.F.; Kham, I.; Chattha, M.U.; Ali, E.F.; Alhammad, B.A.; Jalal, R.; Refay, Y.; Hassan, M. Glycine-betaine induced salinity tolerance in maize by regulating the physiological attributes, antioxidant defense system and ionic homeostasis. *Not. Bot. Horti Agrobot.* **2021**, *49*, 12248. [CrossRef]
26. Yang, Y.; Huang, C.; Ge, Z.; Zhou, B.; Su, G.; Liu, C.; Fei, Y. Exogenous glycine betaine Reduces drought damage by mediating osmotic adjustment and enhancing antioxidant defense in *Phoebe hunanensis*. *Phyton* **2022**, *91*, 129–148. [CrossRef]
27. Hasanuzzaman, M.; Banerjee, A.; Borhannuddin Bhuyan, M.H.M.; Roychoudhury, A.; Al Mahmud, J.; Fujita, M. Targeting glycinebetaine for abiotic stress tolerance in crop plants: Physiological mechanism, molecular interaction and signaling. *Phyton* **2019**, *88*, 185–221. [CrossRef]
28. Allakhverdiev, S.I.; Hayashi, H.; Nishiyama, Y.; Ivanov, A.G.; Aliev, J.A.; Klimov, V.V.; Murata, N.; Carpentier, R. Glycinebetaine protects the D1/D2/Cytb559 complex of photosystem II against photo-induced and heat-induced inactivation. *J. Plant Physiol.* **2003**, *160*, 41–49. [CrossRef]
29. Agboma, P.C.; Jones, M.G.K.; Peltonen-Sainio, P.; Rita, H.; Pehu, E. Exogenous glycinebetaine enhances grain yield of maize, sorghum and wheat grown under two supplementary watering regimes. *J. Agron. Crop Sci.* **1997**, *178*, 29–37. [CrossRef]
30. Dawood, M.G.; Sadak, M.S. Physiological role of glycinebetaine in alleviating the deleterious effects of drought stress on canola plants (*Brassica napus* L.). *Middle East J. Agric. Res.* **2014**, 3, 943–954.

31. Lichtenthaler, H.K. Chlorophylls and carotenoids: Pigments of photosynthetic biomembrane. *Methods Enzymol.* **1987**, *148*, 350–352. [CrossRef]
32. AOAC. *Official Methods of Analysis*, 20th ed.; Association of Official Analytical Chemists: Arlington, VA, USA, 1990.
33. Farouk, S.; Elhindi, K.M.; Alotaibi, M.A. Silicon supplementation mitigates salinity stress on *Ocimum basilicum* L. via improving water balance, ion homeostasis, and antioxidant defense system. *Ecotoxicol. Environ. Saf.* **2020**, *206*, 111396. [CrossRef]
34. Lutts, S.; Kinet, J.M.; Bouharmont, J. NaCl-induced senescence in leaves of rice (*Oryza sativa* L.) cultivars differing in salinity resistance. *Ann. Bot.* **1996**, *78*, 389–398. [CrossRef]
35. Howell, T. Irrigation engineering, evapotranspiration. In *Encyclopedia of Agricultural Science*; Arntzem, C.J., Ritter, E.M., Eds.; Academic Press: New York, NY, USA, 1994.
36. Banon, S.J.; Ochoa, J.; Franco, J.A.; Alarcon, J.J.; Sanchez-Blanco, M.J. Hardening of oleander seedlings by deficit irrigation and low air humidity. *Environ. Exp. Bot.* **2006**, *56*, 36–43. [CrossRef]
37. Blum, A. Osmotic adjustment is a prime drought stress adaptive engine in support of plant production. *Plant Cell Environ.* **2017**, *40*, 4–10. [CrossRef] [PubMed]
38. Shehzadi, A.; Akram, N.A.; Ali, A.; Ashraf, M. Exogenously applied glycinebetaine induced alteration in some key physio_biochemical attributes and plant anatomical features in water stressed oat (*Avena sativa* L.) plants. *J. Arid Land* **2019**, *11*, 292–305. [CrossRef]
39. Lawlor, D.W.; Cornic, G. Photosynthetic carbon assimilation and associated metabolism in relation to water deficits in higher plants. *Plant Cell Environ.* **2002**, *25*, 275–294. [CrossRef]
40. González-Villagra, J.; Rodrigues-Salvador, A.; Nunes-Nesi, A.; Cohen, J.D.; Reyes-Díaz, M. Age-related mechanism and its relationship with secondary metabolism and abscisic acid in *Aristotelia chilensis* plants subjected to drought stress. *Plant Physiol. Biochem.* **2018**, *124*, 136–145. [CrossRef]
41. Tisarum, R.; Theerawitaya, C.; Samphumphung, T.; Takabe, T.; Cha-Um, S. Exogenous foliar application of glycine betaine to alleviate water deficit tolerance in two indica rice genotypes under greenhouse conditions. *Agronomy* **2019**, *9*, 138. [CrossRef]
42. Aldesuquy, H.S.; Ibraheem, F.L.; Ghanem, H.E. Exogenously supplied salicylic acid and trehalose protect growth vigor, chlorophylls and thylakoid membranes of wheat flag leaf from drought-induced damage. *J. Agric. For. Meteorol. Res.* **2018**, *1*, 13–20.
43. Yildirim, E.; Ekinci, M.; Turan, M.; Dursun, A.; Kul, R.; Parlakova, F. Roles of glycine betaine in mitigating deleterious effect of salt stress on lettuce (*Lactuca sativa* L.). *Arch. Agron. Soil Sci.* **2015**, *61*, 1673–1689. [CrossRef]
44. Arteca, R.N. *Plant Growth Substances: Principles and Application*; Chapman and Hall Press: London, UK, 1996.
45. Lai, Q.; Zhiyi, B.; Zhu-Jun, Z.; Qiong-Qiu, Q.; Bi-Zeng, M. Effects of osmotic stress on antioxidant enzymes activities in leaf discs of PSAG12-IPT modified gerbera. *J. Zheijang Univ. Sci.* **2007**, *8*, 458–464. [CrossRef] [PubMed]
46. Haider, M.S.; Zhang, C.; Kurjogi, M.M. Insights into grapevine defense response against drought as revealed by biochemical, physiological and RNA-Seq analysis. *Sci. Rep.* **2017**, *7*, 13134. [CrossRef]
47. Farouk, S.; El-Metwally, I.M. Synergistic responses of drip-irrigated wheat crop to chitosan and/or silicon under different irrigation regimes. *Agric. Water Manag.* **2019**, *226*, 105807. [CrossRef]
48. Farouk, S.; Omar, M.M. Sweet basil growth, physiological and ultrastructural modification, and oxidative defense system under water deficit and silicon forms treatment. *J. Plant Growth Regul.* **2020**, *39*, 1307–1331. [CrossRef]
49. Taiz, L.; Zeiger, E. *Plant Physiology*, 4th ed.; Sinauer Associates Inc Publishers: Sunderland, MA, USA, 2006.
50. Makela, P.; Kärkkäinen, J.; Somersalo, S. Effect of glycinebetaine on chloroplast ultrastructure, chlorophyll and protein content, and RuBPCO activities in tomato grown under drought or salinity. *Biol. Plant.* **2000**, *43*, 471–475. [CrossRef]
51. Farouk, S.; Arafa, S.A.; Nassar, R.M.A. Improving drought tolerance in corn (*Zea mays* L.) by foliar application with salicylic acid. *Int. J. Environ.* **2018**, *7*, 104–123.
52. García-Caparrós, P.; Romero, M.J.; Llanderal, A.; Cermeño, P.; Lao, M.T.; Segura, M.L. Effects of drought stress on biomass, essential oil content, nutritional parameters, and costs of production in six Lamiaceae species. *Water* **2019**, *11*, 573. [CrossRef]
53. Gutierrez-Boem, F.H.; Thomas, G.W. Phosphorus nutrition affects wheat response to water deficit. *Agron. J.* **1998**, *90*, 166–171. [CrossRef]
54. Farahani, H.A.; Valadabadi, S.A.; Daneshian, J.; Shiranirad, A.H.; Khalvati, M.A. Medicinal and aromatic plants farming under drought conditions. *J. Hortic. For.* **2009**, *1*, 86–92.
55. Sarani, M.; Namrudi, M.; Hashemi, S.M.; Raoofi, M.M. The effect of drought stress on chlorophyll content, root growth, glucosinolate and proline in crop plants. *Int. J. Farming Allied Sci. J.* **2014**, *3*, 994–997.
56. Estaji, A.; Kalaji, H.M.; Karimi, H.R.; Roosta, H.R.; Moosavi-Nezhad, S.M. How glycine betaine induces tolerance of cucumber plants to salinity stress? *Photosynthetica* **2019**, *57*, 753–761. [CrossRef]
57. Khoshkharam, M.; Shahrajabian, M.H.; Esfandiary, M. The effects of methanol and amino acid glycine betaine on qualitative characteristics and yield of sugar beet (*Beta vulgaris* L.) cultivars. *Not. Sci. Biol.* **2021**, *13*, 10949. [CrossRef]
58. Rady, M.M.; Boriek, S.H.; El-Mageed, A.; Taia, A.; Seif El-Yazal, M.A.; Ali, E.F.; Hassan, F.A.; Abdelkhalik, A. Exogenous gibberellic acid or dilute bee honey boosts drought stress tolerance in *Vicia faba* by rebalancing osmoprotectants, antioxidants, nutrients, and phytohormones. *Plants* **2021**, *10*, 748. [CrossRef]
59. Genard, H.; Le Saos, J.; Billard, J.P.; Tremolieres, A.; Boucaud, J. Effect of salinity on lipid composition, glycine betaine content and photosynthetic activity in chloroplasts of *Suaeda maritime*. *Plant Physiol. Biochem.* **1991**, *29*, 421–427. [CrossRef]

60. Alasvandyari, F.; Mahdavi, B.; Hosseini, S.M. Glycine betaine affects the antioxidant system and ion accumulation and reduces salinity-induced damage in safflower seedlings. *Arch. Biol. Sci.* **2017**, *69*, 139–147. [CrossRef]
61. Nawaz, M.; Wang, Z. Abscisic acid and glycine betaine mediated tolerance mechanisms under drought stress and recovery in *Axonopus compressus*: A new insight. *Sci. Rep.* **2020**, *10*, 6942. [CrossRef]
62. Nazari, R.; Parsa, S.; Afshari, R.T.; Mahmoodi, S.; Seyyedi, S.M. Salicylic acid priming before and after accelerated aging process increases seedling vigor in aged soybean seed. *J. Crop Improv.* **2020**, *34*, 218–237. [CrossRef]
63. Gong, H.; Zhu, X.; Chen, K.; Wang, S.; Zhang, C. Silicon alleviates oxidative damage of wheat plants in pots under drought. *Plant Sci.* **2005**, *169*, 313–321. [CrossRef]
64. Sheshbahreh, M.J.; Dehnavi, M.M.; Salehi, A.; Bahreininejad, B. Effect of irrigation regimes and nitrogen sources on biomass production, water and nitrogen use efficiency and nutrients uptake in coneflower (*Echinacea purpurea* L.). *Agric Water Manag.* **2019**, *213*, 358–367. [CrossRef]
65. Gholami Zali, A.; Ehsanzadeh, P.; Szumny, A.; Matkowski, A. Genotype-specific response of *Foeniculum vulgare* grain yield and essential oil composition to proline treatment under different irrigation conditions. *Ind. Crops Prod.* **2018**, *124*, 177–185. [CrossRef]
66. Ahmed, N.; Zhang, Y.; Li, K.; Zhou, Y.; Zhang, M.; Li, Z. Exogenous application of glycine betaine improved water use efficiency in winter wheat (*Triticum aestivum* L.) via modulating photosynthetic efficiency and antioxidative capacity under conventional and limited irrigation conditions. *Crop J.* **2019**, *7*, 635–650. [CrossRef]
67. Song, F.B.; Ying, D.J.; Lie, Z.; Kun, H.G.; Yiqing, G. Effect of water stress on maize pollen vigour and filament fertility. *Acta Agron. Sin.* **1998**, *24*, 368–373.
68. Anjum, F.; Yaseen, M.; Rasul, E.; Wahid, A.; Anjum, S. Water stress in barley. I. Effect on chemical composition and chlorophyll content. *Pak. J. Agric. Sci.* **2003**, *40*, 45–49. [CrossRef]
69. Ashraf, M.; Shahbaz, M.; Ali, Q. Drought-induced modulation in growth and mineral nutrients in canola (*Brassica napus* L.). *Pak. J. Bot.* **2013**, *45*, 93–98.
70. Farooq, M.; Wahid, A.; Kobayashi, N.; Fujita, D.; Basra, S.M.A. Plant drought stress: Effects, mechanisms and management. *Agron. Sustain. Dev.* **2009**, *29*, 185–212. [CrossRef]
71. Adak, N. Effects of glycine betaine concentrations on the agronomic characteristics of strawberry grown under deficit irrigation conditions. *Appl. Ecol. Environ. Res.* **2019**, *17*, 3753–3767. [CrossRef]

agriculture

Article

Chemical and Nutritional Characterization of the Different Organs of Taif's Rose (*Rosa damascena* Mill. var. *trigintipetala*) and Possible Recycling of the Solid Distillation Wastes in Taif City, Saudi Arabia

Esmat F. Ali [1,2,*], Hatim M. Al-Yasi [1,2], Ali Majrashi [1,2], Emad A. Farahat [3], Ebrahem M. Eid [4,5] and Tarek M. Galal [1,2]

1 Department of Biology, College of Sciences, Taif University, P.O. Box 11099, Taif 21944, Saudi Arabia
2 High Altitude Research Center, Taif University, P.O. Box 11099, Taif 21944, Saudi Arabia
3 Botany and Microbiology Department, Faculty of Science, Helwan University, Cairo 11790, Egypt
4 Biology Department, College of Science, King Khalid University, P.O. Box 9004, Abha 61321, Saudi Arabia
5 Botany Department, Faculty of Science, Kafrelsheikh University, Kafr El-Sheikh 33516, Egypt
* Correspondence: a.esmat@tu.edu.sa

Citation: Ali, E.F.; Al-Yasi, H.M.; Majrashi, A.; Farahat, E.A.; Eid, E.M.; Galal, T.M. Chemical and Nutritional Characterization of the Different Organs of Taif's Rose (*Rosa damascena* Mill. var. *trigintipetala*) and Possible Recycling of the Solid Distillation Wastes in Taif City, Saudi Arabia. *Agriculture* **2022**, *12*, 1925. https://doi.org/10.3390/agriculture12111925

Academic Editor: Maria Roulia

Received: 5 October 2022
Accepted: 11 November 2022
Published: 16 November 2022

Abstract: The objective of the current study was to examine the chemical composition and biological functions of the various Taif's rose (TR) organs and floral solid distillation wastes (SDW). Additionally, it assessed the SDW's potential use in animal feed and potential health applications. For chemical and biological analyses, the plant stems, leaves, and flowers as well as the SDW of TR were gathered from four farms in the Al-Shafa highland region of Taif, Saudi Arabia. The highest levels of cardiac glycosides, flavonoids, and phenolics were found in the flowers (7.66 mg securiaside g^{-1}, 16.33 mg GAE g^{-1}, and 10.90 mg RUE g^{-1}, respectively), while the highest carbohydrate and alkaloid contents were found in the TR leaves (2.09% and 9.43 mg AE g^{-1}, respectively) with no significant differences from the SDW. Quercetin, apigenin, and rutin flavonoids, as well as isocorydine and boldine alkaloids, were found in larger concentrations in the flowers and floral SDW than in the leaves and stems. The various TR flower extracts were effective against Gram-negative and -positive bacteria but had no effect on fungal strains, but the SDW's methanol extract was only effective against fungi. The plant stem had the highest N, K, and Mg contents (138, 174, and 96.12 mg kg^{-1}, respectively), while the leaves had the highest P and Ca values (6.58 and 173.93 mg kg^{-1}, respectively). The leaves had the highest contents of total carbohydrates and acid detergent fibre (59.85 and 3.93%, respectively), while the stems had the highest total protein and acid detergent fibre (8.66 and 24.17%, respectively), and the SDW had the highest fats and crude fiber (0.57 and 36.52%, respectively). The highest amounts of digestible crude protein, gross energy, and total dissolved nutrients (TDN) (4.52% and 412.61 Mcal kg^{-1}) were found in the plant stem and flowers, respectively. The results of the current experiment showed that the TDN contents of the various organs and the SDW of TR are suitable for mature dry gestating beef cows. It was determined that, in addition to the SDW's potential usage as an ingredient in animal feed, various plant parts and TR's SDW can be utilized for a variety of medical reasons.

Keywords: macronutrients; phytochemistry; distillation wastes; rose flowers; biological activity

1. Introduction

Because of the superior quality of its essential oils, Taif's rose (*Rosa damascena* Mill. var. *trigintipetala*), a Rosaceae plant, is one of the most widely produced commercial commodities [1]. The mature plant (four years or older) is a shrub that grows to a height of about 2.5 m and blooms once a year (in May or June), producing 500–600 flowers [2,3]. TR may be found growing in temperate and subtropical areas between 300 and 2500 m above sea level [4]. It is grown for commerce among other places in Saudi Arabia, Egypt, Turkey, Morocco, Bulgaria, Iran, France, China, and India [5]. In several Taif Governorate locales, it is also one of the significant aromatic plants grown for use in the perfume, medicinal, and culinary sectors [6]. TR has strong antioxidant, antidiabetic, anti-HIV, antibacterial, anti-inflammatory, and cardiotonic properties since it contains a variety of phytochemical components, including alkaloids, phenolic acids, flavonoids, and other phenolic compounds [1,7–10]. It has historically been used to treat menstrual bleeding, inflammation, respiratory issues, depression, constipation, and chest pain [11,12].

TR is thought to be one of the most essential aromatic crops producing highly regarded essential oil [13,14]. The most popular TR products are rose flowers, rose water, essential oils, vegetative wastes, and distillation wastes [10]. Because the yield of rose oil is so low compared to the amount of rose flower utilized, a considerable quantity of rose flowers must be grown in order to produce enough rose water and oil [15]. In total, 4000 kg of rose petals are required for every kilogram of rose oil [16]. Therefore, in order to satisfy the needs, a huge amount of rose flowers are used in the distillation process, resulting in a large amount of rose wastes, which is a concern. Furthermore, according to Galal et al. [17], a lot of rose flowers must be used to make a tiny amount of rose water, and this process generates a lot of semisolid waste that could harm the environment and results in a significant amount of resource waste in addition to environmental pollution [18].

Natural sources of polyphenols, flavor compounds, polysaccharides, proteins, etc. can be found in the solid waste from TR oil distillation [19,20]. Large amounts of trash can be produced if there are insufficient essential oils in the raw materials (flowers, stems, leaves, etc.) [21]. Even though distilleries often discard the wastes, these routine practices may upset the ecological balance in the area and result in the loss of valuable physiologically-active compounds (polyphenols, fragrance molecules, and polysaccharides) that are present in the trash [22]. Distillation wastes can be composted and used as animal feed [3]. A Taif factory produces roughly 7000 L of liquid waste after utilizing 2500 kg of rose flowers to extract the oil. One oil factory may repeat this extraction procedure twice or three times daily for 45 days during the flowering season. Therefore, in addition to the solid distillation wastes (SDW) of the flower leftovers, roughly 63,000–94,500 L of liquid waste are created from 225,000–337,500 kg of rose flowers each season. This is a significant amount of oil distillation wastes, and if they are dumped into the environment from just one factory, they might lead to a number of environmental issues [23]. As a result, recycling these wastes is crucial for maintaining environmental safety. According to the previous studies on TR organs, the leaves and stems can be used for different medicinal purposes [1,10], animal forage and organic fertilizers [17], and phytoremediators [4].

The majority of studies on TR have concentrated on the growth methods, the timing of the harvest, and the physical and chemical qualities of the oil [23]. However, pharmacological investigations have demonstrated the diverse health benefits of TR's vegetative leaves and stems [1] and reproductive flowers [24], which are mostly attributable to their abundant polyphenolic components. Different portions of the TR plant yield a variety of phytochemicals, including flavonoids, glycosides, alkaloids, and other phenolic compounds [1,10,25]. Additionally, macronutrients contribute significantly to the high yield and excellent quality of TR oil production [26]. The flower output and essential oil quality of TR plants depend on their nutritional balance [27]. Galal et al. [17] looked into the chemical make-up of the main macronutrients in the TR pruning wastes. The goal of the current study was to better understand the chemical composition of the various TR organs as well as the solid byproducts of flower distillation and their biological activity. In addition, it assessed how the distillation wastes might be used to improve human health when compared to various plant parts and how their disposal could lessen environmental pollution. Thus, the study investigated the inorganic and organic nutrients, nutritive value, and the phytochemical constituents of the different organs of TR as well as the SDW. Such a study can help in the recycling of TR wastes and their multipurpose utilization including for medicinal use, animal fodder, and organic fertilizers.

2. Materials and Methods

2.1. Plant Sampling

The various TR parts (stem, leaves, and flowers) were gathered from four rose farms in Taif City, Saudi Arabia's Al-Shafa Highland. The residual plant materials and the distillation wastewaters are the two main byproducts of oil distillation that are produced as industrial wastes [28]. The rose petals from each farm were combined with water in a weight-to-volume ratio of 1:10 to extract the oil [29]. Following the extraction process, the solid distillation wastes (SDW) from each farm were collected as triplicates. In addition to the SDW, three composite samples of each plant organ were also collected from each farm (N = 48) during the harvesting season in May 2021 for additional chemical and biological investigations.

2.2. Sample Preparation and Quality Analysis

Before being homogenized in a planetary high-energy mill with a hardened chromium steel vial for plant analysis, three composite samples of TR (leaves, stems, flowers, and SDW) were air-dried at room temperature in the shade. A 250 g sample of plant powder was shaken in 1000 mL of ethanol for 24 h at room temperature using an orbital shaker, and the extract was then filtered through Whatman No 1 filter paper. Through the use of an evaporator, the filtrate was condensed to dryness at 40 °C under reduced pressure. For the examination of alkaloids, phenolic acids, flavonoids, and cardiac glycosides, the extract was kept between 2 and 8 °C. The chemical reagents that were employed included HCl, ethanol, Baljet's solution, picric acid, NaOH), $AlCl_3$, methanol, Folin reagent, $NaHCO_3$, formic acid, acetonitrile, glacial acetic acid, diethylamine, dimethyl sulfoxide, ketoconazole, and gentamicin.

2.3. Phytochemical Analysis

2.3.1. Determination of Soluble Carbohydrates

The total soluble carbohydrates were determined using the anthrone technique [30]. In total, 100 mg of the TR powdered sample were hydrolyzed with 5 mL of 2.5 N HCl in a boiling-water bath for 3 h. Before adding sodium carbonate to neutralize the acid-digested sample, it was cooled to room temperature first. The resultant volume was diluted to 100 mL with distilled water and centrifuged for 15 min at 5000 rpm. The supernatant was then collected in order to calculate the total amount of soluble carbohydrates.

2.3.2. Determination of Cardiac Glycosides

In order to measure cardiac glycosides, the method of Solich et al. [31] and Tofighi et al. [32] was employed. A 10% ethanol extract was combined with 10 mL of freshly-made Baljet's solution (95 mL of 1% picric acid with 5 mL of 10% NaOH) in order to identify cardiac glycosides. A spectrophotometer (CECIL CE 1021) was used to measure the absorbance at 495 nm after the solution had been left to stand for an hour.

2.3.3. Determination of Total Flavonoid Contents (TFC)

The TFC of the sampled plant organs as well as the SDW were calculated using the methods given by Tofighi et al. [32]. The plant sample was extracted using a 20 mL water–ethanol solution at a 60% (v/v) concentration for 60 min at reflux (80 °C; pH = 5.06). The extract was filtered when it had reached room temperature, and the residue was then extracted once more using the same procedures. The volume was increased to 50 mL of water–ethanol solution at 60% (v/v) after the hydroalcoholic extract and the re-extract were combined (stock solution). A portion of the stock solution was transferred to a 10-mL volumetric flask and mixed with methanol to make the stock solution volume (blank solution). A new 10-mL volumetric flask was used, and a second aliquot of the stock solution was placed into it. The flask was then filled with 2% $AlCl_3$ and methanol was used to bring it to volume (test solution). The test solution's absorbance at 430 nm was evaluated after 25 min in comparison to a control solution. The average of three measurements was used to calculate the rutin content of the TFC herbal material. The following formula was used to calculate the flavonoid content (mg g^{-1}) of herbal material (corrected for moisture content): TFC herbal material is calculated as (TFC tested solution 1.25 × 50)/(w-ld), where TFC test solution is the total amount of flavonoids in the test solution (mg mL^{-1}), 1.25 is the dilution factor, 50 is the volume of the stock solution (mL), w is the weight of the herbal material (g), and l d is the herbal material's loss during drying.

2.3.4. Determination of the Total Phenolic Compounds

Using a spectrophotometric technique, the amount of phenolics in the plant's ethanol extract was assessed [32,33]. Amounts of 0.5 mL of ethanol extract, 2.5 mL of 10% Folin reagent Ciocalteu's dissolved in water, and 2.5 mL of 7.5% $NaHCO_3$ made up the reaction mixture. The blank was created using 0.5 mL of methanol, 2.5 mL of water-dissolved Ciocalteu's 10 percent Folin reagent, and 2.5 mL of $NaHCO_3$ at a 7.5 percent concentration. The samples were then incubated for 45 min at 45 °C in a thermostat. Using a spectrophotometer, the absorbance at 765 nm was measured (CECIL CE 1021). For each experiment, samples were prepared in triplicate, and the mean absorbance value was calculated. The same process was used to produce the calibration curve for the standard gallic acid solution. Per gram of dry weight, the amount of phenolics was calculated as milligrams of gallic acid equivalent (GAE) (DW).

2.3.5. Estimation of Phytochemical Compounds Using HPLC

For the separation, identification, and quantification of flavonoids, phenolic acids, and other phenolic chemicals in TR plant samples, HPLC-MS techniques are frequently utilized. The HPLC-MS system (Agilent 1100: Agilent Corp., Palo Alto, CA, USA) is made up of a single quadrupole MS detector with an ion source, a photodiode array detector, a quaternary pump, and UV/Vis detectors (ESI). Using a gradient solvent system with a flow rate of 1.0 mL min^{-1} and a 0.1% formic acid solution, flavonoids were separated in 70 min. They were then detected at 280 nm and identified by ESI-MS [34]. With a gradient mobile phase consisting of water/acetonitrile/glacial acetic acid (980/20/5, *v*/*v*/*v*, pH 2.68) and acetonitrile/glacial acetic acid (1000/5, *v*/*v*), phenolic acid was separated in 60 min [35]. Additionally, alkaloids were examined by HPLC using 0.2% diethylamine and 0.16% formic acid as solvent system A and 0.2% diethylamine and 0.16% formic acid in acetonitrile as solvent system B (0 min, 80:20 (A:B); 5 min, 80:20; 20 min, 60:40; 25 min, 0:100). The column used had a flow rate of 1.0 mL min^{-1} and was a 5 μm, 250 mm × 4.6 mm GraceSmart RP18 (Grace Vydac, Hesperia, CA, USA). At 226 nm, the peaks were discovered.

2.4. Biological Activity

2.4.1. Preparation of Extracts

Since the biological activity of TR leaves and stems had already been established [1], 250 g of plant powder from the flowers and SDW were steeped in 1.5 L of 95% ethanol and methanol before being boiled in both cold (about room temperature) and warm (50 °C) water for 5 days. To ensure a uniform infusion, the mixture was blended every day. After 5 days, Whatman filter paper No. 1 was used to filter the extract. The filtrate was dried in a rotary evaporator at a temperature of 60 °C. The dried extract was stored in sterile glass vials at −20 °C until use [36].

2.4.2. Microorganisms Used

One Gram-positive bacterial strain (*Bacillus subtilis*), two Gram-negative bacterial strains (*Escherichia coli* and *Proteus vulgaris*), and two fungal strains (*Aspergillus fumigatus* and *Candida albicans*) were obtained from Faculty of Science at Al-Azhar University in Cairo, Egypt. In nutrient agar and malt extract, respectively, the bacterial and fungal strains were cultivated.

2.4.3. In Vitro Evaluation of the Antimicrobial Activity

By modifying the agar disc well diffusion method [37], an antimicrobial susceptibility test was carried out. Antimicrobial activity was gauged by measuring the diameter of inhibitory zones. As antibacterial agents, plant extracts were tested on bacterial isolates. All bacterial and fungal isolate inoculum suspensions were dispersed across the surface media. The medium was drilled with holes that were 6 mm in diameter using a 6-mm cork borer. Dimethyl sulfoxide (DMSO) was used to process the dried plant extracts into a final extract with a 10 mg/mL concentration. In total, 100 L of plant extract was placed in the well of each plate. For bacterial growth, the inoculated agar plates were incubated for 24 h at 37 °C and for 48 h at 28 °C. Measurements of the inhibitory zones brought on by active extract components were made after 24–48 h of incubation. The investigations were carried out in duplicate, and an established scale was used to evaluate the inhibitory zone [38]. For fungus, gentamicin (MIC = 4 g mL^{-1}) was utilized as a control therapy, whereas ketoconazole antibiotic (MIC = 100 g mL^{-1}) was employed for bacteria.

2.5. Nutritional Analyses

2.5.1. Mineral Nutrients

Three composite samples of TR plants from each farm (N = 48) were pulverized separately in a metal-free plastic mill, passed through a 2-mm mesh size, and then stored in labeled plastic containers. The samples included leaves, stems, flowers, and SDW. For the measurement of inorganic nutrients, a sulphuric and perchloric acid mixture was used to digest 1 g of each milled plant sample [39]. At the Ecology Laboratory, Faculty of Science, Helwan University, Cairo, Egypt, the determination of N, P, K, Ca, Mg, and Na was performed for the extracts using an Agilent 4210 MP-AES (Microwave Plasma-Atomic Emission Spectrometer, Agilent Inc., Suwon, Korea). The operational processes and instrumental settings were modified in accordance with the user guide provided by the manufacturer. For each element, the final concentrations were given as mg kg^{-1} biomass dry matter.

2.5.2. Organic Nutrients

The ground plant samples were ignited for three hours at 550 °C in a muffle furnace to determine the amount of ash present. The Kjeldahl method [40] was used to assess total nitrogen, and its content was utilized to compute total protein (TP) by multiplying its percentage by 6.25. Allen [39] reported that the total lipids or fats represented by ether extract (EE) were determined by diethyl ether extraction of the plant materials using a Soxhlet extractor. Allen [39] claims that after chemically digesting and solubilizing the other existing components, the crude fibre (CF) is measured gravimetrically. According to Le Houérou [41], the nitrogen-free extract (NFE, or total carbohydrates) was determined in the plant samples as follows: NFE (% DM) = 100 − (TP + CF + fat + ash). According to Goering and Van Soest [42] and Van Soest et al. [43], the fiber portions of cell walls were determined as neutral detergent fiber (NDF), acid detergent fiber (ADF), and acid detergent lignin (ADL).

2.5.3. Nutritional Value

According to NRC [44], the digestible crude protein (DCP) was calculated as follows: DCP percentage (% DCP) = 0.85 CP − 2.5. According to NRC [45], the digestible energy (kcal kg^{-1} DM) was determined as follows: gross energy times 0.76 equals digestible energy (DE). The metabolizable energy (kcal kg^{-1} DM) was determined as follows [44]: digestible energy multiplied by 0.82 equals metabolic energy (ME). According to NRC [44], the following formula was used to determine the net energy (kcal kg^{-1} DM): metabolizable energy times 0.56 equals net energy (NE). The caloric values were expressed in Mcal kg^{-1} DM. The total digestible nutrients (TDN) were computed as follows NRC [44]: % digestible energy divided by 44.3 equals total digestible nutrients (%).

2.6. Statistical Analysis

Following a normality check of the data, one-way analysis of variance (ANOVA I) was used to examine the variations in the chemical composition of the plant's various organs [46]. A post hoc test (Duncan's test) was employed when there were significant differences between the different samples.

3. Results and Discussion

3.1. Phytochemical Analysis

It is crucial to identify phytochemical components in order to evaluate the nutritional and therapeutic value of plants [47]. The TR plants' phytochemical components showed substantial differences ($p < 0.001$) among the various plant organs (Table 1). According to the qualitative phytochemical examination of TR's various organs, carbohydrates, glycosides, alkaloids, flavonoids, and phenolic compounds were found. The same species' petals [48,49], and leaf and stem pruning wastes [1] also revealed similar results. Carbohydrates are substances with a lot of energy that show how much food and energy plants have stored and offer the cellular building blocks for those plants [50]. They are used to monitor a plant's capacity to rebound following pruning [51]. TR leaves were observed to have much higher carbohydrate and alkaloid concentrations than other plants (2.09 percent and 9.43 mg AE g^{-1}, respectively), but lower levels of cardiac glycosides (3.61 mg securiaside g^{-1}). The drought stress of rose plants before and after pruning may be the cause of the elevated carbohydrate content in the leaves [6]. Additionally, the reduced ability to use carbohydrates as an energy source, a component of new cells and tissues, or an osmolyte for the cells, may be the cause of the accumulation of carbohydrates [37].

Table 1. Phytochemical constituents (mean ± SD, n = 12) of the different organs as well as the solid distillation wastes of Taif's rose collected from Al-Shafa highlands in Taif City, Saudi Arabia. Maximum and minimum values are underlined.

Organ	Carbohydrates %	Cardiac Glycosides mg Securiaside g^{-1}	Phenolics mg GAE g^{-1}	Flavonoids mg RUE g^{-1}	Alkaloids mg AE g^{-1}
Leaves	2.09 ± 0.71 a	3.61 ± 0.59 c	9.81 ± 1.79 b	6.33 ± 1.52 b	9.43 ± 1.17 a
Stem	0.85 ± 0.08 b	4.84 ± 0.63 b	5.45 ± 1.55 c	7.46 ± 1.16 b	5.70 ± 1.21 b
Flower	0.66 ± 0.08 b	7.66 ± 0.20 a	16.33 ± 0.10 a	10.90 ± 0.20 a	3.44 ± 0.14 c
Flower wastes	0.56 ± 0.04 b	5.63 ± 0.16 b	14.36 ± 2.11 a	9.80 ± 0.26 a	2.36 ± 0.10 c
F-value	19.73 ***	38.54 ***	55.75 ***	13.98 ***	23.98 ***

Means in the same column followed by different letters are significantly different at $p < 0.05$, according to Duncan's HSD test; ***: $p < 0.001$.

Additionally, the TR flowers contained the most cardiac glycosides (7.66 mg securiaside g^{-1}). This figure was higher than what Galal et al. [1] found in the TR leaf and stem pruning wastes. Cardiac glycosides are a particular class of secondary metabolite that have historically been used to increase cardiac contractile force in people with congestive heart failure or cardiac arrhythmias [52]. Their content was higher than the 2.98–5.69 mg g^{-1} found in TR's leaf and stem pruning wastes [1] but lower than the 9.5–15.2 mg g^{-1} and 9.07–21.09 mg g^{-1} found in *C. procera* and *Aloe* spp. ([52,53], respectively). Moreover, the SDW had the lowest levels of soluble carbohydrates and alkaloid contents (0.56% and 2.36 mg AE g^{-1}, respectively), whereas the stem tissues had the lowest levels of phenolics (5.45 mg GAE g^{-1}).

The TR flowers also had the highest levels of flavonoid and phenolic compounds (16.33 mg GAE g^{-1} and 10.90 mg RUE g^{-1}, respectively), with no discernible changes from the SDW. In the same species, Galal et al. [1] and Liu et al. [18] recorded 12.41 and 386.4 mg g^{-1} of phenolic content in the leaf pruning wastes and flower residue, respectively. These results suggest that the TR flowers and SDW have better pharmacological (antioxidant, anti-ageing, whitening, and anticancer) actions than the TR leaves and stems. Additionally, numerous studies have shown how TR's high phenolic and flavonoid contents contribute to its potent antibacterial and disinfection properties [24]. Moreover, the TR flower's flavonoid concentration was higher than the 9.33 mg g^{-1} found in the same species' stem and leaf pruning wastes [1].

3.2. *HPLC of Phytochemical Compounds*

The HPLC analysis of the flavonoids, and phenolic and alkaloid compounds showed that the flowers and flower SDW of TR plants had lower contents of the separated compounds (except quercetin, apigenin, and rutin flavonoids, and isocorydine, and boldine alkaloids) than the leaves and stems (Table A1). Quercetin, apigenin, luteolin, chrysoeriol, rutin, kaempferol, and avicularin were the seven flavonoid compounds that were isolated and identified (Table A1 and Figure 1). These chemicals have potential antioxidant, anti-inflammatory, and antimicrobial properties as reported by Dahat et al. [54], who discovered that quercetin and its glycoside rutin have been identified in extracts demonstrating nephroprotective characteristics. The highest levels of luteolin, chrysoeriol, and kaempferol were found in the plant stem (21.09, 45.19, and 27.55 mg g^{-1}, respectively), while the highest levels of quercetin, apigenin, and rutin were found in the flowers (20.10, 47.02, and 25.36 mg g^{-1}, respectively), and the highest levels of avicularin were found in the SDW (4.15 mg g^{-1}). Quercetin is a plentiful polyphenolic flavonoid that has a number of health-promoting properties, including strong vasodilators, cancer-preventative properties, anti-inflammatory properties, and advantages for asthma, among many others [55]. Additionally, luteolin and apigenin can lessen the occurrence of mouth sores and provide modest symptom relief as well as limit the viability of leukemic cells, colon and ovarian carcinoma cells, and, in particular, human breast cancer cells [56]. Moreover, kaempferol and avicularin are nontoxic and have potent antioxidant, hepatoprotective, antidiabetic, and anti-inflammatory properties [57].

Five phenolic substances were produced by the HPLC of the extracts of the different TR organs: ellagic acid, catechol, resorcinol, gallic acid, and phloroglucinol (Table A1 and Figure 2). The five compounds were found in plant stems and leaves, while resorcinol and gallic acid and catechol were not found in the SDW or flowers. The greatest content of gallic and ellagic acids (23.53 and 14.90 mg g^{-1}) in TR stems was lower than 37.4 and 50.3 mg g^{-1} recorded in the stem pruning wastes and flower residue recorded by Galal et al. [1] and Marlene et al. [58], respectively, in the same species. Numerous biological functions of gallic acid include antioxidant capabilities, antibacterial activity, anti-inflammatory properties, antiviral and antimutagenic characteristics, and anticancer effects [59]. Additionally, ellagic acid is a crucial compound utilized in the treatment of chronic ulcerative colitis as an anticarcinogenic, multipurpose protector against oxidative stress, and an anti-inflammatory agent [5]. Furthermore, catechol, resorcinol, and phloroglucinol concentrations were highest in the plant leaves (13.70, 11.91, and 6.24 mg g^{-1}, respectively). These compounds were discovered to have possible antimicrobial properties [10].

Alkaloids are physiologically-active substances that are frequently employed as medications and produced by plants as secondary metabolites, but many of these substances are extremely toxic [60]. Six alkaloid compounds were isolated and identified in TR extracts as berbamine, jatrorrhizine, palmatine, reticuline, isocorydine, and boldine (Table A1 and Figure 3). These compounds are widespread in many Chinese medicinal plants and have been shown to have leukocytosis-promoting, antimicrobial, anti-inflammatory, anticancer, and choleretic activities [61]. Galal et al. [1] found comparable results in the same species' pruning leaves and stem. Berbamine, jatrorrhizine, palmatine, and reticuline were all found in the highest concentrations (5.28, 5.83, 3.50, and 8.50 mg g^{-1}, respectively) in the plant stem, whereas isocorydine and boldine were found in the highest concentrations (9.11 and 4.79 mg g^{-1}, respectively) in the flowers. Berbamine, jatrorrhizine, palmatine, and reticuline have antibacterial activity, encourage leukocytosis, and are choleretic substances [62]. However, the long-term use of boldine may cause depression, partial motor aphasia, sound hallucinations, and color hallucinations [63].

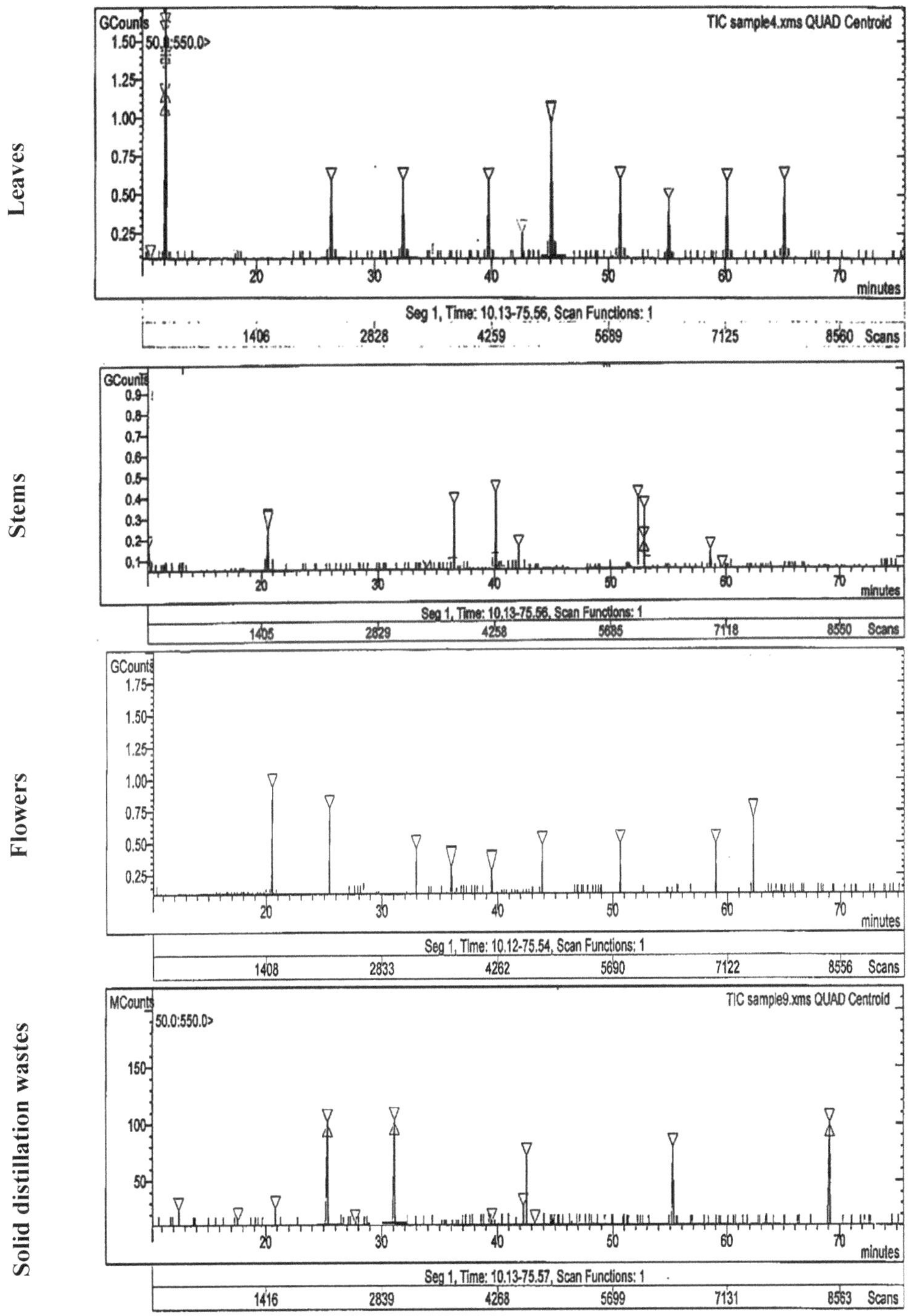

Figure 1. HPLC analysis of the phenolic compounds in the different organs as well as the solid distillation wastes of Taif's rose collected from Al-Shafa highlands in Taif City, Saudi Arabia.

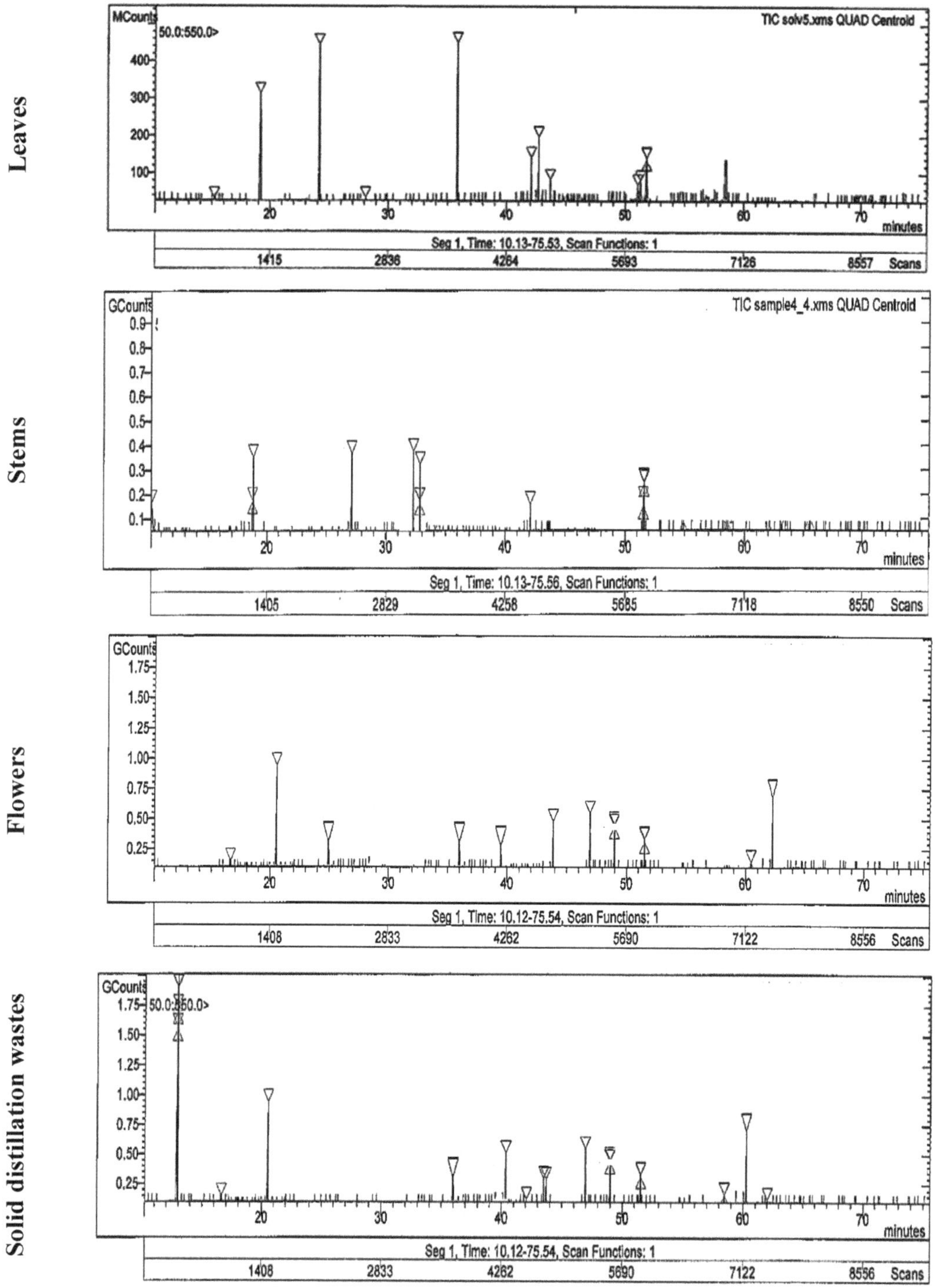

Figure 2. HPLC analysis of the flavonoid compounds in the different organs as well as the solid distillation wastes of Taif's rose collected from Al-Shafa highlands in Taif City, Saudi Arabia.

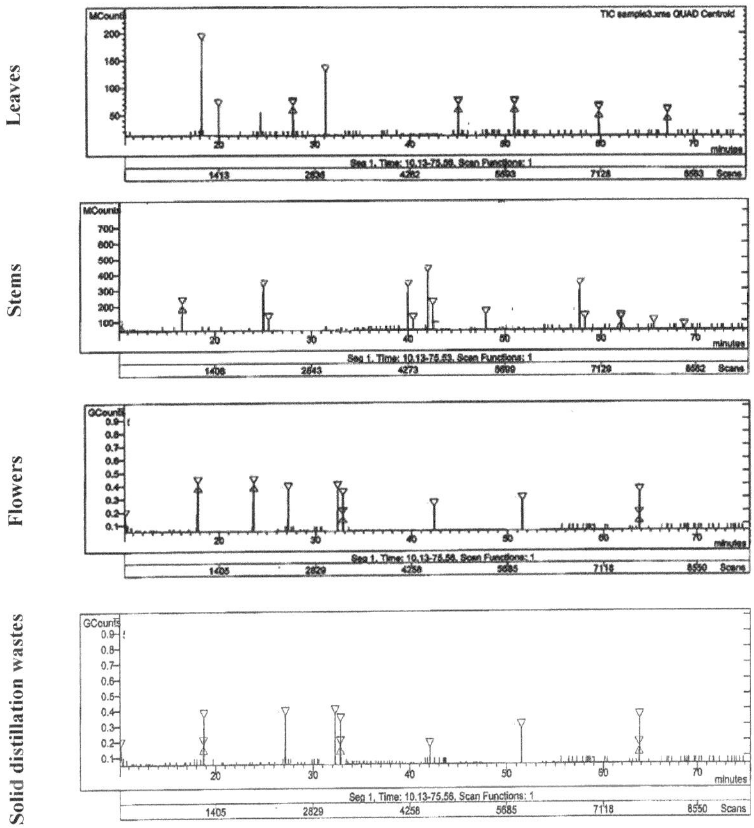

Figure 3. HPLC analysis of the alkaloid compounds in the different organs as well as the solid distillation wastes of Taif's rose collected from Al-Shafa highlands in Taif City, Saudi Arabia.

3.3. Biological Activity

3.3.1. Flower Extracts

The phytogeographical region, the plant component, and the extraction method all affect a plant's antibacterial activity [64]. According to biological activity tests, TR flower extracts were effective against Gram-negative and Gram-positive bacteria but ineffective against fungal strains (Table A2 and Figure 4). In a related investigation, Galal et al. [1] discovered that the leaf pruning wastes of TR showed antibacterial and antifungal activity in the boiling-water extracts but not in the other extracts. Compared to the control (inhibition zones: 26 mm), ethanol and warm-water extracts significantly reduced the sensitivity of *Bacillus subtilis* (14 and 13 mm, respectively). Additionally, *Proteus vulgaris* and *Escherichia coli* were extremely sensitive to most extracts (except for warm water on *E. coli*), with boiling-water extracts having the maximum activity (13 and 18 mm, respectively) when compared to gentamicin (30 and 17 mm, respectively). In comparison to gentamicin antibiotics (control), Galal et al. [1] found that the same bacterial strains were much more sensitive to warm-water extracts of TR leaves and stem pruning wastes. They attributed these findings to the presence of various phytochemical compounds, such as alkaloids, phenolic acids, flavonoids, and other phenolic compounds. Similar findings on *Rosa indica* extracts show antifungal action against *A. fumigatus* strains as well as antibacterial activity against *Proteus* sp. and *E. coli* [65]. The current findings show that the floral extracts had a modest effect on the bacterial strains under investigation, with boiling- and cold-water extracts being the most effective against *P. vulgaris*. It is important to note that *P. vulgaris* was only responsive to TR extracts in the following order:

ethanol > warm water > methanol > boiling and cold water. Norziah et al. [66] claim that because it is nontoxic and beneficial to the environment, using water as an extracting solvent is preferable to using organic solvents. Additionally, water is a good solvent for extracting a significant amount of highly active phenolic and flavonoid chemicals that can be used safely in a variety of food applications [10].

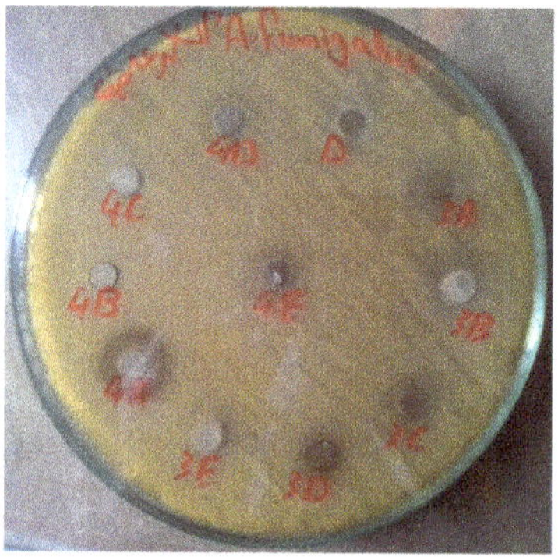

Aspergillus fumigatus

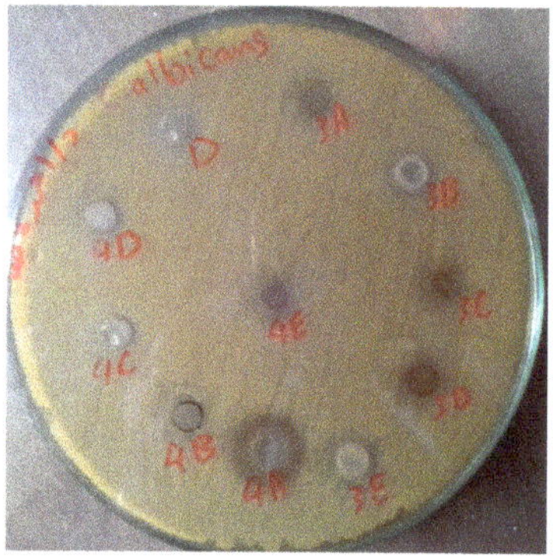

Candida albicans

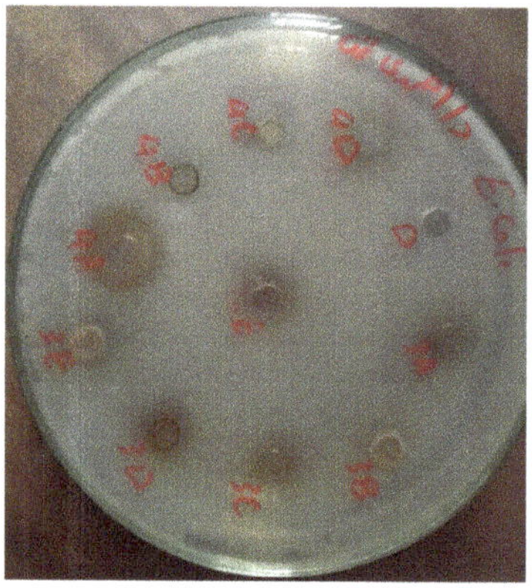

Escherichia coli

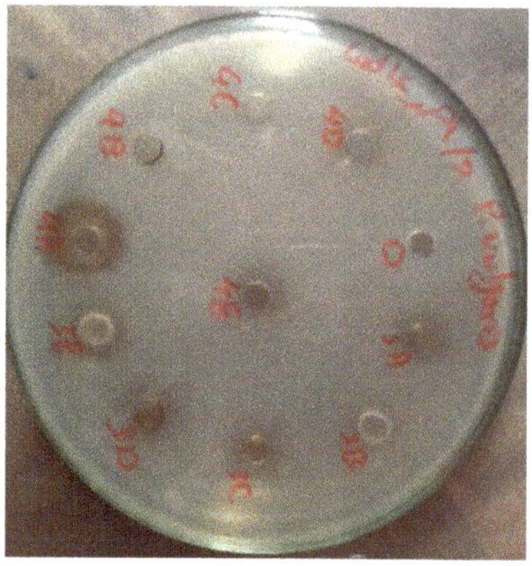

Proteus vulgaris

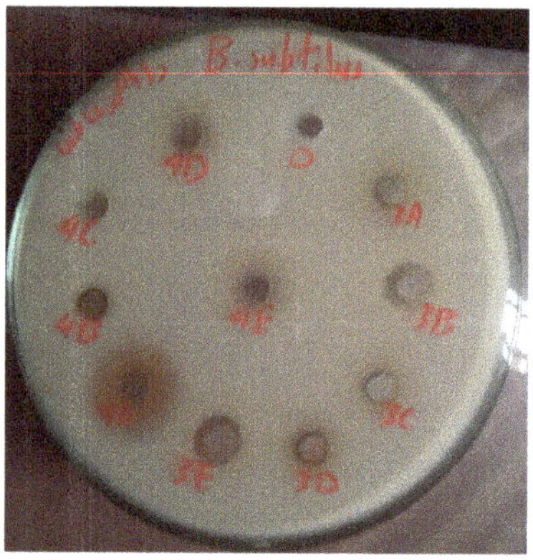

Bacillus subtilis

Figure 4. Antimicrobial activity of the different extracts of Taif's rose. 3: flowers, 4: solid distillation wastes, A: methanol extract, B: ethanol extract, C: boiled water, D: cold water, E: warm water.

3.3.2. Solid Distillation Waste Extracts

The antimicrobial activity data of the SDW extracts of TR showed that the methanol extract was exclusively potent against fungal strains (*Aspergillus fumigatus* and *Candida albicans*) with lower activity (10 and 14 mm, respectively) compared with 17 and 20 mm, respectively, of the ketoconazole (Table A3 and Figure 4). Similarly, the methanol and cold-water extracts of the stem pruning wastes of TR were active against the bacterial and fungal strains [1]. On the other hand, the warm-water extract was only efficient against the investigated bacterial strains with inhibition zones (10, 12, and 18 mm) compared with 26, 30, and 17 mm for gentamicin. It was noticed that the SDW extracts had low efficacy against the investigated fungal and bacterial strains, except for the warmwater extract against *P. vulgaris*. Halawani [23] found that *P. vulgaris* was highly susceptible to all of the TR flower extracts. In contrast, Adom et al. [56] found that the aqueous extract of *P. major* has no antimicrobial activity. Thus, pharmaceutical studies are required to separate, purify, and identify the phytochemical compounds in the ethanolic, methanolic, and water extracts of TR flowers and SDW.

3.4. Nutritional Properties

3.4.1. Mineral Nutrients

Using organic fertilizer made from composted agricultural waste can help reduce the need for chemical fertilizers and nutrient requirements [67]. The statistical analysis (ANOVA I) indicated significant variations ($p < 0.05$) in the concentrations of the analyzed mineral elements (except Ca and Mg) among the different organs of TR (Table 2). The plant stem contributed the highest contents of N, K, and Mg (138.73 ± 16.49, 174.57 ± 17.20, and 96.12 ± 17.01 mg kg^{-1}, respectively), but the lowest Na content (136.15 ± 15.17 mg kg^{-1}). In addition, the plant leaves had the highest contents of P and Ca (6.58 ± 0.54 and 173.93 ± 26.03 mg kg^{-1}, respectively) but the lowest of K and Mg (101.54 ± 10.89 and 85.43 ± 14.90 mg kg^{-1}, respectively). The values of these elements (except K, P, and Ca) were lower than those recorded in the leaf and stem pruning wastes of the same species [17]. Moreover, the highest Na content (198.08 ± 12.34 mg kg^{-1}) was recorded in the flower tissues, which was comparable to 195.18 ± 9.32 mg kg^{-1} recorded in the SDW. The nutritional value of TR plant parts was lower than that found in herbal plant residues and olive mill waste ([68,69], respectively). The current study is in line with the findings of Galal et al. [17], who found that the macronutrient content of TR leaf and stem pruning wastes makes them more promising for composting in order to improve soil quality than flowers and SDW. The various plant organs of TR can be used as additives in organic fertilizers because their nutrient release rate is too slow to meet crop requirements in a timely manner. Based on the observation of Chang et al. [67], those organic fertilizers could not be an absolute replacement for chemical fertilizers.

Table 2. Mineral nutrient content (mean ± SD, $n = 12$) of the different organs of Taif's rose grown on Al-Shafa highlands in Taif City, Saudi Arabia. Maximum and minimum values are underlined.

Organ	N	P	K	Ca	Mg	Na
	mg kg^{-1}					
Leaves	98.01 ± 3.86 ab	6.58 ± 0.54 a	101.54 ± 10.89 b	173.93 ± 26.03	85.43 ± 14.90	167.64 ± 19.09 ab
Stem	138.73 ± 16.49 a	6.53 ± 1.77 a	174.57 ± 17.20 a	165.46 ± 24.28	96.12 ± 17.01	136.15 ± 15.17 b
Flower	71.70 ± 5.31 b	1.49 ± 0.19 b	117.10 ± 2.01 ab	147.54 ± 17.20	91.32 ± 12.01	198.08 ± 12.34 a
Distillation wastes	81.58 ± 6.23 b	2.24 ± 0.17 b	102.13 ± 13.11 b	135.55 ± 13.93	89.83 ± 11.73	195.18 ± 9.32 a
F-value	4.55 *	24.83 ***	5.5 **	0.96	0.22	4.01 *

Means in the same column followed by different letters are significantly different at $p < 0.05$, according to Duncan's HSD test; *: $p < 0.05$, **: $p < 0.01$, ***: $p < 0.001$.

3.4.2. Organic Nutrients

Animal productivity depends on forage quality, which is primarily determined by total protein (TP), crude fiber (CF), digestibility, and other related factors [70]. The statistical analysis (ANOVA I) of the organic nutrients (except CF and NDF) showed significant variation ($p < 0.05$) among the investigated samples of TR plants (Table 3). The plant leaves contributed the highest percentage of total carbohydrates (NFE) and ADL (59.85 and 3.93%, respectively) but the lowest CF content (21.86%). The NFEs of TR leaves, stems, and flowers were comparable to those recorded in the pruning wastes of the same species [17], *Echinochloa stagnina* and *Eichhornia crassipes*, and higher than that of *C. demersum* (33.4%) [71]. The high carbohydrate content is beneficial for giving the rumen microorganisms sufficient energy, which in turn helps lactating cows [72]. Moreover, the highest percentages of TP and ADF (8.66 and 24.17%, respectively) were recorded in the plant stem associated with the lowest ash and fat contents (7.34 and 0.14%). The ADF content of the different investigated organs was lower than that of TR pruning wastes (27.5–45.1%: Galal et al. [17]), *Trifolium alexandrinum* (32.1%: [73]), and wheat straw (46.5–50.8%: [74]). The percentages of NDF and ADF of TR organs were lower than 74.4% and 44.4%, respectively, which were recorded in central Oklahoma [75] and 57.5% and 31.6%, respectively, reported in western Washington [76] for forage intermediate wheatgrass. Favre et al. [77] reported 59.7% NDF and 33.7% ADF in Wisconsin's Kernza intermediate wheatgrass in the similar context.

Table 3. Organic nutrient content (mean ± SD, $n = 12$) of the different organs of Taif's rose grown on Al-Shafa highlands in Taif City, Saudi Arabia. Maximum and minimum values are underlined.

Organ	Total Protein (%)	NFE (%)	Ash (%)	Fat (%)	Crude Fiber (%)	ADF (%)	ADL (%)	NDF (%)
Leaves	6.13 ± 1.69 ab	59.85 ± 9.96 a	11.67 ± 2.69 c	0.50 ± 0.05 ab	21.86 ± 3.48	22.88 ± 7.11 a	3.93 ± 1.12 a	38.21 ± 7.88
Stem	8.66 ± 2.27 a	52.63 ± 13.81 a	7.34 ± 1.75 d	0.14 ± 0.01 c	31.33 ± 3.81	24.17 ± 5.16 a	2.49 ± 0.49 ab	36.51 ± 3.31
Flower	4.71 ± 0.49 b	53.92 ± 5.15 a	18.74 ± 2.34 b	0.36 ± 0.05 b	22.26 ± 2.29	15.54 ± 1.41 ab	1.71 0.09 b	38.51 ± 3.36
Distillation wastes	5.13 ± 0.38 b	28.39 ± 4.78 b	29.39 ± 2.87 a	0.57 ± 0.05 a	36.52 ± 1.58	14.23 ± 1.75 b	3.60 ± 0.31 a	34.56 ± 1.53
F-value	4.31 *	5.28 **	62.75 ***	14.93 ***	2.71	2.96 *	2.85 *	0.46

Means in the same column followed by different letters are significantly different at $p < 0.05$, according to Duncan's HSD test; *: $p < 0.05$, **: $p < 0.01$, ***: $p < 0.001$.

Moreover, the flower SDW with the lowest values of NFE, ADF, and NDF (28.39, 14.23, and 34.56%, respectively) had the highest ash, lipids, and CF contents (29.39, 0.57, and 36.52%, respectively). According to Heneidy and Halmy [78], TP and CF are traditionally used as indicators of the nutritional content of food for grazing animals because TP is used for energy and aids in tissue formation, and CF defines the energy feeding value of the forage [79]. The minimal protein in animal diets ranges from 6 to 12% DM MAFF [80] and, as a result, the TP content of TR's leaves and stems falls within this range but that of the flowers and SDW is lower. In addition, the plant leaves and stems lie within the required range (7–9%) of TP for sheep and gestating cows ([45,81], respectively). CF in plants represents all the cell wall fractions that are resistant to the action of digestive enzymes and includes the insoluble residue of acid hydrolysis and the alkaline one [82]. In the current study, the CF contents of the SDW and stem of TR were slightly higher than the range reported for some known wild forage plants such as *Phragmites australis* (29.9% DM), *Panicum repens* (27.3% DM), and *Cynodon dactylon* (20.5% DM) [83,84]. However, the range of CF contents in the investigated organs of TR was higher than the mean content (20.0%) of temperate grasses [85].

3.4.3. Nutritional Value

The nutritive value of any forage is dependent upon its content of energy-producing nutrients as well as its contents of essential nutrients to the body and mainly depends on high digestible crude protein (DCP) [71]. The investigated parameters of nutritional values (except NE) were significantly varied ($p < 0.05$) among the different organs of TR plants (Table 4). DCP is an important fraction of proteins that are ingested and absorbed by the animal and not excreted in feces [86]. The highest values of DCP and GE (4.52% and 412.61 Mcal kg^{-1}, respectively) and the lowest TDN (56.18%) were recorded in the plant stem, while the highest TDN (59.19%) associated with the lowest DCP (0.86%) were recorded in the plant flowers. The DCP in the present study was relatively low in comparison with TR pruning wastes [17] but was lower than 9% of the main fodder crop *T. alexandrinum* [87]. Additionally, the TDN is defined as the energy content of feeds available to animals after the digestion losses [79]. The present investigation revealed that the different organs as well as the SDW of TR have suitable contents of TDN for mature dry gestating beef cows that require 55–60% [88]. Moreover, the plant leaves had the highest values of DE, ME, and NE (2.68, 2.20, and 1.10 Mcal kg^{-1}, respectively). The SDW had the lowest energy contents represented by DE, ME, NE, and GE (1.23, 1.23, 0.62, and 324.11 Mcal kg^{-1}, respectively). The values of DE and ME of the different living organs of TR were comparable to the 2.65 and 2.17 Mcal kg^{-1} DM values, respectively, recorded in hay of alfalfa (*Medicago sativa*), and higher than the 2.43 and 1.99 Mcal kg^{-1} DM values, respectively, in red clover (*Trifolium pratense*) [45]. However, the values of the GE of the TR stems and leaves were comparable to the 377.02–424.25 Mcal kg^{-1} DM of the pruning wastes [17] and higher than the 389 Mcal kg^{-1} DM reported for *Cynodon dactylon* and 398 Mcal kg^{-1} DM in *Panicum repens* [84].

Table 4. Nutritional value (mean ± SD, n = 12) of the different organs of Taif's rose grown on Al-Shafa highlands in Taif City, Saudi Arabia. Maximum and minimum values are underlined.

Organ	DCP (%)	TDN (%)	DE (Mcal/kg)	ME (Mcal/kg)	NE (Mcal/kg)	GE (Mcal/kg)
Leaves	2.71 ± 0.44 ab	58.28 ± 1.87 ab	<u>2.68 ± 0.37 a</u>	<u>2.20 ± 0.31 a</u>	<u>1.10 ± 0.01 a</u>	385.58 ± 8.24 b
Stem	<u>4.52 ± 0.16 a</u>	<u>56.18 ± 1.82 b</u>	2.61 ± 0.42 a	2.14 ± 0.42 a	1.07 ± 0.03 a	<u>412.61 ± 9.59 a</u>
Flower	<u>0.86 ± 0.05 b</u>	<u>59.19 ± 3.13 a</u>	2.25 ± 0.19 a	1.85 ± 0.16 a	0.93 ± 0.06 a	354.33 ± 6.62 c
Distillation wastes	1.24 ± 0.35 b	59.05 ± 2.32 a	<u>1.23 ± 0.08 b</u>	<u>1.23 ± 0.08 b</u>	<u>0.62 ± 0.04 b</u>	<u>324.11 ± 9.13 d</u>
F-value	4.31 *	5.28 **	62.75 ***	14.93 ***	2.71	2.96 *

Means in the same column followed by different letters are significantly different at $p < 0.05$, according to Duncan's HSD test; *: $p < 0.05$, **: $p < 0.01$, ***: $p < 0.001$.

4. Conclusions

Compared to the flowers and SDW, the leaves and stems of TR contained more carbohydrates and cardiac glycosides but fewer flavonoids and phenolic substances. The largest cardiac glycoside, flavonoid, and phenolic levels were found in the TR flowers, with no discernible difference from the SDW, whereas the highest carbohydrate and alkaloid contents were found in the TR leaves. Quercetin, apigenin, and rutin flavonoids, as well as isocorydine and boldine alkaloids, were found in larger concentrations in the flowers and blossom SDW of TR plants than in the leaves and stems, according to the HPLC analysis of the phytochemical components. The various TR flower extracts were effective against Gram-negative and Gram-positive bacteria but had no effect on fungal strains, while the SDW's methanol extract was exclusively effective against fungi. In contrast to the flowers and SDW, the macronutrient content of TR leaves and stems makes it possible to compost them for improving soil quality. The results of the current investigation show that the TDN contents of the various organs and the SDW of TR are suitable for mature dry gestating beef cows. In addition to the potential use of the SDW as a supplement in animal feed, the various plant organs and the SDW of TR can be utilized for a variety of medicinal purposes. Based on these micro-components as well as the inorganic and organic nutrients and the nutritive value, the different organs of TR can be used in medicine, animal fodder, soil amendment, organic fertilizers, and for other industrial purposes (Table 5).

Table 5. Micro-components of the different organs of Taif's rose and their significance.

Component	Significance
Inorganic nutrients (N, P, Na, K, Ca, Mg)	Soil amendments, organic fertilizers
Organic nutrients (carbohydrates, proteins, fats, fibers)	Animal forage, industrial application
Nutritional components (DCP, TDN, GE, ME, NE)	Animal forage
Secondary metabolites (cardiac glycosides, flavonoids, alkaloids, phenolics)	Medicinal purposes (e.g., cancer-preventative properties, antioxidant, anti-inflammatory, and antimicrobial), pharmaceutical industry.

Author Contributions: Conceptualization, T.M.G., E.A.F., E.F.A. and E.M.E.; methodology, E.A.F., E.F.A., T.M.G., E.M.E., A.M. and H.M.A.-Y.; software, T.M.G. and E.M.E.; validation, E.A.F., E.F.A., T.M.G., E.M.E., A.M. and H.M.A.-Y.; formal analysis, E.A.F., E.F.A., T.M.G., E.M.E. and H.M.A.-Y.; investigation, E.A.F., E.F.A. and A.M.; resources, E.F.A., T.M.G., A.M. and H.M.A.-Y.; data curation, T.M.G., E.F.A., E.M.E., A.M., H.M.A.-Y. and E.A.F.; writing—original draft preparation, E.A.F., E.F.A., T.M.G., E.M.E., A.M., H.M.A.-Y. and E.A.F.; writing—review and editing, E.F.A. and T.M.G.; visualization, E.M.E., A.M., H.M.A.-Y., T.M.G. and E.A.F.; supervision, E.F.A. and T.M.G.; project administration, E.F.A. and T.M.G.; funding acquisition, E.A.F., E.F.A. and T.M.G.; methodology, E.A.F., E.F.A., T.M.G., E.M.E. and A.M. All authors have read and agreed to the published version of the manuscript.

Funding: This research was funded by the High-Altitude Research Centre at Taif University, Taif, Saudi Arabia, under project number 1-442-42.

Institutional Review Board Statement: Not applicable.

Informed Consent Statement: Not applicable.

Data Availability Statement: Not applicable.

Acknowledgments: The authors acknowledge the Dean of Scientific Research at Taif University, Taif, Saudi Arabia, for funding this work through the High-Altitude Research Centre, under project number 1-442-42.

Conflicts of Interest: The authors declare no conflict of interest.

Appendix A

Table A1. HPLC analysis of the phenolic, flavonoid, and alkaloid contents of the different organs as well as the solid distillation wastes of Taif's rose collected from Al-Shafa highlands in Taif City, Saudi Arabia. SDW: solid distillation wastes; ND: not detected.

Chemical Compound		Plant Organ			
		Leaves	Stem	Flowers	SDW
Flavonoids concentration (mg g^{-1})	Quercetin	11.09	12.61	20.10	2.69
	Apigenin	4.59	23.83	47.02	11.25
	Luteolin	10.17	21.09	13.60	ND
	Chrysoeriol	19.20	45.19	ND	6.98
	Rutin	16.87	11.16	25.36	ND
	Kaempferol	21.59	27.55	19.10	9.68
	Avicularin	ND	ND	2.50	4.15
Phenolics concentration (mg g^{-1})	Gallic acid	14.21	23.53	5.41	ND
	Ellagic acid	7.94	14.90	11.69	7.25
	Catechol	13.70	6.59	8.60	ND
	Resorcinol	11.91	7.57	ND	8.79
	Phloroglucinol	6.24	0.93	1.69	3.45
Alkaloids concentration (mg g^{-1})	berbamine	3.03	5.28	ND	ND
	jatrorrhizine	5.10	5.83	5.60	4.26
	palmatine	ND	3.50	ND	3.44
	reticuline	1.88	8.50	5.30	ND
	isocorydine	1.60	4.70	9.11	4.77
	boldine	ND	0.64	4.79	3.56

Table A2. Antimicrobial activity (mm) of the different extracts of Taif's rose flowers on the pathogenic fungal and bacterial strains. NA: no activity.

Extract	**Fungi**		**Bacteria**		
			Gram- + ve Bacteria	**Gram- − ve Bacteria**	
	Aspergillus fumigatus	*Candida albicans*	*Bacillus subtilis*	*Escherichia coli*	*Proteus vulgaris*
	Ketoconazole		Gentamicin	Gentamicin	
Control	17	20	26	30	17
Methanol	NA	NA	NA	10	16
Ethanol	NA	NA	14	11	13
Boiling water	NA	NA	NA	13	18
Cold water	NA	NA	NA	12	18
Warm water	NA	NA	13	NA	14

Table A3. Antimicrobial activity (mm) of the different extracts of Taif's rose solid distillation wastes on the pathogenic fungal and bacterial strains. NA: no activity.

Extract	**Fungi**		**Bacteria**		
			Gram- + ve Bacteria	**Gram- − ve Bacteria**	
	Aspergillus fumigatus	*Candida albicans*	*Bacillus subtilis*	*Escherichia coli*	*Proteus vulgaris*
	Ketoconazole		Gentamicin	Gentamicin	
Control	17	20	26	30	17
Methanol	10	14	10	19	NA
Ethanol	NA	NA	NA	NA	NA
Boiling water	NA	NA	NA	NA	NA
Cold water	NA	NA	NA	14	12
Warm water	NA	NA	10	12	18

References

1. Galal, T.M.; Al-Yasi, H.M.; Fawzy, M.A.; Abdelkader, T.G.; Hamza, R.Z.; Eid, E.M.; Ali, E.F. Evaluation of the Phytochemical and Pharmacological Potential of Taif's Rose (*Rosa damascena* Mill var. *trigintipetala*) for Possible Recycling of Pruning Wastes. *Life* **2022**, *12*, 273. [PubMed]
2. Tsanaktsidis, C.G.; Tamoutsidis, E.; Kasapidis, G.; Itziou, A.; Ntina, E. Preliminary results on attributes of distillation products of the rose *Rosa damascene* as a dynamic and friendly to the environment rural crop. *APCBEE Procedia* **2012**, *1*, 66–73. [CrossRef]
3. Nunes, H.; Miguel, M.G. *Rosa damascena* essential oils: A brief review about chemical composition and biological properties. *Trends Pharmacol. Sci.* **2017**, *1*, 111–128.
4. Galal, T.M.; Majrashi, A.; Al-Yasi, H.M.; Farahat, E.A.; Eid, E.M.; Ali, E.F. Taif's Rose (*Rosa damascena* Mill var. *trigentipetala*) Wastes Are a Potential Candidate for Heavy Metals Remediation from Agricultural Soil. *Agriculture* **2022**, *12*, 1319.
5. Mileva, M.; Ilieva, Y.; Jovtchev, G.; Gateva, S.; Zaharieva, M.M.; Georgieva, A.; Dimitrova, L.; Dobreva, A.; Angelova, T.; Vilhelmova-Ilieva, N.; et al. Rose Flowers—A Delicate Perfume or a Natural Healer? *Biomolecules* **2021**, *11*, 127. [CrossRef]
6. Al-Yasi, H.; Attia, H.; Alamer, K.; Hassan, F.; Ali, E.; Elshazly, S. Impact of drought on growth, photosynthesis, osmotic adjustment, and cell wall elasticity in damask rose. *Plant Physiol. Biochem.* **2020**, *150*, 133–139. [CrossRef]
7. Pal, P.K. Evaluation, genetic diversity, recent development of distillation method, challenges and opportunities of *Rosa damascena*: A review. *J. Essent. Oil Bear. Plants* **2013**, *16*, 1–10. [CrossRef]
8. Labban, L.; Thallaj, N. The medicinal and pharmacological properties of Damascene Rose (*Rosa damascena*): A review. *Int. J. Herb. Med.* **2020**, *8*, 33–37.
9. Ghavam, M.; Afzali, A.; Manconi, M.; Bacchetta, G.; Manca, M.L. Variability in chemical composition and antimicrobial activity of essential oil of *Rosa* × *damascena* Herrm. from mountainous regions of Iran. *Chem. Biol. Technol. Agric.* **2021**, *8*, 22. [CrossRef]
10. Hamza, R.Z.; Al-Yasi, H.M.; Ali, E.F.; Fawzy, M.A.; Abdelkader, T.G.; Galal, T.M. Chemical characterization of Taif rose (*Rosa damascena* Mill var. trigentipetala) waste methanolic extract and its hepatoprotective and antioxidant effects against cadmium chloride ($CdCl_2$)-induced hepatotoxicity and potential anticancer activities against liver cancer cells (HepG2). *Crystals* **2022**, *12*, 460.
11. Moein, M.; Etemadfard, H.; Zarshenas, M.M. Investigation of different Damask rose (*Rosa damascena* Mill.) oil samples from traditional markets in Fars (Iran); focusing on the extraction method. *Trends Pharmacol. Sci.* **2016**, *2*, 51–58.

12. Alizadeh, Z.; Fattahi, M. Essential oil, total phenolic, flavonoids, anthocyanins, carotenoids and antioxidant activity of cultivated Damask Rose (*Rosa damascena*) from Iran: With chemotyping approach concerning morphology and composition. *Sci. Hortic.* **2021**, *288*, 110341. [CrossRef]
13. Rusanov, K.; Kovacheva, N.; Atanassov, A.; Atanassov, I. *Rosa damascena* Mill., the oil-bearing Damask rose: Genetic resources, diversity and perspectives for molecular breeding. *Floricul Ornam Biotech.* **2009**, *3*, 14–20.
14. Thakur, M.; Sharma, S.; Sharma, U.; Kumar, R. Study on effect of pruning time on growth, yield and quality of scented rose (*Rosa damascena* Mill.) varieties under acidic conditions of western Himalayas. *J. Appl. Res. Med. Aromat. Plants* **2019**, *13*, 100202. [CrossRef]
15. Nasir, M.H.; Nadeem, R.; Akhtar, K.; Hanif, M.A.; Khalid, A.M. Efficacy of modified distillation sludge of rose (*Rosa centifolia*) petals for lead (II) and zinc (II) removal from aqueous solutions. *J. Hazard. Mater.* **2007**, *147*, 1006–1014. [CrossRef]
16. Slavov, A.; Kiyohara, H.; Yamada, H. Immunomodulating pectic polysaccharides from waste rose petals of *Rosa damascena* Mill. *Int. J. Biol. Macromol.* **2013**, *59*, 192–200. [CrossRef] [PubMed]
17. Galal, T.M.; Ali, E.F.; Eid, E.M.; Al-Yasi, H.M.; Magrashi, A.; Althobaiti, F.; Farahat, E.A. Evaluating the nutrient contents and nutritive value of Taif's Rose (*Rosa damascena* Mill var. trigintipetala) wastes to be used as animal forage or soil organic fertilizers. *Agriculture* **2022**, *12*, 1481.
18. Liu, W.; Chen, L.; Huang, Y.; Fu, L.; Song, L.; Wang, Y.; Bai, Z.; Meng, F.; Bi, Y. Antioxidation and active constituents analysis of flower residue of *Rosa damascena*. *Chin. Herb. Med.* **2020**, *12*, 336–341. [CrossRef]
19. Slavov, A.; Panchev, I.; Kovacheva, D.; Vasileva, I. Physico-chemical characterization of water-soluble pectic extracts from Rosa damascena, *Calendula officinalis* and *Matricaria chamomilla* wastes. *Food Hydrocoll.* **2016**, *61*, 469–476. [CrossRef]
20. Slavov, A.; Denev, P.; Panchev, I.; Shikov, V.; Nenov, N.; Yantcheva, N.; Vasileva, I. Combined recovery of polysaccharides and polyphenols from *Rosa damascena* wastes. *Ind. Crops Prod.* **2017**, *100*, 85–94. [CrossRef]
21. Slavov, A.; Vasileva, I.; Stefanov, L.; Stoyanova, A. Valorization of wastes from the rose oil industry. *Rev. Environ. Sci. Biotechnol.* **2017**, *16*, 309–325. [CrossRef]
22. Dodevska, T.; Vasileva, I.; Denev, P.; Karashanova, D.; Georgieva, B.; Kovacheva, D.; Yantcheva, N.; Slavov, A. *Rosa damascena* waste mediated synthesis of silver nanoparticles: Characteristics and application for an electrochemical sensing of hydrogen peroxide and vanillin. *Mater. Chem. Phys.* **2019**, *231*, 335–343. [CrossRef]
23. Halawani, E.M. Antimicrobial activity of *Rosa damascena* petals extracts and chemical composition by gas chromatography-mass spectrometry (GC/MS) analysis. *Afr. J. Microbiol. Res.* **2014**, *8*, 2359–2367.
24. Nayebi, N.; Khalili, N.; Kamalinejad, M.; Emtiazy, M. A systematic review of the efficacy and safety of *Rosa damascena* Mill. with an overview on its phytopharmacological properties. *Complement. Ther. Med.* **2017**, *34*, 129–140. [CrossRef] [PubMed]
25. Davoodi, I.; Rahimi, R.; Abdollahi, M.; Farzaei, F.; Farzaei, M.H.; Memariani, Z.; Najafi, F. Promising effect of *Rosa damascena* extract on high-fat diet-induced nonalcoholic fatty liver. *J. Tradit. Complement. Med.* **2017**, *7*, 508–514. [CrossRef]
26. Kumar, R.; Sharma, S.; Kaundal, M.; Sharma, S.; Thakur, M. Response of damask rose (*Rosa damascena* Mill.) to foliar application of magnesium (Mg), copper (Cu) and zinc (Zn) sulphate under western Himalayas. *Ind. Crops Prod.* **2016**, *83*, 596–602. [CrossRef]
27. Pal, P.K.; Mahajan, M. Pruning system and foliar application of $MgSO_4$ alter yield and secondary metabolite profile of *Rosa damascena* under rainfed acidic conditions. *Front. Plant Sci.* **2017**, *8*, 507. [CrossRef]
28. Celano, R.; Piccinelli, A.L.; Pagano, I.; Roscigno, G.; Campone, L.; De Falco, E.; Russo, M.; Rastrelli, L. Oil distillation wastewaters from aromatic herbs as new natural source of antioxidant compounds. *Food Res. Int.* **2017**, *99*, 298–307. [CrossRef]
29. Onursal, E.; Ekinci, K. Co-composting of rose oil processing waste with caged layer manure and straw or sawdust: Effects of carbon source and C/N ratio on decomposition. *Waste Manag. Res.* **2015**, *33*, 332–338. [CrossRef]
30. Sadasivam, S.; Manickam, A. *Biochemical Methods*, 3rd ed.; New Age International Publishers: New Delhi, India, 2008.
31. Solich, P.; Sedliaková, V.; Karlíček, R. Spectrophotometric determination of cardiac glycosides by flow-injection analysis. *Anal. Chim. Acta* **1992**, *269*, 199–203. [CrossRef]
32. Tofighi, Z.; Ghazi, N.; Hadjiakhoondi, A.; Yassa, N. Determination of cardiac glycosides and total phenols in different generations of *Securigera securidaca* suspension culture. *Res. J. Pharmacogn.* **2016**, *3*, 25–31.
33. Singleton, V.L.; Orthofer, R.; Rosa, M.B.T.; Raventós, L. Analysis of total phenols and other oxidation substrates and antioxidants by means of Folin-Ciocalteu reagent. *Methods Enzymol.* **1999**, *299*, 152–178.
34. Schütz, K.; Kammerer, D.R.; Carle, R.; Schieber, A. Characterization of phenolic acids and flavonoids in dandelion (*Taraxacum officinale* WEB. Ex WIGG) root and herb by high performance liquid chromatography/electrospray ionization mass spectrometry. *Rapid Commun. Mass Spectrom.* **2005**, *19*, 179–186. [PubMed]
35. Zheng, W.; Clifford, M.N. Profiling the chlorogenic acids of sweet potato (*Ipomea batatas*) from China. *Food Chem.* **2008**, *106*, 147–152. [CrossRef]
36. Kandil, O.; Radwan, N.M.; Hassan, A.B.; Amer, A.M.; El-Banna, H.A.; Amer, W.M. Extracts and fractions of *Thymus capitatus* exhibit antimicrobial activities. *J. Ethnopharmacol.* **1994**, *44*, 19–24. [CrossRef]
37. Das, M.M.; Deka, D.C. Evaluation of anticancer and antimicrobial activity of arborinine from Glycosmis pentaphylla. *J. Biol. Act. Prod. Nat.* **2017**, *7*, 131–139. [CrossRef]
38. Abd El-Kader, H.A.; Sedde, S.R.; El-Shanawany, A.A. In vitro study of the effect of some medicinal plants on the growth of some dermatophytes. *Assiut Vet. Med. J.* **1995**, *34*, 36–42.
39. Allen, S. *Chemical Analysis of Ecological Materials*; Blackwell Scientific Publications: London, UK, 1989.

40. Chapman, H.D.; Pratt, P.F. *Methods of Analysis for Soil, Plants and Water*; University of California, Division of Agricultural Sciences: Los Angeles, CA, USA, 1961; pp. 233–234.
41. Le Houérou, H.N. Chemical composition and nutritive value of browse in tropical West Africa. In *Browse in Africa*; Le Houérou, M.N., Ed.; ILCA: Addis Ababa, Ethiopia, 1980; pp. 261–289.
42. Goering, H.K.; Van Soest, P.J. Forage fiber analysis (apparatus, reagents, procedures, and some applications). In *Agricultural Handbook 379*; United States Department of Agriculture: Washington, DC, USA, 1970.
43. Van Soest, P.J.; Robertson, J.B.; Lewis, B.A. Methods of dietary fiber, neutral detergent fiber, and non-starch polysaccharides in relation to animal nutrition. *J. Dairy Sci.* **1991**, *74*, 3583–3597. [CrossRef]
44. NRC (National Research Council). *Nutrient Requirements of Beef Cattle*, 7th ed.; updated 2000; The National Academies Press: Washington, DC, USA, 2000.
45. NRC (National Research Council). *Nutrient Requirements of Dairy Cattle*, 8th ed.; NRC (National Research Council): Ottawa, ON, Canada, 2001.
46. SPSS. *IBM SPSS Statistics Version 21.0.*; IBM Corp.: Armonk, NY, USA, 2012.
47. Alzletni, H.; Galal, T.; Khalafallah, A. *The Arable Weed Malva parviflora L., Ecophysiology and Phytochemistry*; Lambert Academic Publishing Gmbh & Co.KG.: Saarbrücken, Germany, 2020; 220p, ISBN 978-3-613-97570-9.
48. Tatke, P.; Satyapal, U.S.; Mahajan, D.C.; Naharwar, V. Phytochemical analysis, In-vitro antioxidant and antimicrobial activities of flower petals of *Rosa damascena*. *Int. J. Pharmacogn. Phytochem. Res.* **2015**, *7*, 246–250.
49. Fathima, S.N.; Murthy, S.V. Cardioprotective effects to chronic administration of *Rosa damascena* petals in isoproterenol induced myocardial infarction: Biochemical, histopathological and ultrastructural studies. *Biomed. Pharmacol. J.* **2019**, *12*, 1155–1166. [CrossRef]
50. Qureshi, M.N.; Stecher, G.; Sultana, T.; Abel, G.; Popp, M.; Bonn, G.K. Determination of carbohydrates in medicinal plants-comparison between TLC, mf-MELDI-MS and GC-MS. *Phytochem. Anal.* **2011**, *22*, 296–302. [CrossRef] [PubMed]
51. Kimbel, J.C.; Carpenter, S.R. Effects of mechanical harvesting on *Myriophyllum spicatum* L. regrowth and carbohydrate allocation to roots and shoots. *Aquat. Bot.* **1981**, *11*, 121–127. [CrossRef]
52. Aseeri, S.A.; Al-Yasi, H.M.; Galal, T.M. *Aloe Species in the Kingdom of Saudi Arabia: Morphological, Phytochemical and Molecular Characterization*; Lambert Academic Publishing Gmbh & Co.KG.: Saarbrücken, Germany, 2020; ISBN 978-620-2-66802-6.
53. El-Bakry, A.A.; Hammad, A.; Galal, T.M.; Ghazi, M.; Rafat, F.A. Polymorphism in *Calotropis procera*: Variation of metabolites in populations from different phytogeographical regions of Egypt. *Rend. Fis. Acc. Lincei* **2014**, *25*, 461–469. [CrossRef]
54. Dahat, Y.; Saha, P.; Mathew, J.T.; Chaudhary, S.K.; Srivastava, A.K.; Kumar, D. Traditional uses, phytochemistry and pharmacological attributes of *Pterocarpus santalinus* and future directions: A review. *J. Ethnopharmacol.* **2021**, *276*, 114127. [CrossRef]
55. Sharma, A.; Gupta, H. Quercetin-a flavonoid. Chron. *Young Sci.* **2010**, *1*, 10–15.
56. Adom, M.B.; Taher, M.; Mutalabisin, M.F.; Amri, M.S.; Abdul Kudos, M.B.; Sulaiman, M.W.; Sengupta, P.; Susanti, D. Chemical constituents and medical benefits of *Plantago major*. *Biomed. Pharmacother.* **2017**, *96*, 348–360. [CrossRef] [PubMed]
57. Marzouk, M.S.; Soliman, F.M.; Shehata, I.A.; Rabee, M.; Fawzy, G.A. Flavonoids and biological activities of *Jussiaea repens*. *Nat. Prod. Res.* **2007**, *21*, 436–443. [CrossRef]
58. Marlene, R.P.; Camila, K.P.; Cesar, M.B.; Evelyn, W.; Tânia, B.C.; Claudriana, L. Gallic acid and dodecyl gallate prevents carbon tetrachloride-induced acute and chronic hepatotoxicity by enhancing hepatic antioxidant status and increasing p53 expression. *Biol. Pharmaceut. Bull. J.* **2017**, *40*, 425–434.
59. Yanni, Y.; Mengyao, W.; Yingjie, H.; Chuankai, L.; Xin, P.; Wen, Z.; Youyi, H. Appropriately raising fermentation temperature beneficial to the increase of antioxidant activity and gallic acid content in *Eurotium cristatum*-fermented loose tea. *LWT-Food Sci. Technol.* **2017**, *82*, 248–254.
60. Algradi, A.M.; Liu, Y.; Yang, B.; Kuang, H. Review on the genus *Brugmansia*: Traditional usage, phytochemistry, pharmacology, and toxicity. *J. Ethnopharmacol.* **2021**, *279*, 113910. [CrossRef]
61. Petruczynik, A. Analysis of alkaloids from different chemical groups by different liquid chromatography methods. *Cent. Eur. J. Chem.* **2012**, *10*, 802–835. [CrossRef]
62. Li, T.S.C. *Chinese and Related North American Herbs: Phytopharmacology and Therapeutic Values*; CRC Press LLC: Boca Raton, FL, USA, 2000; 611p.
63. Duke, J.A. *Duke's Handbook of Medicinal Plants of Latin America*; CRC Press: Boca Raton, FL, USA; Taylor & Francis Group, LLC: Abingdon, UK, 2009; 962p.
64. Metiner, K.; Özkan, O.; Seyyal, A.K. Antibacterial Effects of Ethanol and Acetone Extract of *Plantago major* L. on Gram Positive and Gram Negative Bacteria. *Kafkas Univ. Vet. Fak. Derg.* **2012**, *18*, 503–505.
65. Saeed, R.; Ali, S.; Ullah, H.; Ullah, M.; Hassan, S.; Ahmed, S.; Akhwan, S. Phytochemical analysis and anti-microbial activities of *Rosa indica* collected from Kohat Pakistan. *Am. J. Phytomed. Clin. Ther.* **2014**, *2*, 1370–1377.
66. Norziah, M.H.; Fezea, F.A.; Bhat, R.; Ahmad, M. Effect of extraction solvents on antioxidant and antimicrobial properties of fenugreek seeds (*Trigonella foenum-graecum* L.). *Int. Food Res. J.* **2015**, *22*, 1261–1271.
67. Chang, K.H.; Wu, R.Y.; Chuang, K.C.; Hsieh, T.F.; Chung, R.S. Effects of chemical and organic fertilizers on the growth, flower quality and nutrient uptake of *Anthurium andreanum*, cultivated for cut flower production. *Sci. Hortic.* **2010**, *125*, 434–441. [CrossRef]
68. Khater, E.S.G. Some physical and chemical properties of compost. *Int. J. Waste Resour.* **2015**, *5*, 172. [CrossRef]

69. Roig, A.; Cayuela, M.L.; Sánchez-Monedero, M.A. An overview on olive mill wastes and their valorisation methods. *Waste Manag.* **2006**, *26*, 960–969. [CrossRef]
70. Vasileva, V.; Naydenova, Y.; Stoycheva, I. Nutritive value of forage biomass from sainfoin mixtures. *Saudi J. Biol. Sci.* **2019**, *26*, 942–949. [CrossRef]
71. Shaltout, K.H.; Galal, T.M.; El-Komy, T.M. Evaluation of the nutrient status of some hydrophytes in the water courses of Nile Delta, Egypt. *Ecol. Mediterr.* **2010**, *36*, 77–87. [CrossRef]
72. Batajoo, K.K.; Shaver, R.D. Impact of nonfiber carbohydrate on intake, digestion, and milk production by dairy cows. *J. Dairy Sci.* **1994**, *77*, 1580–1587. [CrossRef]
73. Kholif, A.E.; Gouda, G.A.; Morsy, T.A.; Salem, A.Z.M.; Lopez, S.; Kholif, A.M. *Moringa oleifera* leaf meal as a protein source in lactating goat's diets: Feed intake, digestibility, ruminal fermentation, milk yield and compositio n, and its fatty acids profile. *Small Rumin. Res.* **2015**, *129*, 29–137. [CrossRef]
74. Blümmel, M.; Updahyay, S.R.; Gautam, N.; Barma, N.C.D.; Abdul Hakim, M.; Hussain, M.; Joshi, A.K. Comparative assessment of food-fodder traits in a wide range of wheat germplasm for diverse biophysical target domains in South Asia. *Field Crops Res.* **2019**, *236*, 68–74. [CrossRef]
75. Nelson, M.L.; Finley, J.W.; Scarnecchia, D.L.; Parish, S.M. Diet and forage quality of intermediate wheatgrass managed under continuous and short-duration grazing. *J. Range Manag.* **1989**, *42*, 474–477. [CrossRef]
76. Coleman, S.W.; Rao, S.C.; Volesky, J.D.; Phillips, W.A. Growth and nutritive value of perennial C3 grasses in the southern Great Plains. *Crop Sci.* **2010**, *50*, 1070–1078. [CrossRef]
77. Favre, J.R.; Castiblanco, T.M.; Combs, D.K.; Wattiaux, M.A.; Picasso, V.D. Forage nutritive value and predicted fiber digestibility of Kernza intermediate wheatgrass in monoculture and in mixture with red clover during the first production year. *Anim. Feed. Sci. Technol.* **2019**, *258*, 114298. [CrossRef]
78. Heneidy, S.Z.; Halmy, M.W. The nutritive value and role of *Panicum turgidum* Forssk. in the arid ecosystems of the Egyptian desert. *Acta Bot. Croat.* **2009**, *68*, 127–146.
79. Krachunov, I. Estimation of energy feeding value of forages for ruminants II. Energy prediction through crude fiber content. *J. Mt. Agric. Balk.* **2007**, *10*, 122–134.
80. MAFF (Ministry of Agriculture, Fisheries, and Food). *Energy Allowances and Feeding Systems for Ruminants*; Technical Bulletin 33; Her Majesty's Stationary Office: London, UK, 1975.
81. NRC. *Nutrient Requirements of Domestic Animals: Nutrient Requirement of Cheep*, 6th ed.; Research Council Pamphelets No. 5; National Academic Sciences: Washington, DC, USA, 1985.
82. Wu, G. *Principles of Animal Nutrition*; CRC Press: Boca Raton, FL, USA; London, UK; New York, NY, USA, 2018.
83. El-Kady H Seasonal variation in phytomass and nutrient status of *Phragmites australis* along the water courses in the Middle Delta region. *Taeckholmia* **2002**, *20*, 123–138.
84. Shaltout, K.H.; Galal, T.M.; El-Komy, T.M. Nutrients and heavy metals accumulation in the aboveground biomass of two perennial grasses along the water courses of Nile Delta, Egypt. In Proceedings of the Egyptian Journal of Botany, 3rd International Congress, Helwan, Egypt, 17–18 April 2013.
85. Norton, B.W. Differences between species in forage quality. In *Nutritional Limits to Animal Production from Pastures*; Hacker, J.B., Ed.; CAB: Farnham Royal, UK, 1982; pp. 98–110.
86. Cherian, G. *A Guide to the Principles of Animal Nutrition*; Open textbook library, Oregon State Open Educational Resources: Corvallis, IL, USA, 2020; 163p.
87. Shoukry, M.M. An Actual Vision about the Availability of the Utilization of Water Hyacinth in Feeding Ruminants. In *National Symposium on Water Hyacinth*; Assiut University: Assiut, Egypt, 1992; pp. 75–92.
88. Gill, K.S.; Omokanye, A. Spring triticale varieties forage yield, nutrients composition, and suitability for beef cattle production. *J. Agric. Sci.* **2016**, *8*, 1–14. [CrossRef]

agriculture

MDPI

Article

Properties of Humic Substances in Composts Comprised of Different Organic Source Material

Marge Lanno [1,*], Maris Klavins [2], Oskars Purmalis [2], Merrit Shanskiy [1], Anu Kisand [1] and Mait Kriipsalu [3]

1 Institute of Agricultural and Environmental Sciences, Estonian University of Life Sciences, 51006 Tartu, Estonia
2 Department of Environmental Sciences, University of Latvia, 1586 Riga, Latvia
3 Institute of Forestry and Rural Engineering, Estonian University of Life Sciences, 51006 Tartu, Estonia
* Correspondence: marge.lanno@emu.ee

Abstract: Reusing organic waste as fertilizer is one method to reduce the use of mineral fertilizers and minimize waste disposal in landfills. Regulations have been enacted for the processing of organic waste and for recycling end products, but the humic content of organic fertilizers has been neglected. We studied seven composts with different organic input materials and technologies. Humic substances (HSs) were detected in all composts. The total organic carbon in the HSs constituted 8.7 ± 0.1% (SD)–27.0 ± 0.2% of the compost dry matter. Spectral differences between the studied samples in FTIR spectroscopy could be observed at 1700–1000 cm^{-1}, indicating differences in compost precursor material. The EEM peak, associated with humic acids (HAs), was high in composts containing animal by-products (e.g., fish waste, horse manure, and kitchen biowaste). Kitchen biowaste, also when processed by *Hermetia illucens* larvae and vermicompost, exhibited slower organic material transformation with low humic acid/fulvic acid ratios (<1.60). The results show the importance of source material origin and amendments, which influence the composting process and final products. Our study emphasizes the role of humic substances in the comprehensive evaluation of composts. To maximize the added value of composts, marketing strategies should consider determining the share of humic substances besides the content of organic matter and nutrients.

Keywords: humic acid; fulvic acid; fish waste; horse manure; sewage sludge; *Hermetia illucens* frass; green waste; kitchen biowaste

Citation: Lanno, M.; Klavins, M.; Purmalis, O.; Shanskiy, M.; Kisand, A.; Kriipsalu, M. Properties of Humic Substances in Composts Comprised of Different Organic Source Material. *Agriculture* **2022**, *12*, 1797. https://doi.org/10.3390/agriculture12111797

Academic Editor: Maria Roulia

Received: 4 October 2022
Accepted: 24 October 2022
Published: 28 October 2022

1. Introduction

With rapid urbanization and population growth, global annual waste is expected to increase by 3.4 billion tons by 2050 [1]. It is estimated that, across the European Union (EU), up to 138 million tons of bio-waste is generated annually, of which only 60 million tons is recycled through composting and anaerobic digestion [2]. About 120–130 tons of bio-waste are generated in Estonia annually, of which 30% is recycled. Unsegregated bio-waste is incinerated or disposed in a landfill [3]. Managing this waste to minimize environmental impacts is increasingly important. Waste disposal, especially organic waste disposal in landfills, causes various environmental impacts, including landfill gases, contributing to air pollution and climate warming and leachate contaminating groundwater [4]. Within the framework of a circular economy, EU waste policies (directives 2018/850 and 2018/851) aim to reduce the landfilling of municipal waste to 10% of current levels by 2035 and to increase the re-use and recycling of municipal waste up to 65% [5,6]. This ambitious policy objective is only achievable by implementing organic-waste-management strategies.

The dependence on mineral fertilizers for agriculture can threaten food security if they become less available. Limited reserves of mineral fertilizer raw materials and changes in supply chains due to various environmental, political, and epidemiological crises all affect the price and availability of mineral fertilizers [7,8]. At the same time, the EU is conducting the Green Deal, which aims to reduce greenhouse gas pollution and slow down climate change [8]. To increase self-sufficiency and reduce environmental impacts, the supply of alternative fertilizers could be increased at the local level.

It was stated within EU regulation 2019/1009 that organic fertilizer shall contain organic carbon (C_{org}) and nutrients of a solely biological origin [9]. Composting is a common biological treatment option, suitable for processing organic waste from various sources. Mature compost is a stable and nutrient-rich, humus-like product, representing a valuable source of recovered nutrients. The concentration of nutrients is determined by input materials. The determination of total nutrient contents for fertilizing purposes is regulated, but results do not describe the proportion of nutrients that are readily available to plants [10,11]. The quality of mature compost is described as a combination of two criteria. First, the precautionary criteria (hygiene parameters, impurities, weed seeds, and inorganic pollutants, such as heavy metals) set limits to prevent environmental pollution and the spread of diseases [9]. Second, quality criteria provide information about fertilizing properties, such as nitrogen (N), phosphorus (P), potassium (K) and micronutrients, organic matter content, salinity, and pH [12,13].

Composts tend to be compared to mineral fertilizers, despite having more complex compositions [14,15]. A distinctive feature of compost is the presence of natural organic matter, reflecting a wide range of environmental and biowaste degradation processes. In Europe, an estimated 45% of soils are low in organic matter content, which lowers soil productivity and increases the risk of soil degradation [16]. With climate change, it is estimated that global air temperatures will increase by 2 °C by 2100 compared to pre-industrial levels, which will change environmental conditions and agricultural production [8,17]. Soils with a higher organic matter content are more resistant to drought and climate change [18]. Organic matter is a nutrient reservoir and can retain nutrients in a plant-available form. Compost can also help restore humus in degraded soils [14,18,19]. While compost has various quality standards regarding fertilizing properties and has limits to prevent environmental pollution and the spread of diseases, it is often overlooked that compost adds humic substances (HSs) to the soil [9,20]. HSs, including humic acids (HAs), fulvic acids (FAs), and humin, are part of the soil organic matter (SOM) that forms humus [21]. The formation of fertile soil layers rich in humic substances can take decades in nature, but it takes 6–12 months to produce a humic-acid-like substance by composting [22,23]. HSs are important in soil restoration processes and can be used to promote nutrient uptake, increase soil porosity, enhance nutrient preservation and water-holding capacity, and reduce the abundance of pathogens [24,25]. HS concentration in the compost is correlated with compost maturity and provides added value to the waste material [26,27]. Therefore, the benefits of composts are described by nutrient and HS contents. These natural biostimulants can be an important marketing tool for composts, which are primarily used in agriculture.

The aim of this study is to compare various composts to quantify HSs and determine whether HS content depends on input materials (manure, sewage sludge, vermicompost, fish, food, yard waste, and black soldier frass) or the selected treatment. These results will draw attention to an important feature of compost that has been underappreciated in compost evaluation and marketing processes.

2. Materials and Methods

2.1. Selection of Composts

Seven composts of different source materials were investigated. Composts were produced from low-valued fish (C1), horse manure (C2), green waste (e.g., branches, leaves, plant residues) (C3), sewage sludge (C4), and kitchen biowaste (C5). In addition, *Hermetia illucens* frass (C6) and vermicompost (C7) were selected. All composts were prepared in Estonia by different manufacturers or by the authors. Four composts (C2–C5) were industrially produced and commercially marketed. Composts C3, C4, and C5 were certified according to the Estonian end-of-waste regulation [28], while certification was not required for C2 because manures are not classified as waste. Composts C1, C6, and C7 were experimental (Table 1), and compost amendments and composting technologies were determined by the manufacturer. Straw is a dry and widely available bulking material in Estonia; therefore, it is well-suited for composting wet waste, such as sludge. Shredded wood, another common bulking material, and straw increase porosity, improve air circulation, and adjust C/N ratios (20–30)/1.

Table 1. Characterization of composts. The active period (d) indicates the duration that composts C1–C5 were in outdoor windrows, and for C6–C7, the feeding period of *H. illucens* larvae or *E. fetida*, respectively.

Compost	Target Waste	Amendments	Scale	Certified Quality Regulations	Active Period (d)	Composting Site Location
C1	Fish waste	Straw, peat, inoculum from previous composts	Industrial	No	185	58°28′41.08″ N 24°49′28.66″ E
C2 *	Horse Manure	Straw	Industrial	Yes	90	58°28′41.08″ N 24°49′28.66″ E
C3 *	Sewage sludge	Straw	Industrial	Yes	120	58°13′37.89″ N 26°23′10.17″ E
C4 *	Green waste from municipal areas	None	Industrial	Yes	>180	59°28′14.63″ N 24°54′43.97″ E
C5 *	Kitchen Biowaste	Shredded wood	Industrial	Yes	130	59°27′36.08″ N 25°5′17.18″ E
C6	Cafeteria and grocery store biowaste	None	Pilot	No	51	58°23′30.76″ N 26°41′40.43″ E
C7	Kitchen biowaste	Straw, paper, green waste, biochar	Pilot	No	365	58°23′30.76″ N 26°41′40.43″ E

* Commercially produced compost.

Composts C1–C5 were processed on an industrial scale in outdoor windrows and mixed mechanically with a Backhus 16.30 windrow turner (Eggersmann Anlagenbau GmbH, Bad Oeynhausen, Germany) or an CMC ST 300 windrow turner (Composts Systems GmbH, Wels, Austria). All windrows were approximately 1.5 m high, 3 m wide, and 12 m long and were covered with semi-permeable geotextile KSV 200 (Compost Systems GmbH, Wels, Austria). All composts were maintained in accordance with national biowaste recycling regulations [28,29]. For waste containing animal by-products (ABPRs), such as fish waste and horse manure, the temperature exceeded 70 °C for at least 1 h, as required by EU regulations (EU 142/2011) [10]. To compost biodegradable waste, including kitchen waste and sewage sludge, the temperature had to exceed 55 °C for 10 d or 65 °C for at least 3 d [12,13]. The active composting period was considered complete when the temperature dropped to an ambient level, and after mixing the windrows, no significant temperature increase occurred.

Compost C1 was an experimental treatment made by the authors at an industrial scale. For C1, the targeted waste material, measured as wet weight, was fish by-catch (ABPRs, 22%) amended with structural materials, wheat straw (22%), peat (22%), and inoculated with composts from similar previous batches (34%). Peat and mature compost, which have a small particle size, also served as a binder for fish material [30]. Compost C2 was composed of horse manure (ABPRs, 60%) amended with straw (40%). Compost C3 consisted of municipal sewage sludge (WWTP with a load of 16.733 PE) amended with straw. The ratio of sludge to straw was 1:1 by volume, resulting in a 9:1 ratio by weight. Compost C4 was composed of biowaste from urban green areas. Compost C5 was made of source-separated household biowaste amended with shredded wood (1:1 by volume) and 6% of category III ABPRs from the chicken industry. Treatments C2–C5 were compatible with national compost quality regulations [28].

Treatments C6 and C7 were prepared on a pilot scale. For C6, *H. illucens* larvae were fed for 51 d, with the frass in contact with fresh feed. The frass was sieved off every two weeks during the cycle. Feed consisted of approximately 150 L of raw vegetable and fruit waste. To ensure a suitable moisture content, hot-pressed rapeseed cake, rye/wheat bread, and 2 L of canola oil were used. In addition to vegetable and fruit waste, approximately 90 L of catering waste was used.

For treatment C7, *Eisenia fetida* were fed in quadruplicate three-liter mesocosms for 136 d. In the beginning of the experiment, 20 earthworms were applied in each mesocosm (average mass 4.12 g). The experiment was conducted in a temperature-controlled room at 22 °C. Moisture was controlled weekly using the hand-held W.E.T. Sensor Kit (Delta-T Devices, Cambridge, England). Vermicompost moisture was kept in a range of 75–80%, as recommended by Khan et al. [31]. The organic feed was mixed with 50 g of biochar (Biolan Baltic OÜ, Lavassaare, Estonia), with properties described by Escuer, 2021 [32]. Biochar was added to increase reproduction and create favorable conditions for *E. fetida*, while also enhancing microbial biodegradation and the composting process [31,33]. For amendments, milled straw, garden and kitchen green waste, and shredded paper were used. Feed calculations assumed that individual *E. fetida* consume half of its body mass per day. Later, feeding was adjusted based on earthworm performance in the mesocosms. For the first feeding, 200 g of feed with 50 g of biochar was applied.

2.2. Chemical Analyses

Approximately 250 g of mature compost from three random locations were collected and homogenized from each treatment. For treatments C1–C5, the compost samples were collected 20–30 cm inside the windrow. For treatment C6, *H. illucens* frass was sieved and homogenized before sampling. For treatment C7, subsamples were taken from four replicate mesocosms and mixed to obtain one homogenous sample per mesocosm. Subsamples were collected based on volume, and earthworms and cocoons were separated by hand.

Samples were dried at 105 °C and ball-milled prior to analysis. Elemental analysis (C, N) was conducted using a VarioMAX CNS analyzer (ELEMENTAR, Langenselbold Germany). Potassium (K), calcium (Ca), magnesium (Mg) contents, and phosphorus (P) were determined using the ammonium acetate–lactate (AL) method as described by Egner et al. [34].

Lipids were extracted using 10 g of dried compost sample and 10 mL of $CHCl_3$. The extract was filtered after 24 h through filter paper into a pre-weighed Petri dish. The air-dried lipids were weighed, and their proportion in the compost was calculated.

2.3. Extraction of Humic and Fulvic Acids

For HA extraction, 5 g compost and 50 mL 2% NaOH solution were combined and stirred for 4 h. The suspension was filtered and acidified with 6M HCl under constant stirring until pH = 2 was reached, HA precipitated, and only FA remained in the solution. Freeze-dried HA were used to prepare HA solution (50 mg/L) using 0.1 NaOH [35].

Characterization of isolated humic and fulvic acids were conducted as suggested by Klavins and Purmalis [36]. The ratio of HA/FA was calculated the using total organic carbon (TOC) values in HA and FA solutions. TOC analyses were performed using a Shimadzu TOC-V CSN total carbon analyzer (Shimadzu Corporation, Kyoto, Japan), operating at 720 °C.

2.4. UV–Vis Absorption Spectra of HA

The absorbance of solubilized HA (50 mg/L) in 0.1 M NaOH was measured using a UV–Vis spectrophotometer Shimadzu UV-1800 (Shimadzu Corporation, Kyoto, Japan), with a 1 cm quartz cuvette and a wavelength range of 230–670 nm. Absorbance ratios, E_2/E_3, were calculated using UV–Vis absorption at 250 nm and 365 nm, and E_4/E_6 was calculated from the UV–Vis absorption at 465 nm and 665 nm [37,38].

2.5. Fourier-Transform Infrared Absorption Spectra (FTIR) of HA

To obtain the spectra, 3 mg of HA sample and 200 mg of KBr powder were mixed together, and 30 mg of the mixture was weighed and compressed into a tablet. The FTIR spectra were obtained using a Fourier-transform infrared spectrophotometer Shimadzu IR-Tracer 100 (Shimadzu Corporation, Kyoto, Japan), at an infrared spectrum of 400 to 4000 cm^{-1}, with a resolution 4 cm^{-1}, and by taking 10 scans.

2.6. Excitation Emission Matrix (EEM) Spectroscopy of HA and FA

Extracted HA and FA samples were analyzed using fluorescence spectroscopy TL spectrometer HORIBA Aqualog (Horiba Ltd., Kyoto, Japan). The EEM spectra were measured at an emission wavelength range of 250–600 nm and an excitation wavelength range of 250–600 nm scanning with 5 nm increments. The obtained spectra were evaluated with PARAFAC analysis using MATLAB R2014 a v. 5.3.0.532 software with the DOMFluor and drEEM toolboxes [39].

2.7. Statistical Analysis

Statistical analysis was performed in R [40]. The results of elemental composition, TOC, and lipids results were presented as mean values of triplicate measurements and compared using ANOVA with a post-hoc Tukey's HSD test [41].

3. Results

The nutrient content and pH of the composts is presented in Table 2. Composts C3 and C6 had a remarkably high P concentration, and C5 showed the lowest P and highest N concentration, concurrently. Compost C6 had the lowest Ca and Mg content, while both of these nutrients were high in C2 and C4. The pH of the treatments ranged from 5.77 to 8.10.

Table 2. Elemental composition and pH of the composts. Mean ± standard deviation (SD) of triplicate measurements compared using ANOVA with post-hoc Tukey's HSD test.

Compost	N	C	C/N	Ca	K	Mg	P	pH
	%		ratio		µg/g			
C1	4.3 ± 0.1	65.3 ± 0.3	15.19	493 ± 8	1910 ± 25	80 ± 5	1469 ± 18	5.77
C2	4.5 ± 0.1	62.7 ± 0.3	13.93	1472 ± 20	4700 ± 30	234 ± 5	1122 ± 20	7.40
C3	5.7 ± 01.	60.5 ± 0.3	10.61	546 ± 9	791 ± 12	91 ± 5	2782 ± 20	6.29
C4	3.8 ± 0.1	63.2 ± 0.3	16.63	1224 ± 12	4164 ± 30	145 ± 5	796 ± 12	7.80
C5	8.6 ± 0.1	61.9 ± 0.3	7.20	947 ± 10	6412 ± 35	58 ± 5	120 ± 8	6.85
C6	7.9 ± 0.1	60.5 ± 0.3	7.66	<32	114 ± 8	<3	2432 ± 20	8.10
C7	6.0 ± 0.1	62.8 ± 0.3	10.47	759 ± 8	578 ± 10	120 ± 5	839 ± 12	7.69

The humification of organic material can be estimated from the HA/FA ratio (Table 3). A higher HA/FA ratio corresponded to higher humification rates of the organic material. The highest HA/FA ratios were measured in treatments C2 (7.88) and C4 (7.82), whereas the lowest HA/FA ratios were observed in treatments C6 (1.46) and C5 (1.42). Treatments C6 and C5 had the highest concentrations of lipids (2.1% and 4.2%, respectively).

Table 3. Characterization of HSs in composts. TOC = total organic carbon in humic substances. TOC and lipids are presented as % of compost dry matter. Mean ± standard deviation (SD) of triplicate measurements compared using ANOVA with post-hoc Tukey's HSD test. Connecting letters represent significantly different measurements.

Compost	TOC	HA/FA	E_2/E_3	E_4/E_6	Lipids
	%				%
C1	12.0 ± 0.2 ^{a}	5.70	2.65	7.78	1.5 ± 0.1 ^{b}
C2	27.0 ± 0.2 ^{c}	7.88	2.57	6.95	1.5 ± 0.1 ^{b}
C3	25.7 ± 0.2 ^{c}	4.73	2.49	8.33	1.5 ± 0.1 ^{b}
C4	18.0 ± 0.2 ^{ab}	7.82	2.56	7.34	1.7 ± 0.1 ^{b}
C5	15.3 ± 0.2 ^{a}	1.42	2.38	6.88	4.2 ± 0.1 ^{a}
C6	17.1 ± 0.2 ^{a}	1.46	2.66	6.40	2.1 ± 0.1 ^{b}
C7	8.7 ± 0.1 ^{d}	1.60	2.61	8.59	1.4 ± 0.1 ^{b}

The spectral ratio E_4/E_6 shows the overall aromaticity and condensation of the material and can characterize the size of molecules (Table 3) [37]. E_4/E_6 depends on the source material. The smallest E_4/E_6 ratio, indicating the greatest organic matter polymerization, was observed in compost C6 (6.40), followed by C5 (6.88) and C2 (6.95). The lowest aromaticity was observed in compost C7 (8.59). E_2/E_3 describes not only aromaticity, but also the degradation of original material, including the degradation of phenols [37]. When E_2/E_3 is < 3.5, the content of HA is higher than that of FA [42]. For all composts, the ratio varied in the range 2.38–2.66.

The UV–Vis spectra of HA (Figure 1) monotonically decreased with the increasing wavelength. The absorption peaks of UV–Vis spectra between 255 and 290 nm are typical for aromatic or unsaturated compounds, such as quinones and ketones [36]. These spectra were visible for C1, C2, and C3. Composts C4, C7, and C6 had small shoulders of spectra, and C5 did not exhibit a shoulder of spectra.

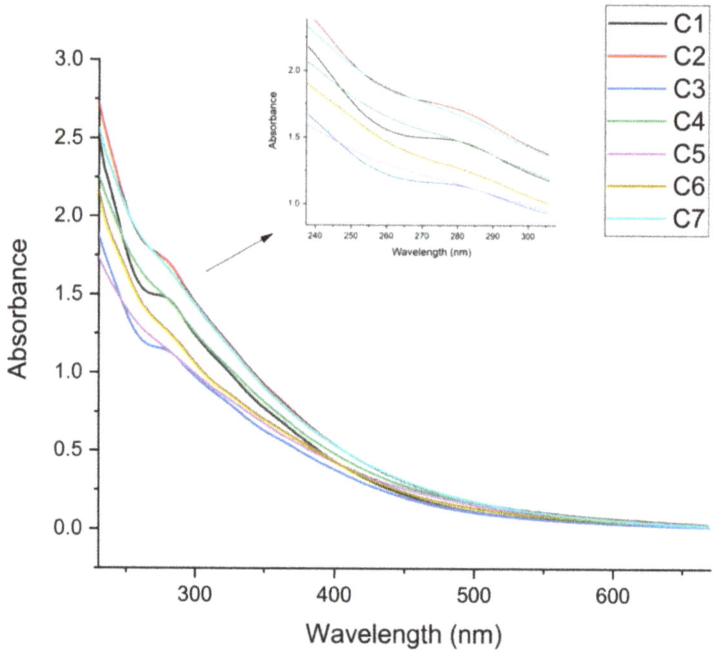

Figure 1. UV–Vis spectra. Compost labels according to the target waste: C1 fish waste; C2 horse manure; C3 sewage sludge; C4 green waste; C5 kitchen biowaste; C6 *H. illucens* frass; and C7 vermicompost.

The FTIR spectra of the analyzed samples had a similar pattern, showing the similarity of the obtained HSs and absorption at 3600–3000 cm^{-1}, representing -OH groups (Figure 2). Methyl groups in aliphatic compounds were visible at 2920 cm^{-1} and were similar in all samples [43]. At 1700 cm^{-1}, the spectra showed the absorption of C=O bonds in carboxyl groups, aldehydes, ketones, and esters, with higher rates in samples C3, C4, C5, and C7. The region at 1637–1610 cm^{-1} shows significantly less signal for C7 and moderately less for C2. Different from other composts, C5 had a small absorption in C-N and N-H bonds in amides, and C7 had none. The region at 1470–1370 cm^{-1} represents carbonyl acids and phenols, and there were visible differences for samples C7 and C5, where the presence of those bands was low. Carboxylic groups and ethers were present in all compost HA samples, with slightly higher amounts for C1 and C4.

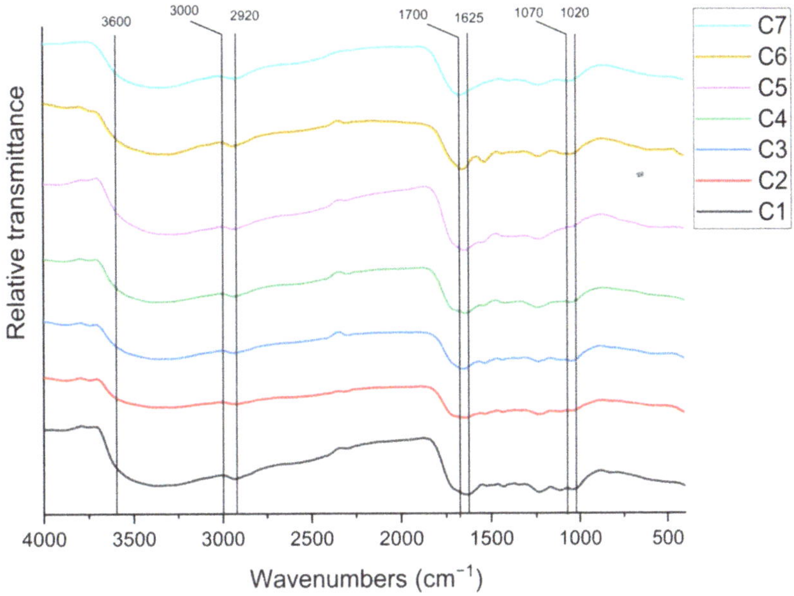

Figure 2. Fourier-transform infrared (FTIR) spectra of humic acids from compost samples. Compost labels according to the target waste: C1 fish waste; C2 horse manure; C3 sewage sludge; C4 green waste; C5 kitchen biowaste; C6 *H. illucens* frass; and C7 vermicompost.

Common, and at the same time similar for all samples, was the presence of C-O in carboxyl groups and ethers. The region at 1070–1020 cm^{-1} describes the C-O bonds in polysaccharides, with noticeably less in the C2 and C5 samples. The presence of detected structural features describes the condensation and conjugation of organic material during composting processes.

The compost EEM fluorescence spectra for FA and HA are shown in Figures 3 and 4. All samples exhibited one FA peak at Ex/Em 340–350/440–450 nm, which is related to the FA visible area and associated with polysaccharides. A very small peak was present at Ex/Em 420–450/520–550 nm, representing condensed aromatic structures [44,45].

Other fractions of humic material, such as HAs, have different compositions and dominant peaks in the EEM. Common peaks vary only by dominant peak and relative height for HA, and they include Ex/Em 350–380/470–510, 430/520, and 460–470/540–560 nm. The peak Ex/Em 350–380/470–510 nm is associated with HA-like substances, which form during the humification of composting organic matter [46]. This peak dominated in samples C1, C2, and C5 and was the second largest peak in samples C4 and C6. Similar to FA, the EEM for HA was Ex/Em 430/520 nm peaks, associated with the degradation and condensation of organic material. Ex/Em 460–470/540–560 nm dominated in samples C4, C5, and C7, which probably indicates the presence of lignin from structural amendments (shredded wood, straw, paper). Treatment C6 differed from other samples, with the core of HA peaking at Ex/Em 370–410/610–630 nm. The origin of these compounds could be fresh plants and fruits added to compost and characterized by low degradation rates and chlorophyll [47].

Figure 5 illustrates the comparative evaluation of studied composts, considering both nutrient and HS indicators. On the PCA biplot, composts C1, C5, and C7 group together, mainly due to low HSs, while C3 shows elevated HSs and P, and C6 is rich in P and N with average HSs.

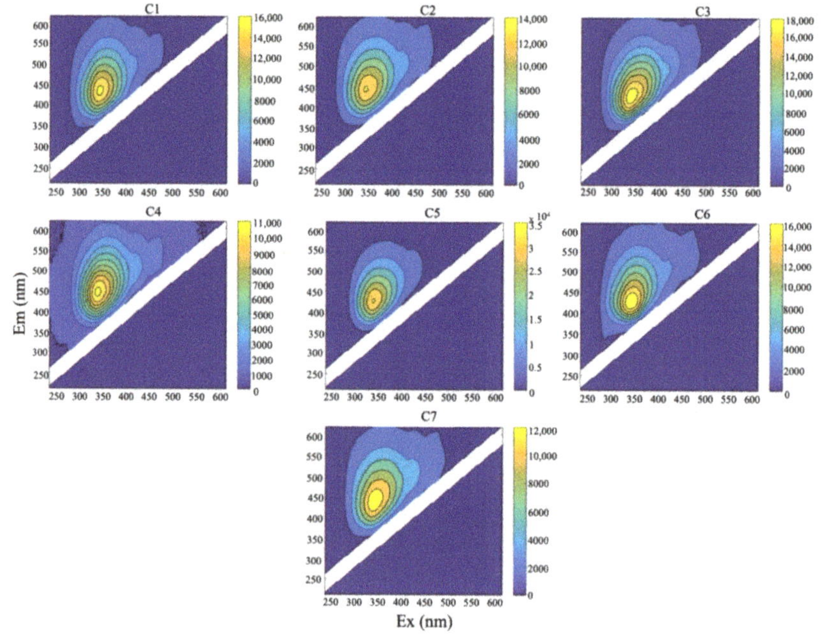

Figure 3. Composition of fulvic acids (FAs) for different treatments. Compost labels according to the target waste: (**C1**) fish waste; (**C2**) horse manure; (**C3**) sewage sludge; (**C4**) green waste; (**C5**) kitchen biowaste; (**C6**) *H. illucens* frass; and (**C7**) vermicompost. The color scale is presented in relative units.

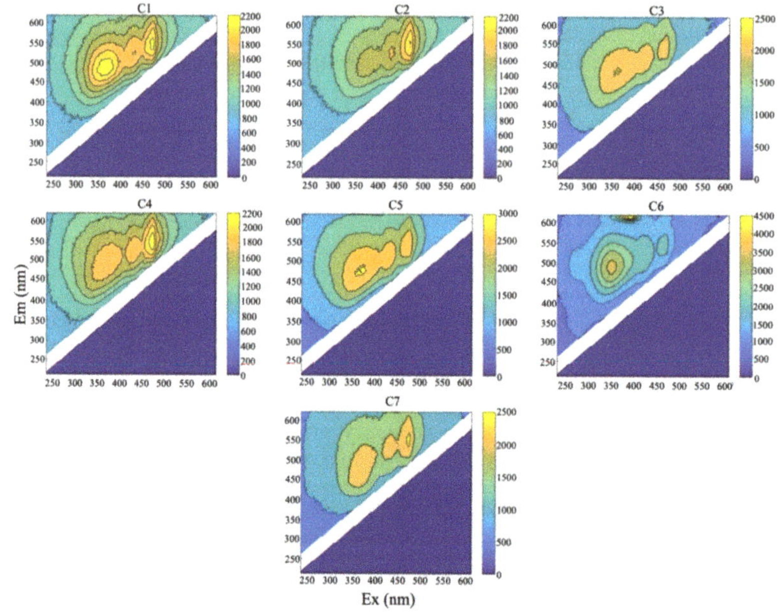

Figure 4. Composition of humic acids (HAs) of different treatments. Compost labels according to the target waste: (**C1**) fish waste; (**C2**) horse manure; (**C3**) sewage sludge; (**C4**) green waste; (**C5**) kitchen biowaste; (**C6**) *H. illucens* frass; and (**C7**) vermicompost. The color scale is presented in relative units.

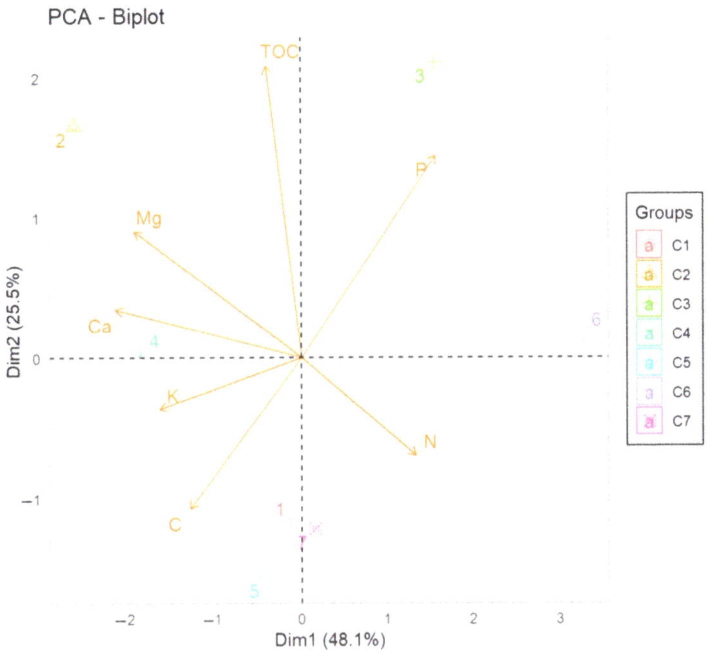

Figure 5. Principal component analysis (PCA) of composts based on nutrient and humic substances' (HSs) TOC concentrations. Axis 1 explains 48.1% of the variance, and axis 2 explains an additional 25.5% of the variance. Compost labels according to the target waste: C1 fish waste; C2 horse manure; C3 sewage sludge; C4 green waste; C5 kitchen biowaste; C6 *H. illucens* frass; and C7 vermicompost.

4. Discussion

The selected composts varied in input material, selection of treatment, and the active composting time (d). These features are known to affect the concentration and type of HSs in the final product [25,48,49].

In the treatment of organic waste, the concentration of humus-like substances and the degree of humification of hydrophilic components increases, and the structure of the substance becomes more stable [50]. HSs are biostimulants that enhance nutrition efficiency, abiotic stress tolerance, and/or crop quality traits [9,51]. All compost treatments in the present study contained HSs, including HA and FA. While all treatments presented an FA peak for polysaccharides, they differed more relative to the HA peak, showing that input materials directly influence the core structure of HA in mature compost. As shown previously, various input materials contain different amounts of basic components that form HSs [25,50]. Organic waste, rich in polyphenols, amino acids, and reducing sugars can promote overall HS production [52].

During composting, the concentration of FAs decreased, whereas the concentration of HAs increases. HAs are generated in the last stage of composting; thus, an increased concentration of HAs indicates a higher maturity of the compost. Compost is considered mature if the HA/FA ratio is higher than 1 [15,25,53,54], and the HA/FA ratio is also affected by the type of organic material. For example, the lowest HA/FA was in treatment C5, which contained a high quantity of food waste. Amino acids and sugars, e.g., polysaccharides, break down easily and are first to form FA, and then HA [55]. The highest HA/FA ratio was in treatment C2. After fermentation in animal guts, horse manure contained a high proportion of lignin from hay and grass, aromatic compounds, and older material. Lignin and aromatic compound remnants could be the core of the initial HA, and less FAs are formed [23,26].

The chemical structures of HSs were similar in the compost treatments. As shown previously, the UV–Vis spectra of hydrophilic components decreased monotonically, and the observed spectra were typical of aromatic or unsaturated compounds, such as the conjugation of quinones and ketones [50,56].

Treatments C1, C2, and C5, which contained different ABPRs (e.g., fish waste, horse manure, chicken body parts), had a dominant HA peak, reflecting the HA-like substances formed during humification. ABPRs are rich in nutrients and add microorganisms to the composting mixture. In addition to the source materials, the content of HSs depends on the amounts of nutrients (e.g., C, N) and the presence of microorganisms [57]. Humification is more intense and achieved more quickly if composting piles are inoculated with microorganisms. Composts inoculated with microorganisms stimulated the degradation of hemicellulose, cellulose, and lignin by 28%, 21%, and 25%, respectively [58]. The greater the amount of amino acids and reducing sugars produced during degradation, the higher the positive effect on HS synthesis [52].

Treatments C4, C5, and C7 had a dominant peak in the area representing lignin-like structures. These treatments contained large quantities of green waste (e.g., branches, leaves), shredded wood, straw, and paper, which are rich in cellulose and lignin. Selected bulking material affects the composting duration and product quality, as material rich in cellulose and lignin can influence the humification process [25]. These kind of materials decompose more slowly; therefore, they can slow down composting processes and humification [59].

Treatment C6 differed from other composts made from more traditional composting technologies. C6 contained frass from *H. illucens* larvae activity, as well as from chitin and food residues during feeding, and the core structure of HA differed compared to other treatments. Although *H. illucens* larvae frass can be used as an organic fertilizer, the fresh frass might have contained phytotoxins, which decrease plant growth [60]. To increase the concentration of HA and the stability of the frass, it is suggested to post-compost the frass [60,61].

It has been shown, that *H. illucens* treatment can increase the HA/FA ratio and improve humification [62]. Compared to other treatments, C6 had the lowest E_4/E_6 ratio, indicating a more advanced humification. However, the HA/FA ratio was quite low, reflecting a high FA concentration. *H. illucens* can also suppress pathogens and is therefore suitable for processing manure in addition to food waste [60,63].

The highest aromaticity was in compost C7, indicating that by vermicomposting, HA is a more aggregated macromolecule with a more incomplete humification processes [64]. However, it has been proposed that there is no direct relationship between aromaticity and humification, and vermicompost can be used as a soil improver [50].

The quality of composts is reflected by the content of both HSs and nutrients. Although some composts may have a lower nutrient concentration, they may still be a valuable source of HSs. The total value of compost is thus reflected by the nutrient content, the disclosure of which is currently required by regulations (Appendix A) [9,13]. Depending on the main purpose of compost application, the effects of added nutrients or added HSs can be the primary advantage of using compost. Including HS data for composts yields a better comprehensive evaluation. Composts C2 and C4 were grouped together on the PCA bioplot, considering only their nutrient content, but elevated HSs in compost C2 signified an advantage over C4 (Figure 5). Composts with a high proportion of HSs, but a relatively low nutrient content, can provide a maximum fertilizing effect when used together with mineral fertilizers. Thus, soil health and nutrient content are ensured to improve crop yield [65,66].

Although different treatments varied in HS concentrations, both HA and FA act as biostimulants and promote plant growth [67,68]. Compared to chemical fertilizers, composts increase the content of soil-available nitrogen (N) and soil organic carbon (SOC) and improve soil fertility [69]. Furthermore, using composts can decrease eutrophication risks caused by the excessive use of chemical fertilizers and the subsequent discharge into surface waters [25]. Because all composts included HSs, their use provides an advantage over using mineral fertilizers alone.

5. Conclusions

A variety of organic materials can be processed with different treatments (windrow composting, insect frass formation, vermicomposting, etc.) and durations, which all influence the properties of the final product. The present research indicates that, regardless of the input organic waste and composting technology, all studied organic fertilizers are a valuable source of HSs, including HA and FA. The presence of HSs is an advantage over mineral fertilizers alone. All samples exhibited a fluorescence EEM peak of FA associated with polysaccharides. The HA peak dominated in composts comprised of animal by-products, such as fish waste, horse manure, and kitchen biowaste. The core structure of humic acids in *H. illucens* larvae frass differed from other treatments.

Compost quality was assessed commercially by nutrient content and contamination criteria, without considering HSs. To improve soil health and fertility, the concentration and characteristics of HSs could serve as a comprehensive indicator for compost quality evaluation. While HSs act as important biostimulants, their presence and features affect compost's fertilizing value. In the present study, we showed the importance of HS content and properties in a comprehensive evaluation of compost quality.

Author Contributions: Conceptualization, M.L., M.K. (Maris Klavins), O.P., M.S., A.K. and M.K. (Mait Kriipsalu); methodology, M.K. (Maris Klavins) and O.P.; software, O.P.; validation, M.K. (Maris Klavins) and O.P.; formal analysis, O.P.; investigation, M.K. (Maris Klavins) and O.P.; resources, M.K. (Maris Klavins), M.S. and M.K. (Mait Kriipsalu); data curation, M.L. and O.P.; writing—original draft preparation, M.L. and M.K. (Mait Kriipsalu); writing—review and editing, O.P., M.S., A.K. and M.K. (Mait Kriipsalu); visualization, M.L. and O.P.; supervision, M.K. (Mait Kriipsalu); project administration, M.K. (Mait Kriipsalu); funding acquisition, M.K. (Mait Kriipsalu). All authors have read and agreed to the published version of the manuscript.

Funding: This research was funded by the Estonian University of Life Sciences project PM180261MIMV "From bio-waste to high value-added products", the European Regional Development Fund under project number 1.1.1.2/VIAA/1/16/001, post-doctoral research project number 1.1.1.2/VIAA/4/20/723 and the Doctoral School of Earth Sciences and Ecology, supported by the European Union, European Regional Development Fund (Estonian University of Life Sciences ASTRA project "Value-chain based bio-economy").

Institutional Review Board Statement: Not applicable.

Data Availability Statement: Data is available upon request from corresponding author M.L.

Acknowledgments: We would like to thank Mark J. McCarthy for language editing and Jordi Escuer Gatius for contributing to data visualization.

Conflicts of Interest: The authors declare no conflict of interest.

Appendix A

Table A1. "Indicators to be measured and presented." from the Regulation of the Republic of Estonia. "Requirements for production of compost from biodegradable waste" RT I, 18.12.2020, 23 2021. issued by the Minister of the Environment.

Indicator	Parameter	Measurement Unit	Assessment Result
Improvement of soil qualities	Organic substance (OM)	[% in dry matter]	≥15% presentation
	Total nitrogen (N)	[% in dry matter]	presentation
Fertilising qualities	Total phosphorus (P)	[% in dry matter]	presentation
	Total potassium (K)	[% in dry matter]	presentation
	Maximum size of particles	[mm]	presentation
	Overall density	[g/L]	presentation
Material qualities	Water content	[g/L]	
	Salinity/conductivity	[mS/m]	presentation
	pH value		presentation

References

1. Kaza, S.; Yao, L.; Bhada-Tata, P.; Van Woerden, F. *What a Waste 2.0: A Global Snapshot of Solid Waste Management to 2050*; World Bank Publications: Washington, DC, USA, 2018.
2. Interreg Europe. *The Biowaste Management Challenge. A Policy Brief from the Policy Learning Platform on Environment and Resource Efficiency 2021*; Interreg Europe: Lille, France, 2021.
3. Ministry of the Environment Estonian National Waste Management Plan. *Riigi Jäätmekava 2014–2020 (Pikendatud Kuni 2022 Lõpuni)*; Ministry of the Environment Estonian: Tallinn, Estonia, 2014; p. 94.
4. Nanda, S.; Berruti, F. Municipal Solid Waste Management and Landfilling Technologies: A Review. *Environ. Chem. Lett.* **2021**, *19*, 1433–1456. [CrossRef]
5. European Parliament Directive (EU). *2018/850 of the European Parliament and of the Council of 30 May 2018 amending Directive 1999/31/EC on the Landfill of Waste (Text with EEA Relevance) 2018*; European Parliament Directive (EU): Strasbourg, France, 2018.
6. European Parliament Directive (EU). *2018/851 of the European Parliament and of the Council of 30 May 2018 amending Directive 2008/98/EC on Waste (Text with EEA Relevance) 2018*; European Parliament Directive (EU): Strasbourg, France, 2018.
7. Smol, M. Transition to Circular Economy in the Fertilizer Sector—Analysis of Recommended Directions and End-Users' Perception of Waste-Based Products in Poland. *Energies* **2021**, *14*, 4312. [CrossRef]
8. The European Green Deal. *European Comission Communication from the Commission to the European Parliament, the European Council, the Council, The European Economic and Social Committee and the Committee of the Regions*; The European Green Deal: Brussels, Belgium, 2019.
9. European Parliament Regulation (EU). *2019/1009 of the European Parliament and of the Council of 5 June 2019 Laying down Rules on the Making Available on the Market of EU Fertilising Products and Amending Regulations (EC) No 1069/2009 and (EC) No 1107/2009 and repealing Regulation (EC) No 2003/2003 2019*; European Parliament Directive (EU): Strasbourg, France, 2019.
10. Lanno, M.; Kriipsalu, M.; Shanskiy, M.; Silm, M.; Kisand, A. Distribution of Phosphorus Forms Depends on Compost Source Material. *Resources* **2021**, *10*, 102. [CrossRef]
11. Wei, Y.; Zhao, Y.; Xi, B.; Wei, Z.; Li, X.; Cao, Z. Changes in Phosphorus Fractions during Organic Wastes Composting from Different Sources. *Bioresour. Technol.* **2015**, *189*, 349–356. [CrossRef] [PubMed]
12. Siebert, S. *Quality Requirements and Quality Assurance of Digestion Residuals in Germany*; Bundesgütegemeinschaft Kompost t e.V.: Nuremberg, Germany, 2008.
13. Siebert, S.; Auweele, W.V. *ECN-Qas European Quality Assurance Scheme for Compost and Digestate; Quality Manual*; European Compost Network ECN e.V.: Bochum, Germany, 2014; ISBN 978-3-00-047599-3.
14. Bhattacharya, S.S.; Kim, K.-H.; Das, S.; Uchimiya, M.; Jeon, B.H.; Kwon, E.; Szulejko, J.E. A Review on the Role of Organic Inputs in Maintaining the Soil Carbon Pool of the Terrestrial Ecosystem. *J. Environ. Manage.* **2016**, *167*, 214–227. [CrossRef]
15. Ben Mbarek, H.; Ben Mahmoud, I.; Chaker, R.; Rigane, H.; Maktouf, S.; Arous, A.; Soua, N.; Khlifi, M.; Gargouri, K. Change of Soil Quality Based on Humic Acid with Date Palm Compost Incorporation. *Int. J. Recycl. Org. Waste Agric.* **2019**, *8*, 317–324. [CrossRef]
16. European Commission; European Environment Agency. *The State of Soil in Europe: A Contribution of the JRC to the European Environment Agency's Environment State and Outlook Report—SOER 2010*; Publications Office of the European Union: Luxembourg, 2012.
17. Gao, Y.; Gao, X.; Zhang, X. The 2 °C Global Temperature Target and the Evolution of the Long-Term Goal of Addressing Climate Change—From the United Nations Framework Convention on Climate Change to the Paris Agreement. *Engineering* **2017**, *3*, 272–278. [CrossRef]
18. Bot, A.; Benites, J. *The Importance of Soil Organic Matter: Key to Drought-Resistant Soil and Sustained Food Production*; Food & Agriculture Org.: Rome, Italy, 2005; Volume 80.
19. Lakhdar, A.; Rabhi, M.; Ghnaya, T.; Montemurro, F.; Jedidi, N.; Abdelly, C. Effectiveness of Compost Use in Salt-Affected Soil. *J. Hazard. Mater.* **2009**, *171*, 29–37. [CrossRef]
20. Nguyen, T.H.; Shindo, H. Quantitative and Qualitative Changes of Humus in Whole Soils and Their Particle Size Fractions as Influenced by Different Levels of Compost Application. *Agric. Sci.* **2011**, *2*, 1–8. [CrossRef]
21. Dou, S.; Shan, J.; Song, X.; Cao, R.; Wu, M.; Li, C.; Guan, S. Are Humic Substances Soil Microbial Residues or Unique Synthesized Compounds? A Perspective on Their Distinctiveness. *Pedosphere* **2020**, *30*, 159–167. [CrossRef]
22. Von Lützow, M.; Kögel-Knabner, I.; Ekschmitt, K.; Flessa, H.; Guggenberger, G.; Matzner, E.; Marschner, B. SOM Fractionation Methods: Relevance to Functional Pools and to Stabilization Mechanisms. *Soil Biol. Biochem.* **2007**, *39*, 2183–2207. [CrossRef]
23. Harrison, R.B. *Composting and Formation of Humic Substances*; Elsevier Science, Encyclopedia of Ecology: Amsterdam, Netherlands, 2008; pp. 713–719.
24. Gerke, J. Concepts and Misconceptions of Humic Substances as the Stable Part of Soil Organic Matter: A Review. *Agronomy* **2018**, *8*, 76. [CrossRef]
25. Guo, X.; Liu, H.; Wu, S. Humic Substances Developed during Organic Waste Composting: Formation Mechanisms, Structural Properties, and Agronomic Functions. *Sci. Total Environ.* **2019**, *662*, 501–510. [CrossRef]
26. Zhou, Y.; Selvam, A.; Wong, J.W.C. Evaluation of Humic Substances during Co-Composting of Food Waste, Sawdust and Chinese Medicinal Herbal Residues. *Bioresour. Technol.* **2014**, *168*, 229–234. [CrossRef]
27. Zingaretti, D.; Lombardi, F.; Baciocchi, R. Soluble Organic Substances Extracted from Compost as Amendments for Fenton-like Oxidation of Contaminated Sites. *Sci. Total Environ.* **2018**, *619–620*, 1366–1374. [CrossRef]

28. Minister of the Environment Regulation of the Republic of Estonia. *"Requirements for Production of Compost from Biodegradable Waste" RT I, 18.12.2020, 23 2021*; Environment Regulation of the Republic of Estionia: Tallinn, Estonia, 2021.
29. Minister of the Environment Regulation of the Republic of Estonia. *Haljastuses, Rekultiveerimisel Ja Põllumajanduses Kasutatava Reoveesette Kvaliteedi Piirväärtused Ning Kasutamise Nõuded. RT I, 06.08.2019, 7 2019*; Environment Regulation of the Republic of Estionia: Tallinn, Estonia, 2019.
30. Lanno, M.; Silm, M.; Shanskiy, M.; Kisand, A.; Orupõld, K.; Kriipsalu, M. Open Windrow Composting of Fish Waste in Estonia. *EMU DSpace* **2020**, *18*, 428. [CrossRef]
31. Khan, M.B.; Cui, X.; Jilani, G.; Tang, L.; Lu, M.; Cao, X.; Sahito, Z.A.; Hamid, Y.; Hussain, B.; Yang, X.; et al. New Insight into the Impact of Biochar during Vermi-Stabilization of Divergent Biowastes: Literature Synthesis and Research Pursuits. *Chemosphere* **2020**, *238*, 124679. [CrossRef]
32. Escuer, O.; Karp, K.; Escuer-Gatius, J.; Raave, H.; Teppand, T.; Shanskiy, M. Hardwood Biochar as an Alternative to Reduce Peat Use for Seed Germination and Growth of *Tagetes patula*. *Acta Agric. Scand. Sect. B Soil Plant Sci.* **2021**, *71*, 408–421. [CrossRef]
33. Malińska, K.; Zabochnicka-Świątek, M.; Cáceres, R.; Marfà, O. The Effect of Precomposted Sewage Sludge Mixture Amended with Biochar on the Growth and Reproduction of Eisenia Fetida during Laboratory Vermicomposting. *Ecol. Eng.* **2016**, *90*, 35–41. [CrossRef]
34. Egner, H.; Riehm, H.; Domingo, W.R. Investigations on the Chemical Soil Analysis as a Basis for Assessing the Soil Nutrient Status. II: Chemical Extraction Methods for Phosphorus and Potassium Determination. *K. Lantbrukshügskolans Ann.* **1960**, *26*, 199–215.
35. Tan, K.H. *Soil Sampling, Preparation, and Analysis*; CRC Press: Boca Raton, FL, USA, 1995.
36. Klavins, M.; Purmalis, O. Properties and Structure of Raised Bog Peat Humic Acids. *J. Mol. Struct.* **2013**, *1050*, 103–113. [CrossRef]
37. Chen, Y.; Senesi, N.; Schnitzer, M. Information Provided on Humic Substances by E4/E6 Ratios. *Soil Sci. Soc. Am. J.* **1977**, *41*, 352–358. [CrossRef]
38. Peuravuori, J.; Pihjala, K. Molecular Size Distribution and Spectroscopic Properties of Aquatic Humic Substances. *Anal. Chim. Acta* **1997**, *337*, 133–149. [CrossRef]
39. Murphy, K.R.; Stedmon, C.A.; Graeber, D.; Bro, R. Fluorescence Spectroscopy and Multi-Way Techniques. *PARAFAC Anal. Methods* **2013**, *5*, 6557–6566. [CrossRef]
40. R Core Team. *R: A Language and Environment for Statistical Computing. R Foundation for Statistical Computing*; R Core Team: Vienna, Austria, 2020.
41. Mendiburu, F.D. Agricolae: Statistical Procedures for Agricultural Research, R Package Version 1.2-3. 2015. Available online: http://cran.r-project.org/package=agricolae (accessed on 3 October 2022).
42. Minero, C.; Lauri, V.; Falletti, G.; Maurino, V.; Pelizzetti, E.; Vione, D. Spectrophotometric Characterisation of Surface Lakewater Samples: Implications for the Quantification of Nitrate and the Properties of Dissolved Organic Matter. *Ann. Chim.* **2007**, *97*, 1107–1116. [CrossRef]
43. Verillo, M.; Salzano, M.; Cozzolino, V.; Spaccini, R.; Piccolo, A. Bioactivity and Antimicrobial Properties of Chemically Characterized Compost Teas from Different Green Composts. *Waste Manag.* **2021**, *120*, 98–107. [CrossRef]
44. Wei, Z.; Zhao, X.; Zhu, C.; Xi, B.; Zhao, Y.; Yu, X. Assessment of Humification Degree of Dissolved Organic Matter from Different Composts Using Fluorescence Spectroscopy Technology. *Chemosphere* **2014**, *95*, 261–267. [CrossRef]
45. Valencia, S.; Marin, J.M.; Restrepo, G.; Frimmel, F.H. Application of Excitation–Emission Fluorescence Matrices and UV/Vis Absorption to Monitoring the Photocatalytic Degradation of Commercial Humic Acid. *Sci. Total Environ.* **2013**, *442*, 207–214. [CrossRef]
46. Da Silva, L.S.; Constantino, I.C.; Bento, L.R.; Tadini, A.M.; Bisinoti, M.C.; Boscolo, M.; Ferreira, O.P. Humic Extracts from Hydrochar and Amazonian Anthrosol: Molecular Features and Metal Binding Properties Using EEM-PARAFAC and 2D FTIR Correlation Analyses. *Chemosphere* **2020**, *256*, 127110. [CrossRef]
47. Richard, C.; Guyot, G.; Trubetskaya, O.; Grigatti, M.; Cavani, L. Fluorescence Analysis of Humic-Like Substances Extracted from Composts: Influence of Composting Time and Fractionation. *Environ. Chem. Lett.* **2009**, *7*, 61–65. [CrossRef]
48. Liang, J.; Zhao, C.; Cheng, Y.; Lei, T.; Wu, N.; Wang, X.; Xu, X. Dissolved Organic Matter Features in Black Soldier Fly Larvae Frass Derived From Two Types of Meat and Bone Meals with Straw Amendments. *Waste Biomass Valorization* **2022**, 1–12. [CrossRef]
49. Aguiar, N.O.; Olivares, F.L.; Novotny, E.H.; Dobbss, L.B.; Balmori, D.M.; Santos-Júnior, L.G.; Chagas, J.G.; Façanha, A.R.; Canellas, L.P. Bioactivity of Humic Acids Isolated from Vermicomposts at Different Maturation Stages. *Plant Soil* **2013**, *362*, 161–174. [CrossRef]
50. Dores-Silva, P.R.; Landgraf, M.D.; Rezende, M.O.O. Humification Process in Different Kinds of Organic Residue by Composting and Vermicomposting: Have Microbioreactors Really Accelerated the Process? *Environ. Sci. Pollut. Res.* **2018**, *25*, 17490–17498. [CrossRef]
51. Mosa, A.; Taha, A.; Elsaeid, M. Agro-Environmental Applications of Humic Substances: A Critical Review. *Egypt. J. Soil Sci.* **2020**, *60*, 211–229. [CrossRef]
52. Wu, J.; Zhao, Y.; Qi, H.; Zhao, X.; Yang, T.; Du, Y.; Zhang, H.; Wei, Z. Identifying the Key Factors that Affect the Formation of Humic Substance during Different Materials Composting. *Bioresour. Technol.* **2017**, *244*, 1193–1196. [CrossRef]
53. Raj, D.; Antil, R.S. Evaluation of Maturity and Stability Parameters of Composts Prepared from Agro-Industrial Wastes. *Bioresour. Technol.* **2011**, *102*, 2868–2873. [CrossRef]
54. Silva, M.E.; Lemos, L.T.; Cunha-Queda, A.C.; Nunes, O.C. Co-Composting of Poultry Manure with Low Quantities of Carbon-Rich Materials. *Waste Manag. Res. J. Sustain. Circ. Econ.* **2009**, *27*, 119–128. [CrossRef]
55. Stevenson, F.J. *Humus Chemistry: Genesis, Composition, Reactions*; John Wiley & Sons: New Jersey, NJ, USA, 1994.
56. Zilong, Z.; Yu, X. Changes in the Spectral Characteristics of Hydrophilic Fraction in Compost Organic Matter. *J. Phys. Conf. Ser.* **2021**, *2011*, 012069. [CrossRef]

57. Sutton, R.; Sposito, G. Molecular Structure in Soil Humic Substances: The New View. *Environ. Sci. Technol.* **2005**, *39*, 9009–9015. [CrossRef]
58. Jurado, M.M.; Suárez-Estrella, F.; López, M.J.; Vargas-García, M.C.; López-González, J.A.; Moreno, J. Enhanced Turnover of Organic Matter Fractions by Microbial Stimulation during Lignocellulosic Waste Composting. *Bioresour. Technol.* **2015**, *186*, 15–24. [CrossRef]
59. Kulikowska, D.; Sindrewicz, S. Effect of Barley Straw and Coniferous Bark on Humification Process during Sewage Sludge Composting. *Waste Manag.* **2018**, *79*, 207–213. [CrossRef] [PubMed]
60. Lopes, I.G.; Yong, J.W.; Lalander, C. Frass Derived from Black Soldier Fly Larvae Treatment of Biodegradable Wastes. A Critical Review and Future Perspectives. *Waste Manag.* **2022**, *142*, 65–76. [CrossRef] [PubMed]
61. Song, S.; Ee, A.W.L.; Tan, J.K.N.; Cheong, J.C.; Chiam, Z.; Arora, S.; Lam, W.N.; Tan, H.T.W. Upcycling Food Waste Using Black Soldier Fly Larvae: Effects of Further Composting on Frass Quality, Fertilising Effect and Its Global Warming Potential. *J. Clean. Prod.* **2021**, *288*, 125664. [CrossRef]
62. Liu, T.; Awasthi, M.K.; Awasthi, S.K.; Zhang, Y.; Zhang, Z. Impact of the Addition of Black Soldier Fly Larvae on Humification and Speciation of Trace Elements during Manure Composting. *Ind. Crops Prod.* **2020**, *154*, 112657. [CrossRef]
63. Phooi, C.L.; Azman, E.A.; Ismail, R. Do it Yourself: Humic Acid. *Pertanika J. Trop. Agric. Sci.* **2022**, *45*, 547–564. [CrossRef]
64. Campitelli, P.; Ceppi, S. Effects of Composting Technologies on the Chemical and Physicochemical Properties of Humic Acids. *Geoderma* **2008**, *144*, 325–333. [CrossRef]
65. Iqbal, J.; Sarwar, G.; Shah, S.H.; Sabah, N.-U.; Tahir, M.A.; Muhammad, S.; Zeeshan Manzoor, M.; Zafar, A.; Shehzad, I. Evaluating the Combined Effect of Compost and Mineral Fertilizers on Soil Health, Growth and Mineral Acquisition in Maize (*Zea mays* L.). *Pak. J. Bot.* **2022**, *54*, 1793–1801. [CrossRef]
66. Issoufa, B.B.; Ibrahim, A.; Abaidoo, R.C.; Ewusi-Mensah, N. Combined Use of Millet Glume-Derived Compost and Mineral Fertilizer Enhances Soil Microbial Biomass and Pearl Millet Yields in a Low-Input Millet Cropping System in Niger. *Arch. Agron. Soil Sci.* **2019**, *65*, 1107–1119. [CrossRef]
67. Klavins, M.; Grandovska, S.; Obuka, V.; Ievinsh, G. Comparative Study of Biostimulant Properties of Industrially and Experimentally Produced Humic Substances. *Agronomy* **2021**, *11*, 1250. [CrossRef]
68. Savy, D.; Brostaux, Y.; Cozzolino, V.; Delaplace, P.; Jardin, P.D.; Piccolo, A. Quantitative Structure-Activity Relationship of Humic-Like Biostimulants Derived from Agro-Industrial Byproducts and Energy Crops. *Front. Plant Sci.* **2020**, *11*, 581. [CrossRef]
69. Hu, C.; Xia, X.; Chen, Y.; Han, X. Soil Carbon and Nitrogen Sequestration and Crop Growth as Influenced by Long-Term Application of Effective Microorganism Compost. *Chil. J. Agric. Res.* **2018**, *78*, 13–22. [CrossRef]

agriculture

MDPI

Article

Biochar-Based Fertilizer Enhances the Production Capacity and Economic Benefit of Open-Field Eggplant in the Karst Region of Southwest China

Meng Zhang, Yanling Liu, Quanquan Wei, Lingling Liu, Xiaofeng Gu and Jiulan Gou *

Institute of Soil and Fertilizer, Guizhou Academy of Agricultural Sciences, Guiyang 550006, China
* Correspondence: gzgoujiulan@163.com

Abstract: Biochar as an amendment has been widely applied to enhance crop productivity and improve soil quality. However, the effect of biochar-based fertilizer (BF) on the production capacity and economic benefits of open-field eggplant in the karst region remains unclear. A field experiment was carried out in the karst region of Southwest China from 2020 to 2021 to study the ameliorative roles of different application rates (1875, 2250, 2625, and 3000 kg ha^{-1}) of BF on the fresh yield, quality, fertilizer utilization, and economic benefits of fresh eggplant. The results show that BF increased the yield of fresh eggplant by 3.65–13.76% (2020) and 23.40–49.04% (2021) compared to the traditional fertilization practice (TFP). The application of BF reduced the nitrate content and increased the vitamin C (VC) and soluble sugar content of the fruits, which is beneficial for improving the quality of eggplant fruits. Meanwhile, the application of BF not only increased the nutrient uptake of the eggplant but also significantly improved the fertilizer utilization rates, especially the agronomic efficiency (AE) and recovery efficiency (RE). Moreover, BF could also significantly increase the output value and net income of fresh eggplant, which can help farmers increase their income. In conclusion, a BF application rate of 2544–2625 kg ha^{-1} could be used to improve the yield, fertilizer efficiency, and economic benefits of open-field eggplant and is recommended for managing agricultural production in the karst region of Southwest China.

Keywords: economic potential; fertilizer management strategies; karst ecosystem; utilization efficiency; yield and nutritional value

Citation: Zhang, M.; Liu, Y.; Wei, Q.; Liu, L.; Gu, X.; Gou, J. Biochar-Based Fertilizer Enhances the Production Capacity and Economic Benefit of Open-Field Eggplant in the Karst Region of Southwest China. *Agriculture* **2022**, *12*, 1388. https://doi.org/10.3390/agriculture12091388

Academic Editor: Maria Roulia

Received: 16 August 2022
Accepted: 1 September 2022
Published: 4 September 2022

1. Introduction

Karst landform is one of the main geomorphic types worldwide. Due to the low rate of soil formation and high leakage, there is no transition layer of weathered parent material between the rock interface and soil, and thus, the ecological environment is extremely fragile [1,2]. China has the widest distribution and largest area of karst landforms, mainly across Guangxi, Guizhou, and Yunnan provinces, accounting for one-third of the total. The exposed rock area in the karst region is large, and the phenomenon of rocky desertification is serious. Simultaneously, the soil in the karst region is easily affected by human activities and the natural environmental processes, which greatly restricts agricultural production herein [3,4].

Biochar is prepared by the pyrolysis of biomass materials under low oxygen or hypoxia conditions [5]. Due to the large specific surface area and rich pore structure, biochar retains soil nutrients and reduces their loss [6]. The surface of biochar has abundant negative charges and a high charge density, resulting in the absorption and attachment of polar or non-polar organic compounds and inorganic ions, such as NH_4^+- and NO_3^-- in water, soil, and sediment [7,8]. Therefore, the application of biochar in farmland improves soil quality and carbon fixation, reduces the contents of nitrogen and phosphorus, and improves crop yield [9,10]. Biochar is made by the carbonization of agricultural wastes, and its nutrient content is low; thus, its application alone is insufficient to meet the needs of crop growth [11,12]. However, increasing the crop yield using biochar may require long-term accumulation, and its application for a short period may not increase the yield significantly [13,14]. Therefore, the application of biochar for agricultural production has some limitations.

In recent years, with adjustments to agricultural structure and the development of ecologically circular agriculture in China with the goal to recycle agricultural wastes, the development and application of new high-efficiency fertilizers have been research hotspots in the field of plant nutrition and fertilizer. Biochar-based fertilizers (BF), as a new type of environmental protection fertilizer, have received traction in the fields of agriculture and environmental protection [15,16]. BFs are made by mixing biochar with organic or inorganic fertilizers; these can be used as a slow-release nutrient carrier [17]. BF not only confers the advantage of soil improvement but also continuously supplies nutrients in combination with the fertilizer requirements of crops, thereby reducing nutrient loss and achieving the dual functions of slow release and carbon fixation [18,19]. As a slow-release fertilizer, BF can maintain high nutrient levels in the soil for a long time, which is attributed to the fact that biochar retains nitrogen, phosphorus, potassium, and other nutrients [20,21]. The interaction of biochar with nitrogen, phosphorus, and potassium is through electrostatic interaction, complexation, and mineralization [22]. Biochar can be combined with NH_4^+-N through electrostatic adsorption, thus reducing the leaching of ammonium ions and improving the utilization efficiency of nitrogen fertilizers [23]. Biochar also complexes with urea to form biochar-based nitrogen fertilizer, and the amino groups on the surface of urea react with the carboxylic anhydride on the biochar for fixation [24]. Additionally, biochar can absorb and fix the ammonia released during the decomposition of urea, which can further prevent the loss of nitrogen fertilizer [25]. When biochar is mixed with phosphorus fertilizer, the former helps capture P-complexing metallic ions, thus reducing the chances of P-fixation [26]. Biochar can also fix K^+, resulting in the formation of biochar-based potassium fertilizer through $\pi-$cation bond action, which helps reduce the rate of loss and improve the utilization rate of potassium fertilizers [27]. Many studies have shown that following BF application in the soil, it oxidizes and decomposes, ultimately leading to nutrient fixation on the biochar, which is gradually released and absorbed by crops [28]. Accumulating evidence shows that BF has good application prospects in improving the physical and chemical properties of soil, reducing chemical fertilizer input, and promoting crop growth [29,30].

Open-field eggplant is an important summer vegetable in the karst region of Guizhou province. However, due to long-term and continuous cropping, insufficient organic fertilizer input, soil carbon/nitrogen (C/N) imbalance, and reduction in soil fertility, the yield, quality, fertilizer utilization, and economic benefits of cultivating eggplant have hardly improved [31,32]. In the present study, distillers grains were used as raw material to prepare biochar through carbonization at a high temperature. Subsequently, the biochar and chemical fertilizers were mixed to a certain proportion to obtain BF. A two-year field experiment was conducted to evaluate the ameliorative roles of BF on the biological and economic benefits of cultivating eggplant in the karst region of Southwest China. In this study, a new technique of fertilizer application using distillers grains was explored, and the findings may provide a reference for the agricultural utility of BF in the karst region of Southwest China.

2. Materials and Methods

2.1. Site Description and Experimental Materials

The field experiment was conducted in Xishan town (27°4′42″ N, 106°41′56″ E) in Xifeng County, Guiyang City, Guizhou Province, China in 2020 and 2021. The soil type of the experimental region was as follows: yellow, zonal with a high aluminization intensity, formed under perennial, humid, and bioclimatic conditions in the subtropical zone. Due to intense leaching caused by the perennial humidity, the exchangeable base content was only 20%, and therefore, the base was extremely unsaturated. The nutrient content of the topsoil of the test field was measured; its pH was 5.74, the soil organic matter (SOM) content was 44.58 g kg^{-1}, the total nitrogen (TN) was 0.52 g kg^{-1}, and the available phosphorus (AP) and available potassium (AK) contents were 30.31 mg kg^{-1} and 107.49 mg kg^{-1}, respectively. The eggplant variety used in the experiment was 'Ruibao 3'.

The raw materials of the biochar were distillers grains, which comprise biomass waste generated from the production process of distilled spirits (Kweichow Moutai (Group) Circular Economy Industrial Investment and Development Co., Ltd., Zunyi, China). The biochar was prepared by the oxygen-limited cracking method in a biomass carbonization furnace (SSDP-5000-A, Jiangsu Huaian Huadian Environmental Protection Machinery Manufacturing Co., Ltd., Huaian, China). Briefly, we obtained appropriate amounts of distillers grain samples and put them in the equipment, followed by blowing in N_2 for 5–10 min to exhaust the excess air in the furnace. The sample was pyrolyzed at 550 °C for 2 h. After cooling, it was passed through a 100-mesh sieve and placed in the shade until subsequent experiments. The nutrient content of the distillers grain biochar was measured. Its pH was 9.05, the soil organic carbon (SOC) was 265.88 g kg^{-1}, and the total nitrogen (TN), phosphorus (TP), and potassium (TK) contents were 47.51, 11.43, and 25.33 g kg^{-1}, respectively. The structural characteristics of the distillers grain biochar were measured. The specific surface area (SSA) was 2.12 m^2 g^{-1}, the single point adsorption total pore volume (SPATPV) was 2.95×10^{-3} m^3 g^{-1}, and the average pore size (APS) was 5.55 nm. According to the local experience of eggplant cultivation, the suitable proportion of $N/P_2O_5:/K_2O$ for growth was 5:3:7. To meet these nutritional requirements, the biochar-based fertilizer was blended with fertilizer and biochar, comprising distillers grains biochar (16%), urea (30%), mono-ammonium phosphate (12%), potassium sulfate (40%), and solid binder (2%). The BF was prepared using a flat grinding extrusion granulator (SKJ-120, Shanghai Jiale Electromechanical Group Co., Ltd., Shanghai, China). Compound fertilizer (N, 15%; P_2O_5, 15%, and K_2O, 15%; Guizhou Xiyang Industrial Co., Ltd., Guiyang, China) and organic fertilizer (comprising mainly cow dung, organic matter $\geq$ 50%, and N + P_2O_5 + $K_2O \geq 5\%$; Guizhou Dibao Co., Ltd., Guiyang, China) were also used.

2.2. Experimental Design and Management

The experiment consisted of six treatments, and each treatment was repeated thrice according to a randomized complete block design (RCBD). The treatments included no fertilizer (CK), traditional fertilization practice (TFP, based on the local practices, 3000 kg ha^{-1} of compound fertilizer, 450.00 kg ha^{-1} N, 450.00 kg ha^{-1} P_2O_5, 450.00 kg ha^{-1} K_2O, and the ratio of the basal to the top dressing was 50:50), BF application at 1875 kg ha^{-1} (BF1, N 281.25 kg ha^{-1}, P_2O_5 112.5 kg ha^{-1}, K_2O 375.00 kg ha^{-1}), BF application at 2250 kg ha^{-1} (BF2, N 337.50 kg ha^{-1}, P_2O_5 135.00 kg ha^{-1}, K_2O 450.00 kg ha^{-1}), BF application at 2625 kg ha^{-1} (BF3, N 393.75 kg ha^{-1}, P_2O_5 157.50 kg ha^{-1}, K_2O 252.00 kg ha^{-1}), and BF application at 3000 kg ha^{-1} (BF4, N 450.00 kg ha^{-1}, P_2O_5 180.00 kg ha^{-1}, K_2O 600.00 kg ha^{-1}). According to the local planting traditions, organic fertilizer at 1500 kg ha^{-1} was used in all treatment groups except for the CK treatment. Table 1 shows the type and amount of fertilizer used for each treatment.

Table 1. Fertilizer amounts of different treatments.

Treatments	Basal Dressing Fertilizer ($kg\ ha^{-1}$)			Top Dressing Fertilizer ($kg\ ha^{-1}$)
	BF	Compound Fertilizer	Organic Fertilizer	Compound Fertilizer
CK	—	—	—	—
TFP	—	1500	1500	1500
BF1	1875	—	1500	—
BF2	2250	—	1500	—
BF3	2625	—	1500	—
BF4	3000	—	1500	—

All the basal dressing fertilizers were spread in the soil simultaneously before planting eggplant seeds and mixed with the topsoil. Eggplant seedlings were transplanted after fertilizer application at a planting density of 18,000 plants ha−1 (with a plant spacing of 70 cm and row spacing of 80 cm). The area of each plot was 56.00 m2 (7.0 m × 8.0 m). After the eggplant seedlings were transplanted, for 30 days, 1500 kg ha−1 of compound fertilizer was applied to the plants in the TFP treatment group. A unified management mode was adopted for all eggplants throughout the growing season to reduce interference due to external factors.

2.3. Sampling and Measurement

2.3.1. Soil Sample Collection and Determination Method

Soil samples between depths of 0 and 20 cm were collected from 10 randomly selected spots on the main experimental area before fertilization. The soil samples were composited and air-dried, ground, and passed through 1 mm and 0.149 mm sieves for determining their physicochemical characteristics. The physical and chemical properties of soil were determined according to the methods described by Bao [33]. The soil pH was measured using a 1:2.5 extraction mixture (soil/water, *w*/*v*) with a pH meter (FE20K, Mettler Toledo, Zurich, Switzerland). The soil organic matter (SOM) was determined using the potassium dichromate volumetric–external heating method. The TN was determined using the semi-micro Kjeldahl method. The AP was determined by extraction using hydrochloric acid combined with ammonium fluoride and assessed by molybdenum antimony anti-colorimetry. The soil AK content was determined by extraction using ammonium acetate and assessed using a flame photometer (FP640, Shanghai Aopu Analytical Instrument Co., Ltd., Shanghai, China).

2.3.2. Plant Sample Collection and Determination Method

At maturity, six plants from each experimental plot were sampled before the final harvest, which was used to test the plant's nutrition and fruit quality. The eggplant plants were divided into two parts, namely the stem-leaf and the fruit, and dried to a constant weight at 60 °C after heating at 105 °C for 30 min. All dried samples were ground and passed through a 0.25 mm sieve and digested in a mixture of concentrated H_2SO_4 and H_2O_2 to determine the concentrations of N, P, and K. The TN concentration was determined by the Kjeldahl nitrogen method; the TP concentration was determined by vanadium molybdenum yellow colorimetry, and the TK concentration using a flame photometer (FP640, Shanghai Aopu Analytical Instrument Co., Ltd., Shanghai, China) [33]. For the determination of nitrate content, 2 g of fresh fruits were taken, to which 10 mL of deionized water was added, and the sample was placed in a boiling water bath for 30 min. The extraction solution (0.1 mL) was taken and mixed with 0.4 mL of 5% salicylic acid-sulfuric acid solution at 25 °C for 20 min. Then, 9.5 mL of 8% NaOH was added to the solution, and the absorbance was measured on a visible spectrophotometer (UV-3600i Plus, Shimadzu, Tokyo, Japan). The VC content was determined by high-performance liquid chromatography (HPLC, LC-2040, Nexera-i, Shimadzu, Japan) after grinding, centrifuging, and filtering with 10 mL 0.2% metaphosphoric acid. For assessing the content of soluble sugar, 0.2 g of fresh leaves were taken, and 10 mL of distilled water was added to the extract in a boiling water bath for 30 min. The extract (0.5 mL) was taken in a test tube, to which 0.5 mL ethyl anthrone and 5 mL sulfuric acid were added. The test tube was placed in a boiling water bath and incubated for 1 min. Subsequently, it was taken out and naturally cooled to room temperature. The absorbance was measured at 630 nm. The content of soluble sugar was determined by anthrone colorimetry. Briefly, 0.6 g of fresh fruit was used for extraction, which was performed twice in 3 mL of 80% ethanol for 60 min. The extracts were then mixed and filtered, and the alcohol was removed by evaporation. The anthrone reagent was added to samples, and after thorough shaking, the absorption of the samples was measured at 625 nm.

2.3.3. Eggplant Yield

When the fresh eggplants were harvested, the yield of each subplot was determined by the actual harvest. According to the growth status of the plants, harvesting was performed multiple times, and the total yield was recorded.

2.4. Calculations and Statistical Analysis

The following parameters and models were calculated according to the method of Zhang et al. [34].

2.4.1. Fertilizer Utilization

$$AE = \frac{Y_F - Y_{CK}}{NI}$$

$$RE = \frac{N_F - N_{CK}}{NI}$$

where AE stands for agronomic efficiency (kg kg^{-1}), RE stands for recovery efficiency (%), Y_F denotes the fresh yield of the fertilization treatment (kg ha^{-1}), Y_{CK} denotes the fresh yield of the CK treatment (kg ha^{-1}), N_F denotes the nutrient accumulation of the fertilization treatment (kg ha^{-1}), N_{CK} denotes the nutrient accumulation of the CK treatment (kg ha^{-1}), and NI denotes the nutrient input of fertilization treatment (kg ha^{-1}).

2.4.2. Economic Benefits

$$OV = Y \times UP$$

$$IOV = OV_F - OV_{CK}$$

$$NET = OV - FI$$

where OV stands for the output value (USD ha^{-1}), Y stands for the fresh yield (kg ha^{-1}), UP stands for the unit price of fresh eggplant (USD kg^{-1}), IOV stands for the increased output value (USD ha^{-1}), OV_F stands for the output value of the fertilization treatment (USD ha^{-1}), OV_{CK} stands for the output value of the CK treatment (USD ha^{-1}), NET stands for the net income (USD ha^{-1}), and FI stands for the fertilizer input (USD ha^{-1}). In the calculation of economic benefits, the unit price of fresh eggplant was USD 0.4471 kg^{-1}, and the BF, compound fertilizer, and organic fertilizer were 0.4471, 0.2980, and 0.1490 USD kg^{-1}, respectively.

2.4.3. Linear Plus Platform Model

In the experiment, a linear plus platform model was used to fit the response of the yield of fresh eggplant to the BF application rate, so that the best application amount of BF on fresh eggplant could be calculated.

The calculation equation was:

$$Y = AX + B\ (X \leq C);\ Y = P\ (X > C)$$

where Y stands for the yield of fresh eggplant (kg ha^{-1}), X stands for the BF application rate (kg ha^{-1}), A stands for the slope, B stands for the intercept, C stands for the intersection of the line and the platform, and P stands for the maximum yield (kg ha^{-1}).

2.5. Statistical Analysis

Microsoft Excel 2007 and SPSS 20.0 (SPSS Inc., Chicago, IL, USA) were used for data processing and statistical analysis. The significance of the differences between the soil and the plant indicators was measured by one-way ANOVA. The significance of the differences was tested using Duncan's new compound extreme difference method, and the significance level was set as $\alpha = 0.05$. All the figures were constructed with Origin 12.0 (Origin Lab Corporation, Northampton, MA, USA).

3. Results

3.1. Yield of Fresh Eggplants

The TFP and BF treatments both significantly enhanced the yield of fresh eggplants, with an increase of 4905–9081 kg ha^{-1} in 2020 and 11,036–25,018 kg ha^{-1} in 2021 compared to the CK treatment group (Figure 1). The yield of fresh eggplant in the BF group increased by 1109–4176 kg ha^{-1} in 2020 and 6672–13,982 kg ha^{-1} in 2021 compared to the TFP treatment group, and the increasing rates were 3.65–13.76% in 2020 and 23.40–49.04% in 2021. The fresh yield following the BF3 treatment was the highest in 2020 and 2021, at 34,514 and 42,419 kg ha^{-1}, respectively. The optimal application rate of BF for eggplant in the open field was estimated (Figure 2). The results show that the maximum yield of fresh eggplant was obtained with a BF application of 2625 kg ha^{-1} in 2020; the rate was 2544 kg ha^{-1} in 2021.

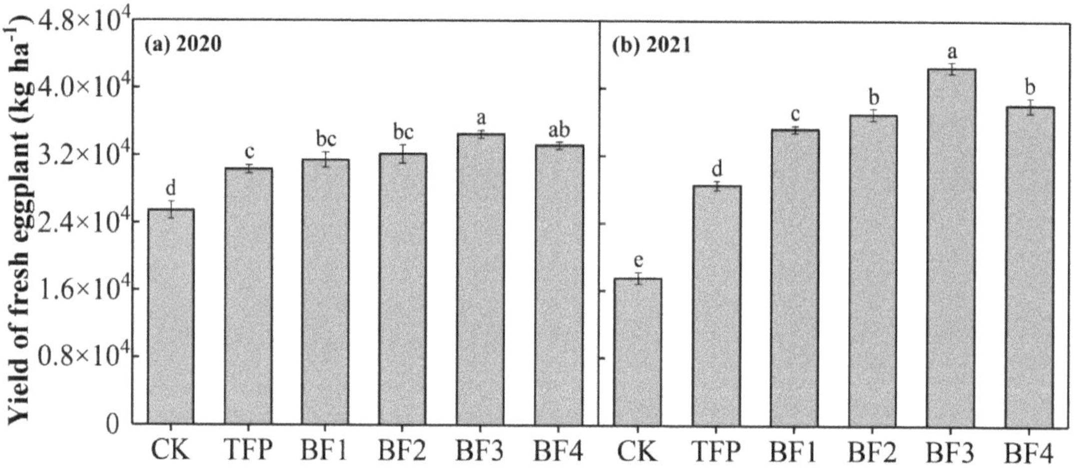

Figure 1. Effects of BF on the yield of fresh eggplant in 2020 (**a**) and 2021 (**b**). CK—no fertilizer; TFP—traditional fertilization practice; BF1—1875 kg ha^{-1} of biochar-based fertilizer (BF); BF2—2250 kg ha^{-1} of BF; BF3—2625 kg ha^{-1} of BF; BF4—3000 kg ha^{-1} of BF. Different lowercase letters denote significant differences among treatment means at the $\alpha = 0.05$ level using Duncan's MRT method.

3.2. Fruit Quality of the Eggplant

Indicators of the fresh fruit quality of eggplant are listed in Table 2. BF application significantly reduced the content of nitrate in fresh eggplant fruits, while the VC and soluble sugar were enhanced. BF reduced the content of nitrate in fresh eggplant fruits by 10.14–17.16% in 2020 and 19.02–30.18% in 2021 compared to the TFP treatment group, and the nitrate content in the BF4 treatment group was the lowest. The contents of VC and soluble sugar in fresh eggplant fruits increased by 10.28–22.52% and 6.79–12.20% in 2020, and by 15.17–40.68% and 11.58–21.00% in 2021, respectively.

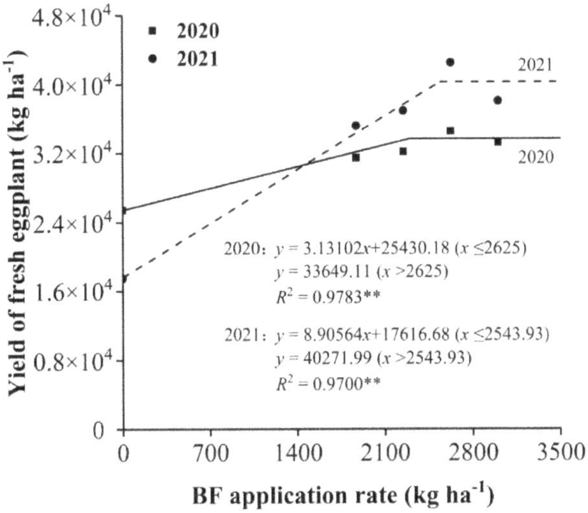

Figure 2. Effects of BF application rate on the yield of fresh eggplant. ** represents significant at α = 0.01 probability level.

Table 2. Effects of BF on quality of fresh eggplant in 2020 and 2021.

Year	Treatments	Nitrate (mg kg^{-1})	VC (mg kg^{-1})	Soluble Sugar (mg kg^{-1})
2020	CK	107.18 ± 4.78 a	99.65 ± 4.44 c	155.24 ± 5.13 b
	TFP	101.13 ± 4.51 a	108.32 ± 4.83 c	161.69 ± 6.43 b
	BF1	90.88 ± 3.61 b	119.45 ± 3.07 b	172.67 ± 4.50 a
	BF2	88.62 ± 3.95 bc	128.16 ± 5.71 ab	180.73 ± 4.08 a
	BF3	85.08 ± 2.33 bc	132.71 ± 5.92 a	181.41 ± 5.14 a
	BF4	83.78 ± 2.17 c	124.31 ± 5.54 ab	176.42 ± 5.70 a
2021	CK	104.97 ± 4.91 a	94.01 ± 3.08 e	153.45 ± 4.99 c
	TFP	104.88 ± 6.16 a	105.44 ± 2.25 d	158.78 ± 8.16 c
	BF1	84.93 ± 3.01 b	121.43 ± 2.09 c	177.16 ± 4.79 b
	BF2	78.40 ± 2.41 bc	138.94 ± 6.60 b	186.19 ± 4.83 ab
	BF3	76.80 ± 3.13 c	148.33 ± 3.98 a	192.12 ± 5.44 a
	BF4	73.23 ± 2.89 c	143.25 ± 5.06 ab	189.82 ± 2.23 a

Note: Different lowercase letters denote significant differences among treatment means at the α = 0.05 level using Duncan's MRT method. The results are presented as the mean value ± standard error.

3.3. Nutrients Accumulation

The application of BF affected the nutrient accumulation in eggplant, and the N, P, and K contents in the eggplant plants are shown in Figure 3. The application of fertilizer improved the accumulation of N, P, and K by 62.93–157.44%, 54.07–92.68%, and 110.04–225.41%, respectively, in 2020 and 58.05–269.51%, 60.07–172.91%, and 118.05–339.71% in 2021 compared to the CK treatment group. Furthermore, the N, P, and K accumulation following BF application increased by 19.77–58.01%, 0.70–25.05%, and 16.11–54.93% in 2020, and 43.94–133.78%, 18.42–70.50%, and 35.54–101.65% in 2021, respectively, compared to the TFP treatment group. The N, P, and K accumulation rates were the highest in the BF3 treatment group, at 222.30, 69.08, and 379.20 kg ha^{-1}, in 2020, and 248.35, 84.74, and 445.21 kg ha^{-1}, in 2021, respectively.

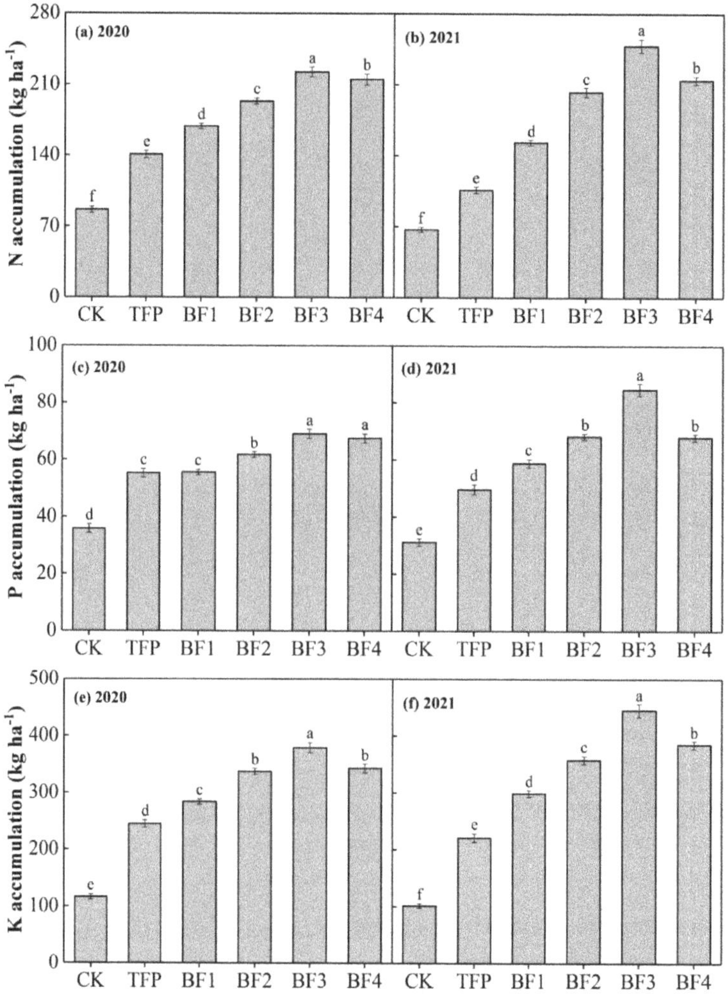

Figure 3. Effects of BF on the N, P and K accumulation in 2020 (**a**,**c**,**e**) and 2021 (**b**,**d**,**f**). CK—no fertilizer; TFP—traditional fertilization practice; BF1—1875 kg ha^{-1} of biochar-based fertilizer (BF); BF2—2250 kg ha^{-1} of BF; BF3—2625 kg ha^{-1} of BF; BF4—3000 kg ha^{-1} of BF. Different lowercase letters denote significant differences among treatment means at the $\alpha = 0.05$ level using Duncan's MRT method.

3.4. Fertilizer Utilization

Table 3 shows the AE and RE of different fertilization treatment groups. Compared to the TFP treatment group, the AE_N, AE_P, and AE_K following BF treatment increased by 58.35–111.56%, 295.96–428.99%, and 18.81–58.72% in 2020, and 86.47–159.03%, 366.12–547.57%, and 39.83–94.25% in 2021, respectively. The AE_N, AE_P, and AE_K in the BF3 treatment group were the highest, at 23.06, 57.66, and 17.30 kg kg^{-1} in 2020, and 63.54, 158.85, and 47.65 kg kg^{-1} in 2021, respectively. Similarly, the RE_N, RE_P, and RE_K following BF treatment increased by 136.67–185.84%, 307.89–389.33%, and 32.50–75.61% in 2020, and 251.44–430.57%, 393.98–721.45%, and 77.97–146.69% in 2021, respectively, compared to the TFP treatment group. The RE_N, RE_P, and RE_K in the BF3 treatment group were also the highest, at 34.53%, 21.09%, and 50.03% in 2020, and 46.00%, 34.09%, and 65.52% in 2021, respectively.

Table 3. Effects of BF on agronomic efficiency and recovery efficiency in 2020 and 2021.

Year	Treatments	AE (kg kg^{-1})			RE (%)		
		AE_N	AE_P	AE_K	RE_N	RE_P	RE_K
2020	CK	—	—	—	—	—	—
	TFP	10.90 ± 3.37 b	10.90 ± 3.37 b	10.90 ± 3.37 b	12.08 ± 1.03 d	4.31 ± 0.44 c	28.49 ± 1.68 d
	BF1	21.38 ± 4.62 a	53.46 ± 11.56 a	16.04 ± 3.47 ab	29.21 ± 0.15 c	17.58 ± 0.53 b	44.71 ± 0.16 b
	BF2	19.95 ± 4.63 a	49.86 ± 11.58 a	14.96 ± 3.47 ab	31.78 ± 1.80 b	19.22 ± 1.87 ab	49.09 ± 2.09 a
	BF3	23.06 ± 0.18 a	57.66 ± 5.44 a	17.30 ± 1.63 a	34.53 ± 0.89 a	21.09 ± 0.81 a	50.03 ± 1.35 a
	BF4	17.26 ± 2.21 ab	43.16 ± 5.52 a	12.95 ± 1.66 ab	28.59 ± 1.03 c	17.60 ± 0.93 b	37.75 ± 1.27 c
2021	CK	—	—	—	—	—	—
	TFP	24.53 ± 2.19 d	24.53 ± 2.19 d	24.53 ± 2.19 d	8.67 ± 1.10 d	4.15 ± 0.53 e	26.56 ± 2.12 e
	BF1	62.96 ± 3.78 a	157.41 ± 9.46 a	47.22 ± 2.84 a	30.47 ± 1.53 c	24.72 ± 1.68 c	52.80 ± 2.01 c
	BF2	57.66 ± 1.63 b	144.15 ± 4.07 b	43.25 ± 1.22 b	40.20 ± 2.05 b	27.52 ± 1.68 b	57.04 ± 2.30 b
	BF3	63.54 ± 2.42 a	158.85 ± 6.05 a	47.65 ± 1.82 a	46.00 ± 1.22 a	34.09 ± 0.95 a	65.52 ± 1.73 a
	BF4	45.74 ± 1.70 c	114.34 ± 4.25 c	34.30 ± 1.27 c	32.75 ± 0.79 c	20.50 ± 0.49 d	47.27 ± 0.91 d

Note: Different lowercase letters denote significant differences among treatment means at the $\alpha = 0.05$ level using Duncan's MRT method. The results are presented as the mean value ± standard error.

3.5. Economic Benefits

The application of BF affected the economic benefits of fresh eggplant (Table 4). The output of fresh eggplant in the BF treatment group increased by 495–1867 USD ha^{-1} in 2020 and 2983–6251 USD ha^{-1} in 2021 compared to the TFP treatment group. The corresponding rates of increase were 3.65–13.76% in 2020 and 23.40–49.04% in 2021. After the cost of the fertilizers was deducted, the net income from fresh eggplants treated with BF increased by 535–1588 USD ha^{-1} in 2020 and 3039–5972 USD ha^{-1} in 2021, and the increase rates were 4.30–12.76% in 2020 and 26.13–51.35% in 2021. The net income in the BF3 treatment group was the highest in both years.

Table 4. Effects of BF on economic benefits of eggplant in 2020 and 2021.

Year	Treatments	Output Value (USD ha^{-1})	Increased Output Value with Fertilizer (USD ha^{-1})	Fertilizer Inputs (USD ha^{-1})	Net Incomes (USD ha^{-1})
2020	CK	11,372 ± 457 d	—	—	11,372 ± 457 d
	TFP	13,565 ± 221 c	2193 ± 677 b	1118	12,447 ± 221 c
	BF1	14,060 ± 398 bc	2688 ± 582 b	1062	12,998 ± 398 bc
	BF2	14,381 ± 485 b	3009 ± 699 b	1230	13,151 ± 485 b
	BF3	15,432 ± 211 a	4060 ± 383 a	1397	14,035 ± 211 a
	BF4	14,547 ± 428 b	3175 ± 240 ab	1565	12,982 ± 428 b
2021	CK	7812 ± 314 e	—	—	7812 ± 314 e
	TFP	12,747 ± 252 d	4935 ± 440 d	1118	11,629 ± 252 d
	BF1	15,730 ± 201 c	7918 ± 476 c	1062	14,668 ± 201 c
	BF2	16,513 ± 318 b	8701 ± 245 b	1230	15,283 ± 318 b
	BF3	18,998 ± 296 a	11,186 ± 426 a	1397	17,601 ± 296 a
	BF4	17,014 ± 400 b	9202 ± 342 b	1565	15,449 ± 400 b

Note: Different lowercase letters denote significant differences among treatment means at the $\alpha = 0.05$ level using Duncan's MRT method. The results are presented as the mean value ± standard error.

4. Discussion

4.1. Developmental Potential of Biochar-Based Fertilizer

Biochar has obvious advantages for carbon sequestration and emission reduction, water and fertilizer preservation, and soil improvement, thus addressing the difficulties in sustainable agricultural development, environmental protection, and governance [35,36]. However, biochar application has some limitations for agricultural production at present due to the following fundamental reasons: (i) the application amount of biochar and input costs are high, (ii) the application method of biochar is controversial, and (iii) the economic benefit and output following biochar application is unclear [37–40]. Biochar-based fertilizer is produced through the secondary processing of biochar and other mineral fertilizers. Thus, the granulation effect and the quantity ratio effects are not only exerted but also temporally and spatially consistent, which is beneficial for reducing agricultural production costs and improving the commercial utility of biochar [38]. Therefore, biochar-based fertilizer may be a new developmental direction in agriculture.

4.2. Biochar-Based Fertilizer for Improving Yield and Quality of Crops

The advantages of BF in improving crop production and soil environmental quality have been confirmed [41,42]. The present study showed that the yield and quality of fresh eggplant following BF treatment improved significantly (Figure 1, Table 2). Due to its loose and porous characteristics, biochar can improve the physical properties and soil porosity of clayey yellow soil in the karst region [31]. Moreover, the characteristics of biochar with a large SSA and high adsorption capacity make it possible to absorb fertilizer nutrients through pore closure and surface adsorption following granulation with chemical fertilizers, thus delaying the fertilizer efficiency of BF [43,44]. This view was confirmed in the study of Chew et al. [45] in another experiment. They pointed out that the nutrient release rate of chemical fertilizer was much higher than that of BF, and most of the released nutrients were leached and lost in the early stage of crop growth. However, the nutrient release rate of BF was slow, which can provide the nutrients needed by plants in the later stage of growth, promote the accumulation of fruit nutrients, and improve the quality of crops [45]. Melo et al. [46] observed that BF was able to further contribute to an increase in productivity beyond that of conventional fertilizers, especially when involving N fertilizers. This is mainly attributed to the following: (i) BF increases the photosynthetic rate of crop leaves, which is caused by BF reducing the limitation of photosynthesis by non-stomatal factors, thus increasing the accumulation of carbohydrates, and (ii) increases in the specific root surface area, root branching, and fine roots under field conditions in soils result in higher crop productivity [47]. Biochar not only contains several elements such as N, P, and K but also contains abundant mineral nutrients, including Ca, Mg, and Zn, contributing to a balanced nutrient supply [48]. Meanwhile, the mineral elements in biochar could promote the synthesis of related enzymes in plants, thus promoting an increase in the VC and soluble sugar contents in fruits and reducing the content of acids in fruits [49]. Notably, biochar could affect the processes in the N cycle in the agricultural ecosystem, which may reduce the excessive nitrate uptake by plants [50,51].

4.3. The Functions of BF in Improving Fertilizer Utilization Rate and Economic Benefits

Based on the results of this study, the AE and RE of nutrients following BF treatment increased significantly, indicating that BF could significantly improve the utilization rates of fertilizers in the karst region, especially N and K fertilizers (Table 3). Studies have shown that biochar in BF can adsorb nitrogen (NH_4^+ or NO_3^-) in soil and fertilizer on its surface with a relatively large number of exchange ions and active carboxyl, hydroxyl, and other functional groups, thus preventing nitrogen from leaching downward into the subsoil, reducing fixed and gaseous losses of nitrogen, and improving the efficiency of nitrogen utilization [52,53]. Moreover, biochar can stimulate the activity of bacteria related to nitrogen, and owing to its porosity and large surface area, it can provide a habitat for microorganisms, which is conducive to adsorbing microorganisms, ultimately affecting the processes of the nitrogen cycle in the soil system [54]. Biochar application can provide a carbon source for soil nitrogen-fixing microorganisms, which can also promote their growth [55]. Some studies have shown a significant positive correlation between the abundance of soil nitrogen-fixing bacteria and the activity of nitrogen-fixing enzymes and the soil carbon content. An increase in the SOC content is crucial for biochar to promote biological nitrogen fixation and improve nitrogen utilization efficiency [55]. Some studies have shown that K on the surface of biochar can be quickly released for absorption and utilization by plants after application into the soil, which is related to the high availability of

K in biochar [56,57]. The carboxyl functional groups in biochar can improve the adsorption capacity of soil for cations, thus increasing the probability of K^+ entering the soil lattice, which is conducive to improving the utilization rate of potassium [58,59]. However, the increase in AE and RE may be the result of the inherent nutrient composition of the biochar. Therefore, during the preparation of biochar-based fertilizers, nitrogen and potassium can replace chemical fertilizers in the future.

The economic benefit is an important index to measure the increase in farmers' output and income. The results show that the net income following BF treatments increased by 4.30–12.76% in 2020 and 26.13–51.35% in 2021 (Table 4), similar to the results of previous studies [60,61]. At present, most research on biochar focuses on soil and environmental effects but studies on economic benefits are relatively scarce. This is mainly because the production and transportation costs of biochar are relatively high; its large-scale application will lead to higher production costs in the early stage, and the economic benefits cannot be increased rapidly [62,63]. However, the application of BF not only ensures a balanced supply of various nutrients but also improves the crop yield and quality following a reduction in the amount of fertilizer, thus achieving a comprehensive utilization of fertilizer and agricultural wastes; this has obvious ecological and economic benefits [64,65]. The optimal application amount of BF was estimated using a linear model, and the results show that the application of 2544–2625 kg ha^{-1} was optimal (Figure 2), which can be popularized in agricultural production in the karst region. Furthermore, future research needs to focus on the exploration of new types of BF to improve its nutrient-controlled release performance to the maximum. In addition, functional biochar-based fertilizers should be prepared according to the soil status and production demands of different regions.

5. Conclusions

In conclusion, biochar-based fertilizer significantly improved the yield, quality, fertilizer utilization, and economic benefits of eggplant cultivation in open fields in the karst region of Southwest China. The application of biochar-based fertilizer is a nutrient-efficient management strategy for open-field eggplant cultivation in Southwest China, which increases the production capacity and economic benefits while reducing nutrient leakage. Although there have been many reports on the impact of biochar-based fertilizer on crop growth, studies on the underlying mechanism of action are lacking. Therefore, more fieldwork should be conducted to provide experimental support for the application of biochar-based fertilizer.

Author Contributions: Conceptualization, M.Z., Q.W. and J.G.; data curation, M.Z. and X.G.; formal analysis, M.Z. and Y.L.; supervision, L.L. and J.G.; writing—original draft, M.Z.; writing—review and editing, M.Z. and J.G. All authors have read and agreed to the published version of the manuscript.

Funding: This research was funded by the National Natural Science Foundation of China (No. 31860594 and No. 32060302), the Subsidy Project from the NSFC of Guizhou Academy of Agricultural Sciences ([2021]32) and the Science and Technology Planning Project of Guizhou Province ([2020]1Y087).

Institutional Review Board Statement: Not applicable.

Informed Consent Statement: Not applicable.

Data Availability Statement: Not applicable.

Conflicts of Interest: The authors declare no conflict of interest.

References

1. Zhang, J.; Zhang, Z.; Huang, X.; Cui, M.; Wu, X.; Liu, H. Adsorption and desorption characteristics of metal(oid)s in the yellow soils of a typical karst area, southwest China. *Soil Sediment Contam. Int. J.* **2022**, *31*, 834–854. [CrossRef]
2. Yu, P.; Li, Y.; Liu, S.; Liu, J.; Ding, Z.; Ma, M.; Tang, X. Afforestation influences soil organic carbon and its fractions associated with aggregates in a karst region of Southwest China. *Sci. Total Environ.* **2021**, *814*, 152710. [CrossRef] [PubMed]

3. Cao, Y.; Xiong, K.N.; Lv, X.X.; Zhang, H.Z. Effects of different vegetation restoration modes on soil microbial quantity and biomass in a degraded karst ecosystem in south China. *Eurasian Soil Sci.* **2021**, *54*, 2009–2021.
4. Xiao, L.; Zhang, W.; Hu, P.; Xiao, D.; Yang, R.; Ye, Y.; Wang, K. The formation of large macroaggregates induces soil organic carbon sequestration in short-term cropland restoration in a typical karst area. *Sci. Total Environ.* **2021**, *801*, 149588. [CrossRef] [PubMed]
5. Gustavo, C.; Antonio, R.; Ana, M.; Marcelo, P.; Andrés, M.P.S.; Sebastián, M.; Rodrigo, M. Effects of biochar amendment on wheat production, mycorrhizal status, soil microbial community, and properties of an andisol in southern Chile. *Field Crops Res.* **2021**, *273*, 108306. [CrossRef]
6. Usevičiūtė, L.; Baltrėnaitė-Gedienė, E.; Feizienė, D. The combined effect of biochar and mineral fertilizer on triticale yield, soil properties under different tillage systems. *Plants* **2022**, *11*, 111. [CrossRef]
7. Li, W.; Yao, L.; Lng, M.; Zhang, H.; Zhang, Y.; Wen, X.; Mo, F.; Han, J.; Liao, Y. Biochar reduced soil N losses and improved N balance in a rainfed winter-wheat cropping agroecosystem. *Appl. Soil Ecol.* **2022**, *179*, 104591. [CrossRef]
8. Jabborova, D.; Annapurna, K.; Azimov, A.; Tyagi, S.; Pengani, K.R.; Sharma, P.; Vikram, K.V.; Poczai, P.; Nasif, O.; Ansari, M.J.; et al. Co-inoculation of biochar and arbuscular mycorrhizae for growth promotion and nutrient fortification in soybean under drought conditions. *Front. Plant Sci.* **2022**, *13*, 947547. [CrossRef]
9. Fungo, B.; Lehmann, J.; Kalbitz, K.; Thiongo, M.; Okeyo, I.; Tenywa, M.; Neufeldt, H. Aggregate size distribution in a biochar-amended tropical ultisol under conventional hand-hoe tillage. *Soil Tillage Res.* **2017**, *165*, 190–197. [CrossRef]
10. Kang, Y.G.; Chun, J.H.; Lee, J.H.; Park, S.Y.; Luyima, D.; Oh, T.K.; Yun, Y.U. Effects of varying rates of nitrogen and biochar pH on NH_3 emissions and agronomic performance of Chinese cabbage (*Brassica rapa* ssp. pekinensis). *Agronomy* **2022**, *12*, 61. [CrossRef]
11. Egamberdieva, D.; Ma, H.; Reckling, M.; Omari, R.A.; Wirth, S.; Bellingrath-Kimura, S.D. Interactive effects of biochar, nitrogen, and phosphorous on the symbiotic performance, growth, and nutrient uptake of soybean (*Glycine max* L.). *Agronomy* **2022**, *12*, 27. [CrossRef]
12. Manzoor, S.; Habib-Ur-Rahman, M.; Haider, G.; Ghafoor, I.; Ahmad, S.; Afzal, M.; Nawaz, F.; Iqbal, R.; Yasin, M.; Tanveer-ul-Haq Danish, S.; et al. Biochar and slow-releasing nitrogen fertilizers improved growth, nitrogen use, yield, and fiber quality of cotton under arid climatic conditions. *Environ. Sci. Pollut. Res.* **2021**, *29*, 13742–13755. [CrossRef] [PubMed]
13. Liao, W.; Zhang, X.; Ke, S.; Shao, J.; Yang, H.; Zhang, S.; Chen, H. Effect of different biomass species and pyrolysis temperatures on heavy metal adsorption, stability and economy of biochar. *Ind. Crops Prod.* **2022**, *186*, 115238. [CrossRef]
14. Kochanek, J.; Soo, R.M.; Martinez, C.; Dakuidreketi, A.; Mudge, A.M. Biochar for intensification of plant-related industries to meet productivity, sustainability and economic goals: A review. *Resour. Conserv. Recycl.* **2022**, *179*, 106109. [CrossRef]
15. Agarwal, H.; Kashyap, V.H.; Mishra, A.; Bordoloi, S.; Singh, P.K.; Joshi, N.C. Biochar-based fertilizers and their applications in plant growth promotion and protection. *3 Biotech* **2022**, *12*, 136. [CrossRef]
16. Yin, D.; Yang, X.; Wang, H.; Guo, X.; Wang, S.; Wang, Z.; Ding, G.; Yang, G.; Zhang, J.; Jin, L.; et al. Effects of chemical-based fertilizer replacement with biochar-based fertilizer on albic soil nutrient content and maize yield. *Open Life Sci.* **2022**, *17*, 517–528. [CrossRef]
17. Gao, T.; Zhu, Q.; Zhou, Z.; Wu, Y.; Xue, J. Effects of biochar-based fertilizers on energy characteristics and growth of black locust seedlings. *Sustainability* **2022**, *14*, 5045. [CrossRef]
18. Wen, P.; Wu, Z.S.; Han, Y.J.; Cravotto, G.; Wang, J.; Ye, B.C. Microwave-assisted synthesis of a novel biochar-based slow-release nitrogen fertilizer with enhanced water-retention capacity. *ACS Sustain. Chem. Eng.* **2017**, *5*, 7374–7382. [CrossRef]
19. Guo, L.; Yu, H.; Niu, W.; Kharbach, M. Biochar promotes nitrogen transformation and tomato yield by regulating nitrogen-related microorganisms in tomato cultivation soil. *Agronomy* **2021**, *11*, 381. [CrossRef]
20. Grafmüller, J.; Schmidt, H.P.; Kray, D.; Hagemann, N. Root-zone amendments of biochar-based fertilizers: Yield increases of white cabbage in temperate climate. *Horticulturae* **2022**, *8*, 307. [CrossRef]
21. Rasse, D.P.; Weldon, S.; Joner, E.J.; Joseph, S.; Kammann, C.I.; Liu, X.; O-Toole, A.; Pan, G.; Kocatürk, S.N.P. Enhancing plant N uptake with biochar-based fertilizers: Limitation of sorption and prospects. *Plant Soil* **2022**, *475*, 213–236. [CrossRef]
22. Wang, C.; Luo, D.; Zhang, X.; Huang, R.; Cao, Y.; Liu, G.; Zhang, Y.; Wang, H. Biochar-based slow-release of fertilizers for sustainable agriculture: A mini review. *Environ. Sci. Ecotechnol.* **2022**, *10*, 100167. [CrossRef]
23. Gao, Y.; Fang, Z.; Van Zwieten, L.; Bolan, N.; Dong, D.; Quin, B.F.; Meng, J.; Li, F.; Wu, F.; Wang, H.; et al. A critical review of biochar-based nitrogen fertilizers and their effects on crop production and the environment. *Biochar* **2022**, *4*, 36. [CrossRef]
24. Gamage, A.; Basnayake, B.; De Costa, J.; Merah, O. Effects of rice husk biochar coated urea and anaerobically digested rice straw compost on the soil fertility, and cyclic effect of phosphorus. *Plants* **2022**, *11*, 75. [CrossRef]
25. Melo, L.C.A.; Lehmann, J.; Carneiro, J.S.; Camps-Arbestain, M. Biochar-based fertilizer effects on crop productivity: A meta-analysis. *Plant Soil* **2022**, *472*, 45–58. [CrossRef]
26. Dib, S.R.; Senesi, G.S.; Gomes Neto, J.A.; Ribeiro, C.A.; Ferreira, E.C. Phosphorous determination in biochar-based fertilizers by spark dischargelaser-induced breakdown spectroscopy. *Chemosensors* **2021**, *9*, 337. [CrossRef]
27. Piash, M.I.; Iwabuchi, K.; Itoh, T. Synthesizing biochar-based fertilizer with sustained phosphorus and potassium release: Co-pyrolysis of nutrient-rich chicken manure and Ca-bentonite. *Sci. Total Environ.* **2022**, *822*, 153509. [CrossRef]
28. Roy, A.; Chaturvedi, S.; Singh, S.V.; Kasivelu, G.; Dhyani, V.C.; Pyne, S. Preparation and evaluation of two enriched biochar-based fertilizers for nutrient release kinetics and agronomic effectiveness in direct-seeded rice. *Biomass Convers. Biorefinery* **2022**, 1–12. [CrossRef]

29. Huang, Y.; Chen, Y.; Wen Dian Zhao, P.; Li, F.; Li Lei Du, R.; Shi, H.; Deng, T.; Du, Y. Biochar-based molybdenum slow-release fertilizer enhances nitrogen assimilation in Chinese flowering cabbage (*Brassica parachinensis*). *Chemosphere* **2022**, *303*, 134663. [CrossRef]
30. Meyer zu Drewer, J.; Köster, M.; Abdulai, I.; Rötter, R.P.; Hagemann, N.; Schmidt, H.P. Impact of different methods of root-zone application of biochar-based fertilizers on young cocoa plants: Insights from a pot-trial. *Horticulturae* **2022**, *8*, 328. [CrossRef]
31. Jiang, M.; Li, C.; Gao, W.; Cai, K.; Tang, Y.; Cheng, J. Comparison of long-term effects of biochar application on soil organic carbon and its fractions in two ecological sites in karst regions. *Geoderma Reg.* **2021**, *28*, e00477. [CrossRef]
32. Zhang, Y.; Li, Y.; Liu, Y.; Huang, X.; Zhang, W.; Jiang, T. Responses of soil labile organic carbon and carbon management index to different long-term fertilization treatments in a typical yellow soil region. *Eurasian Soil Sci.* **2021**, *54*, 605–618. [CrossRef]
33. Bao, S.D. *Soil Agricultural-Chemical Analysis*; China Agricultural Press: Beijing, China, 2000.
34. Zhang, M.; Gou, J.; Wei, Q.; Liu, Y.; Qin, S. Effects of different biochar-based fertilizers on the biological properties and economic benefits of pod pepper (*Capsicum annuum var. frutescens* L.). *Appl. Ecol. Environ. Res.* **2021**, *19*, 2829–2841. [CrossRef]
35. Qayyum, M.F.; Haider, G.; Raza, M.A.; Mohamed, A.K.S.H.; Rizwan, M.; El-Sheikh, M.A.; Alyemeni, M.N.; Ali, S. Straw-based biochar mediated potassium availability and increased growth and yield of cotton (*Gossypium hirsutum* L.). *J. Saudi Chem. Soc.* **2020**, *24*, 963–973. [CrossRef]
36. Liu, X.; Wang, H.; Liu, C.; Sun, B.; Zheng, J.; Bian, R.; Drosos, M.; Zhang, X.; Li, L.; Pan, G. Biochar increases maize yield by promoting root growth in the rainfed region. *Arch. Agron. Soil Sci.* **2020**, *67*, 1411–1424. [CrossRef]
37. Mohammadi, A.; Khoshnevisan, B.; Venkatesh, G.; Eskandari, S. A critical review on advancement and challenges of biochar application in paddy fields: Environmental and life cycle cost analysis. *Processes* **2020**, *10*, 1275. [CrossRef]
38. Lustosa, F.J.F.; Carneiro, J.S.D.S.; Barbosa, C.F.; Lima, K.P.; Leite, A.D.A.; Melo, L.C.A. Aging of biochar-based fertilizers in soil: Effects on phosphorus pools and availability to urochloa brizantha grass. *Sci. Total Environ.* **2020**, *709*, 136028. [CrossRef]
39. Li, Y.; Hu, S.; Chen, J.; Müller, K.; Li, Y.; Fu, W.; Lin, Z.; Wang, H. Effects of biochar application in forest ecosystems on soil properties and greenhouse gas emissions: A review. *J. Soils Sediments* **2018**, *18*, 546–563. [CrossRef]
40. Li, Z.; Song, Z.; Singh, B.P.; Wang, H. The impact of crop residue biochars on silicon and nutrient cycles in croplands. *Sci. Total Environ.* **2019**, *659*, 673–680. [CrossRef]
41. Da Silva Carneiro, J.S.; Andrade, I.C.R.; Nardis, B.O.; Barbosa, C.F.; Filho, J.F.L.; Melo, L.C.A. Long-term effect of biochar-based fertilizers application in tropical soil: Agronomic efficiency and phosphorus availability. *Sci. Total Environ.* **2021**, *760*, 143955. [CrossRef]
42. Sutradhar, I.; Jackson-deGraffenried, M.; Akter, S.; McMahon, S.A.; Waid, J.L.; Schmidt, H.P.; Wendt, A.S.; Gabrysch, S. Introducing urine-enriched biochar-based fertilizer for vegetable production: Acceptability and results from rural Bangladesh. *Environ. Dev. Sustain.* **2021**, *23*, 12954–12975. [CrossRef]
43. Kotuš, T.; Šimanský, V.; Drgo ˇnová, K.; Illéš, M.; Wójcik-Gront, E.; Balashov, E.; Buchkina, N.; Aydın, E.; Horák, J. Combination of biochar with N-fertilizer affects properties of soil and N_2O emissions in maize crop. *Agronomy* **2022**, *12*, 1314. [CrossRef]
44. Liu, M.; Linna, C.; Ma, S.; Ma, Q.; Guo, J.; Wang, F.; Wang, L. Effects of biochar with inorganic and organic fertilizers on agronomic traits and nutrient absorption of soybean and fertility and microbes in purple soil. *Front. Plant Sci.* **2022**, *13*, 871021. [CrossRef] [PubMed]
45. Chew, J.; Joseph, S.; Chen, G.; Zhang, Y.; Zhu, L.; Liu, M.; Taherymoosavi, S.; Munroe, P.; Mitchell, D.R.G.; Pan, G.; et al. Biochar-based fertiliser enhances nutrient uptake and transport in rice seedlings. *Sci. Total Environ.* **2022**, *826*, 154174. [CrossRef]
46. Sim, D.H.H.; Tan, I.A.W.; Lim, L.L.P.; Hameed, B.H. Encapsulated biochar-based sustained release fertilizer for precision agriculture: A review. *J. Clean. Prod.* **2021**, *303*, 127018. [CrossRef]
47. Liao, J.; Liu, X.; Hu, A.; Song, H.; Chen, X.; Zhang, Z. Effects of biochar-based controlled release nitrogen fertilizer on nitrogen-use efficiency of oilseed rape (*Brassica napus* L.). *Sci. Rep.* **2020**, *10*, 11063. [CrossRef]
48. Khan, M.A.; Basir, A.; Fahad, S.; Adnan, M.; Saleem, M.H.; Iqbal, A.; Amanullah Al-Huqail, A.A.; Alosaimi, A.A.; Saud, S.; Liu, K.; et al. Biochar optimizes wheat quality, yield, and nitrogen acquisition in low fertile calcareous soil treated with organic and mineral nitrogen fertilizers. *Front. Plant Sci.* **2022**, *13*, 879788. [CrossRef]
49. Wang, F.; Wang, X.; Song, N. Biochar and vermicompost improve the soil properties and the yield and quality of cucumber (*Cucumis sativus* L.) grown in plastic shed soil continuously cropped for different years. *Agric. Ecosyst. Environ.* **2021**, *315*, 107425. [CrossRef]
50. Wu, D.; Zhang, W.; Xiu, L.; Sun, Y.; Gu, W.; Wang, Y.; Zhang, H.; Chen, W. Soybean yield response of biochar-regulated soil properties and root growth strategy. *Agronomy* **2022**, *12*, 1412. [CrossRef]
51. Tartaglia, M.; Arena, S.; Scaloni, A.; Marra, M.; Rocco, M. Biochar administration to san marzano tomato plants cultivated under lowinput farming increases growth, fruit yield, and affects gene expression. *Front. Plant Sci.* **2020**, *11*, 1281. [CrossRef]
52. Khajavi-Shojaei, S.; Moezzi, A.; Norouzi Masir, M.; Taghavi, M. Synthesis modified biochar-based slow-release nitrogen fertilizer increases nitrogen use efficiency and corn (*Zea mays* L.) growth. *Biomass Convers. Biorefinery* **2020**, *56*, 1–9. [CrossRef]
53. Zheng, J.; Luan, L.; Luo, Y.; Fan, J.; Xu, Q.; Sun, B.; Jiang, Y. Biochar and lime amendments promote soil nitrification and nitrogen use efficiency by differentially mediating ammonia-oxidizer community in an acidic soil. *Appl. Soil Ecol.* **2022**, *180*, 104619. [CrossRef]
54. Li, Y.; Yang, Y.; Sun, Y.; Jiao, Y.; Wang, Y.; Chen, H.; Wu, Q.; Chi, D. Using nitrogen-loaded biochar for soil improvement to decrease applied nitrogen and stabilize rice yield under alternate wet-dry irrigation. *Soil Tillage Res.* **2022**, *223*, 105493. [CrossRef]

55. Chen, K.; Peng, J.; Li, J.; Yang, Q.; Zhan, X.; Liu, N.; Han, X. Stabilization of soil aggregate and organic matter under the application of three organic resources and biochar-based compound fertilizer. *J. Soils Sediments* **2020**, *20*, 3633–3643. [CrossRef]
56. Liao, X.; Mao, S.; Chen, Y.; Zhang, J.; Müller, C.; Malghani, S. Combined effects of biochar and biogas slurry on soil nitrogen transformation rates and N_2O emission in a subtropical poplar plantation. *Sci. Total Environ.* **2022**, *848*, 157766. [CrossRef] [PubMed]
57. Najafi-Ghiri, M.; Boostani, H.R.; Hardie, A.G. Investigation of biochars application on potassium forms and dynamics in a calcareous soil under different moisture conditions. *Arch. Agron. Soil Sci.* **2022**, *68*, 325–339. [CrossRef]
58. Wu, X.; Wang, D.; Riaz, M.; Zhang, L.; Jiang, C. Investigating the effect of biochar on the potential of increasing cotton yield, potassium efficiency and soil environment. *Ecotoxicol. Environ. Saf.* **2019**, *182*, 109451. [CrossRef]
59. Farrar, M.B.; Wallace, H.M.; Xu, C.Y.; Joseph, S.; Nguyen, T.T.N.; Dunn, P.K.; Bai, S.H. Biochar compound fertilizers increase plant potassium uptake 2 years after application without additional organic fertilizer. *Environ. Sci. Pollut. Res.* **2021**, *29*, 7170–7184. [CrossRef]
60. Guo, L.; Yu, H.; Kharbach, M.; Zhang, W.; Wang, J.; Niu, W. Biochar improves soil-tomato plant, tomato production, and economic benefits under reduced nitrogen application in northwestern China. *Plants* **2021**, *10*, 759. [CrossRef]
61. Zhang, M.; Liu, Y.; Wei, Q.; Liu, L.; Gou, J. Biochar-based fertiliser improved the yield, quality and fertiliser utilisation of open field tomato in karst mountainous area. *Plant Soil Environ.* **2022**, *68*, 163–172. [CrossRef]
62. Han, M.; Zhao Qing Li Wei Ciais, P.; Wang, Y.P.; Goll, D.S.; Zhu, L.; Zhao, Z.; Wang, J.; Wei, Y.; Wu, F. Global soil organic carbon changes and economic revenues with biochar application. *GCB Bioenergy* **2021**, *14*, 364–377. [CrossRef]
63. Rogers, P.M.; Fridahl, M.; Yanda, P.; Hansson, A.; Pauline, N.; Haikola, S. Socio-economic determinants for biochar deployment in the southern highlands of Tanzania. *Energies* **2022**, *15*, 144. [CrossRef]
64. Li, Y.; Cheng, J.; Lee, X.; Chen, Y.; Gao, W.; Pan, W.; Tang, Y. Effects of biochar-based fertilizers on nutrient leaching in a tobacco-planting soil. *Acta Geochim.* **2018**, *38*, 1–7. [CrossRef]
65. Gwenzi, W.; Nyambishi, T.J.; Chaukura, N.; Mapope, N. Synthesis and nutrient release patterns of a biochar-based N-P-K slow-release fertilizer. *Int. J. Environ. Sci. Technol.* **2018**, *15*, 405–414. [CrossRef]

agriculture

MDPI

Article

Taif's Rose (*Rosa damascena* Mill var. *trigentipetala*) Wastes Are a Potential Candidate for Heavy Metals Remediation from Agricultural Soil

Tarek M. Galal [1,2], Ali Majrashi [1,2], Hatim M. Al-Yasi [1,2], Emad A. Farahat [3,*], Ebrahem M. Eid [4,5] and Esmat F. Ali [1,2,*]

1 Department of Biology, College of Sciences, Taif University, P.O. Box 11099, Taif 21944, Saudi Arabia
2 High Altitude Research Center, Taif University, P.O. Box 11099, Taif 21944, Saudi Arabia
3 Botany and Microbiology Department, Faculty of Science, Helwan University, Cairo 11790, Egypt
4 Botany Department, Faculty of Science, Kafrelsheikh University, Kafr El-Sheikh 33516, Egypt
5 Biology Department, College of Science, King Khalid University, P.O. Box 9004, Abha 61321, Saudi Arabia
* Correspondence: emad23_1999@yahoo.com (E.A.F.); a.esmat@tu.edu.sa (E.F.A.)

Abstract: The current study examines the bioaccumulation potential of Taif rose shrubs by analyzing the shrubs' wastes. f. At Al-Shafa Highland, four farms with plants of different ages were chosen to collect soil samples and vegetative waste (leaves and stems) for morphological and chemical analysis. The tallest stem and largest crown diameter (184.2 and 243.5 cm, respectively) were found in the oldest (20-year-old) shrubs, which also produced the highest biomass of pruning wastes of stems and leaves (3.9 and 1.3 t/ha, respectively). The 10-year-old shrubs gathered the highest concentration of Co and Pb (1.74 and 7.34 mg kg^{-1}) in the stem and the highest Fe, Mn, and Ni (18.55, 18.60, and 9.05 mg kg^{-1}) in the leaves, while the youngest plants (4 years) accumulated the highest Cr and Zn (0.83 and 13.44 mg kg^{-1}) in their leaves. The highest contents of Cd, Cr, Cu, Fe, Mn, Pb, and Zn were found in the oldest Taif rose stem (34.94, 1.16, 36.29, 49.32, 51.22, 24.76, and 32.51 g ha^{-1}), while the highest contents of Co and Ni were found in the stems of plants that were 10 and 12 years old (3.21 and 9.54 g ha^{-1}, respectively). The Taif rose's stem and leaves can absorb the majority of heavy metals that have been studied with BAF values greater than one. Significant relationships between various heavy metals in the soil and the same in the stems (Al, Co, and Pb) and leaves (Co, Fe, Mn, Ni, and Pb) of Taif roses have been observed. According to the current findings, the Taif rose is a promising viable and safe crop for heavy metals phytoremediation if it is grown in polluted soil because there is little to no risk of contamination in the use of its end products, high biomass of pruning wastes, and high efficiency of heavy metal removal.

Keywords: Taif's rose; aromatic plants; phytoremediation; soil amendments; vegetative wastes

Citation: Galal, T.M.; Majrashi, A.; Al-Yasi, H.M.; Farahat, E.A.; Eid, E.M.; Ali, E.F. Taif's Rose (*Rosa damascena* Mill var. *trigentipetala*) Wastes Are a Potential Candidate for Heavy Metals Remediation from Agricultural Soil. *Agriculture* **2022**, *12*, 1319. https://doi.org/10.3390/agriculture12091319

Academic Editors: Maria Roulia and Wanting Ling

Received: 31 July 2022
Accepted: 24 August 2022
Published: 26 August 2022

1. Introduction

Large amounts of waste biomass are produced by numerous biomass-based operations in a variety of industries including agriculture, forestry, and the biotechnology sector [1]. Instead of causing issues with the environment, economy, or ecology, turning these wastes into useful goods will sustainably contribute to the preservation of natural resources and the ecosystem [2,3]. Aromatic plants hold an adjunct position to food crops for phytoremediation due to their specialized application for the manufacture of essential oils, cosmetics, personal care products, etc., and the fact that they are not directly connected to the food chain [4]. It has frequently been claimed that growing non-edible commercial aromatic crops at heavy metal-contaminated locations would be both profitable and practical [5]. Essential oil, which is the main by-product of aromatic crops, is primarily used for non-edible purposes such as the production of soaps and detergents and the preparation of insect repellents, cosmetics, and perfumes; as a result, they can be thought of as a potential choice for minimizing food chain contamination [6].

Roses are significant aromatic plants that are grown all over the world due to the demand from florists, the necessity for perfumes and fragrances, and their use in medicine [7]. The Rosaceae family plant *Rosa damascena* Mill var. *trigentipetala*, often known as Taif's rose, is cultivated extensively in Bulgaria, Turkey, Saudi Arabia, Egypt, Russia, India, and China [8]. It is a tall shrub that may grow to a height of 2.5 m, and only blooms once per year (in May or June) producing 500–600 flowers when fully grown [9]. At elevations between 300 and 2500 m, Taif's rose thrives in temperate and subtropical climates [10]. Additionally, it is one of the lovely and fragrant plants that are raised for use in the culinary, medical, and fragrance industries in several Taif governorate places [11]. There are around 860 Taif's rose farms in the Taif governorate, ranging in size from large to medium to small [7]. The trash produced by these farms may come from industrial oil distillation processes as well as agricultural rose bush trimming [12]. A very small amount is used for vegetative propagation, but the vast majority is dried and burned, creating health dangers to nearby populations as well as environmental problems such as air and soil pollution [13]. Additionally, a great number of rose flowers must be used to produce a small amount of rose water, and this procedure creates a significant amount of semisolid debris that could pollute the environment [14].

Heavy metal contamination of the soil has been caused by the prolonged use of fertilizers, the application of sewage sludge, industrial waste, and improper irrigation of agricultural regions [15,16]. Due to the inherent characteristics of the soil resulting from various parent materials, or differing management of fields within and across farms, smallholder farms, such as Taif's rose farms, are known to be spatially heterogeneous in terms of soil fertility [17]. The production of crops on agricultural land is greatly influenced by the nutritional components' bioavailability [18,19]. If the soils do not contain the proper amounts of nutrient components, farmers must integrate nutrients from external sources such as fertilizers and soil conditioners [20]. However, adding external fertilizer or nutrients to the soil speeds up plants' uptake of harmful heavy metals [21–23]. Land use poses a threat to the environment and public safety by introducing dangerous pollutants into the soil, groundwater, and food chain [24]. The buildup of heavy metals in agricultural products has become a significant issue that has disastrous repercussions on consumer health [25]. Low quantities of heavy metals in agricultural soils may also be deposited in crop plants [18]. Heavy metals are well-known dangerous inorganic contaminants due to their environmental persistence, bioaccumulation, and toxicity to plants [26].

To mitigate soil heavy metals contamination, it is crucial to look into low-cost, eco-friendly technology made from agricultural waste [27,28]. With or without chemical changes, biosorbents made from agro-wastes such as Taif's rose wastes that are found in nature have been employed [29]. While bioaccumulation needs the presence of living organisms and is completed in the later phases of biosorption, biosorption is a metabolically passive process that is typically carried out utilizing non-living microbiological or biological materials (such as agricultural wastes) [30]. It will take substantial research to fully utilize the phytoremediation capability of aromatic plants for the successful management of heavy metal-contaminated locations, which could result in "Green Scented Technology" [5]. The ultimate goal of the current research is to investigate the potential of Taif's rose plants for heavy metal bioaccumulation and its impact on the growth performance of these plants. Such research may aid in determining the value of using soil additives, such as irrigation with treated wastewater or sewage sludge in Taif's rose farms.

2. Materials and Methods

2.1. Plant Sampling

To gather materials for an investigation of the potential to accumulate heavy metals, four Taif's rose farms at Al-Shafa highland, Taif Province, Saudi Arabia, were chosen. This investigation was carried out in December 2020. The ages of the farms F1, F2, F3, and F4 were 10, 12, 20, and 4 years, respectively. The soils of the study farms are sandy. The height and diameter of 10 rose plants of varied sizes were chosen from each farm. After that, bushes were clipped until they were between 80 and 90 cm tall. The fresh wastes (stems and leaves) were collected and weighed to determine their fresh biomass (kg individual^{-1}), and then the average individual weight was multiplied by the number of individuals per farm to calculate the total fresh biomass as kg ha^{-1}. The samples were then dried in an oven at 65 °C until they reached a consistent weight for chemical analysis.

2.2. Plant Analysis

Three composite samples of oven-dried Taif's rose plant leaves and stems were homogenized by grinding them individually in a metal-free plastic mill, passing through a 2 mm mesh size, and then stored in plastic containers with labels. To conduct chemical studies, a sample of 1 g of each plant part was digested in 20 mL of a tri-acid combination of HNO_3:H_2SO_4:$HClO_4$ (5:1:1, $v/v/v$) until a translucent colour developed (2000). Using an atomic absorption photometer (Shimadzu AA-6200), the concentrations of Al, Cd, Co, Cr, Cu, Fe, Mn, Ni, Pb, and Zn were measured following the recommended procedures of Allen [31]. The instrument settings and operational circumstances were carried out in line with the instructions provided by the manufacturers. Heavy metal digestion and measurement were performed three times. Heavy metal digestion and their measurement were performed in triplicates. Additionally, the heavy metal concentrations were multiplied by the pruning waste biomass of the relevant plant part to determine the amount of heavy metals in the stem and leaves (g DM m^{-2}).

2.3. Soil Sampling and Analysis

From each farm under study, three composite soil samples ($N = 3 \times 4 = 12$) were taken from profiles that included the topsoil at a depth of 0–50 cm. The soil samples were air-dried in the lab before being sieved through a 2 mm sieve and placed in paper bags for additional chemical analysis. Soil–water extracts (1:5, w:v) were prepared for chemical analysis. A glass electrode pH meter (Model 9107 BN, ORION type) and a multi-range Cryson-HI8734 electrical conductivity meter (Crison Instruments, S.A., Barcelona, Spain) were used to measure soil pH and EC, respectively. Soil samples were digested using the acid digestion method adopted by Wade et al. [32] to determine the soil heavy metals (Al, Cd, Co, Cr, Cu, Fe, Mn, Ni, Pb, and Zn) using the same methods of plant analysis [31]. The concentrations of soil elements were expressed in mg kg^{-1} dry weight.

2.4. Data Analysis

The concentration of heavy metals in the stem and leaves of Taif's rose plants and the chemical properties of the soil were compared using the simple linear correlation coefficient (r) formula. A relevant measure to assess a plant's capacity to collect heavy metals from the soil is the bioaccumulation factor (BAF). According to Eid et al. [19], the BAF was calculated as follows: The heavy metal concentrations (mg kg^{-1}) in the soil, stem, and leaves, respectively, are represented by C_{soil}, C_{stem}, and C_{leaf} in the formulas $BAF_{stem} = C_{stem}/C_{soil}$ and $BAF_{leaf} = C_{leaf}/C_{soil}$. The significant differences in the examined growth characteristics, bioaccumulation factors, as well as the soil chemical variables of Taif's rose plants across the investigated farms were determined using a one-way analysis of variance (ANOVA I) test and Tukey's HSD test. The significant variations in the levels of heavy metals in the various plant parts and farms were examined using a two-way analysis of variance (ANOVA II) test. SPSS software was used to conduct the statistical analysis [33].

3. Results

3.1. Soil Properties

The findings of the soil analysis revealed a considerable difference in all parameters among the study farms ($p < 0.05$) (Figure 1). Farm 2 (F2: 10 years old) was found to have the lowest soil pH (6.3), the highest EC (0.87 dS m^{-1}), and the highest concentrations of Cr, Cu,

Fe, Mn, Ni, Pb, Zn, Al, and Co (0.76, 0.59, 0.99, 0.67, 0.66, 1.15, 1.36, 0.26, and 0.36 mg kg^{-1}, respectively). In contrast, the soil on the newest farm (F1: 4 years old) had the greatest pH value (8.4) along with the lowest EC (0.61 dS m^{-1}) and the lowest concentrations of Cu, Pb, Zn, and Cd (0.14, 0.05, 0.06, and 0.23 mg kg^{-1}, respectively). Moreover, the oldest farm's (F4: 20 years old) soil had the highest Cd content (0.82 mg kg^{-1}), but the lowest Fe and Al (0.07 and 0.06 mg kg^{-1}). However, the lowest Cr, Mn, Ni, and Co contents (0.12, 0.15, 0.18, and 0.02 mg kg^{-1}) were recorded in farm 3 (F3: 12 years old).

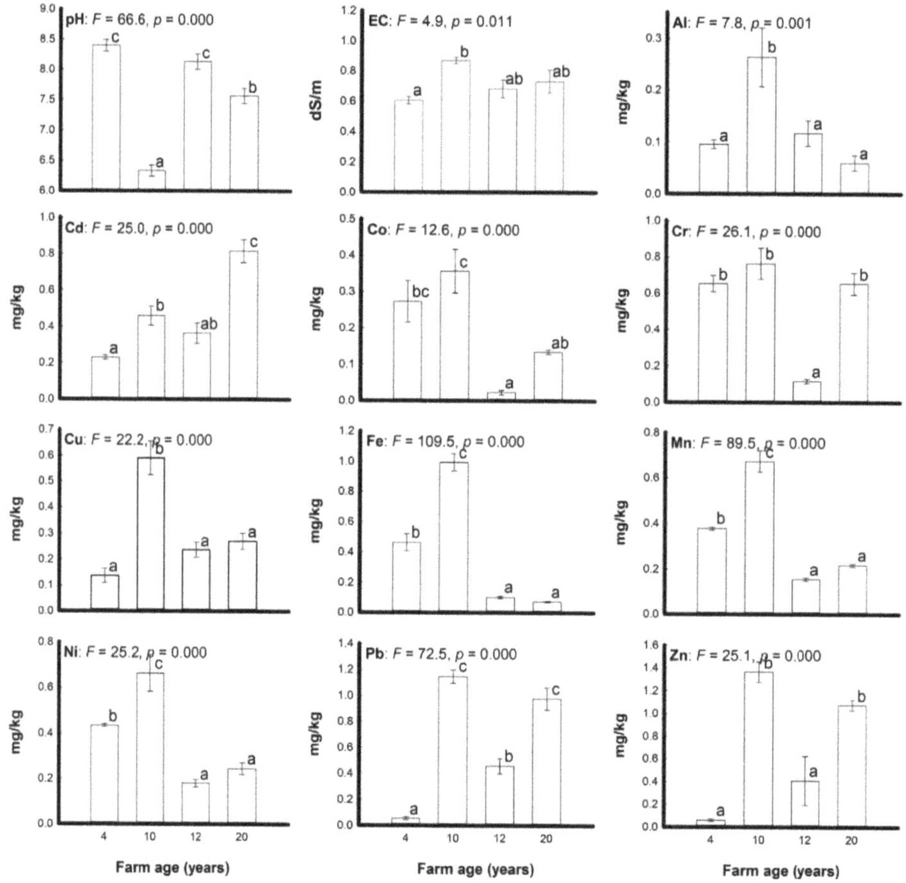

Figure 1. Mean soil characteristics of the sampled farms support the growth of Taif's rose populations in the mountainous area of Taif City, Saudi Arabia. The standard errors of the means (n = 6) are indicated by vertical bars. F-values demonstrate the one-way analysis of variance (ANOVA), degrees of freedom (df) = 3. Means with different letters are significantly different at $p < 0.05$ according to Tukey's HSD test.

3.2. Plant Growth Properties

Data on plant morphology and growth parameters revealed that as plants aged, their height and diameter as well as their biomass from pruning wastes grew dramatically (Figure 2). The stem height and crown diameter of the oldest (20-year-old) shrubs were the largest (184.2 and 243.5 cm, respectively), while the crown diameter of the youngest (4-year-old) shrubs was the lowest (110.5 and 112.7 cm). Therefore, the oldest shrubs produced the largest biomass of the standing crop biomass of stem, leaves, and the total aboveground biomass (AGB) (3.5, 1.2, and 4.7 kg/shrub), as well as the highest pruning waste biomass represented by the stem, leaf, and AGB (3.9, 1.3, and 5.2 t ha^{-1}, respectively). In addition, the youngest shrubs had the lowest biomass of any plant organ.

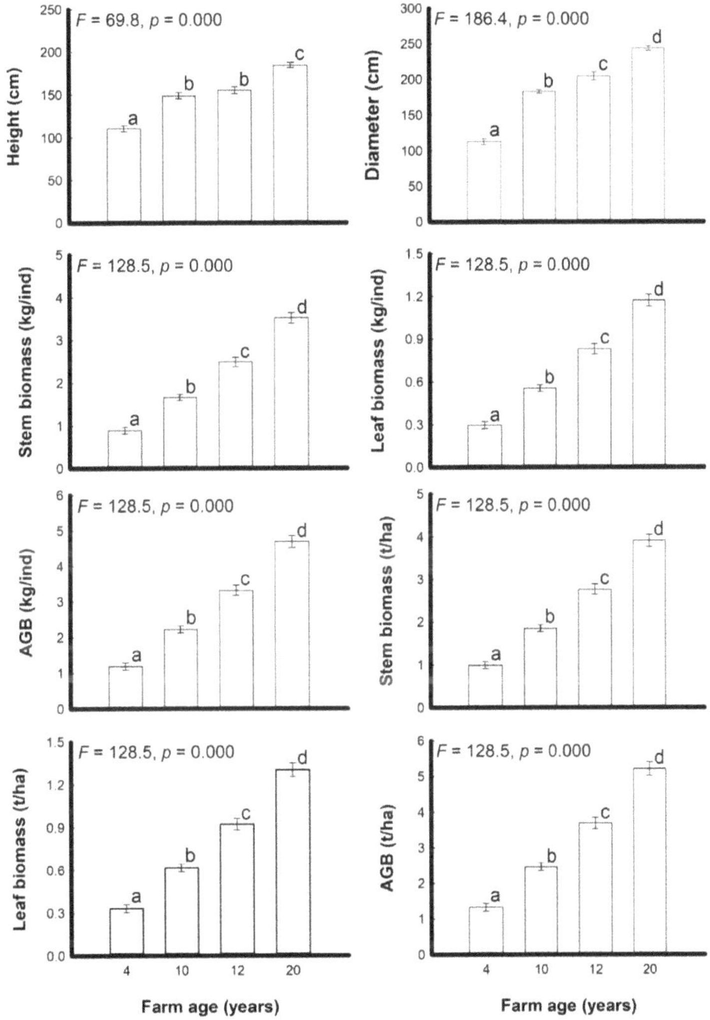

Figure 2. Mean morphological and biomass variables of Taif's rose populations grown on four farms in the mountainous area of Taif City in Saudi Arabia. The standard errors of the means (n = 10) are indicated by vertical bars. F-values demonstrate the one-way analysis of variance (ANOVA), degrees of freedom (df) = 3. Means with different letters are significantly different at $p < 0.05$ according to Tukey's HSD test.

3.3. Heavy Metals Concentration

According to data on heavy metal concentration, the leaves of Taif's rose shrubs collected larger concentrations of Al, Cr, Cu, Fe, Mn, Ni, and Zn, but the stem accumulated higher concentrations of Cd, Co, and Pb (Figure 3). Additionally, the concentration of heavy metals in the various plant organs varies significantly throughout time. The Cr and Zn concentrations in the leaves of the youngest (4-year-old) plants were highest (0.83 and 13.44 mg kg^{-1}), and the Cu, Mn, and Pb concentrations in the stem were the lowest (3.19, 2.48, and 1.63 mg kg^{-1}). The highest concentrations of Co and Pb (1.74 and 7.34 mg kg^{-1})

and the lowest concentrations of Ni and Zn (1.12 and 5.50 mg kg^{-1}) were found in the stems of shrubs that were 10 years old, while the highest concentrations of Fe, Mn, and Ni (18.55, 18.60, and 9.05 mg kg^{-1}) were found in the leaves. Additionally, 12-year-old shrubs had the highest leaf Cu (10.29 mg kg^{-1}) and the lowest leaf Fe (6.11 mg kg^{-1}) as well as Al and Co in the stem (0.10 and 0.06 mg kg^{-1}). The 20-year plant's leaves collected the highest levels of Al and the lowest levels of Cd and Cr (1.16, 2.70, and 0.17 mg kg^{-1}, respectively), whereas the plant stem contained the highest levels of Cd (8.94 mg kg^{-1}).

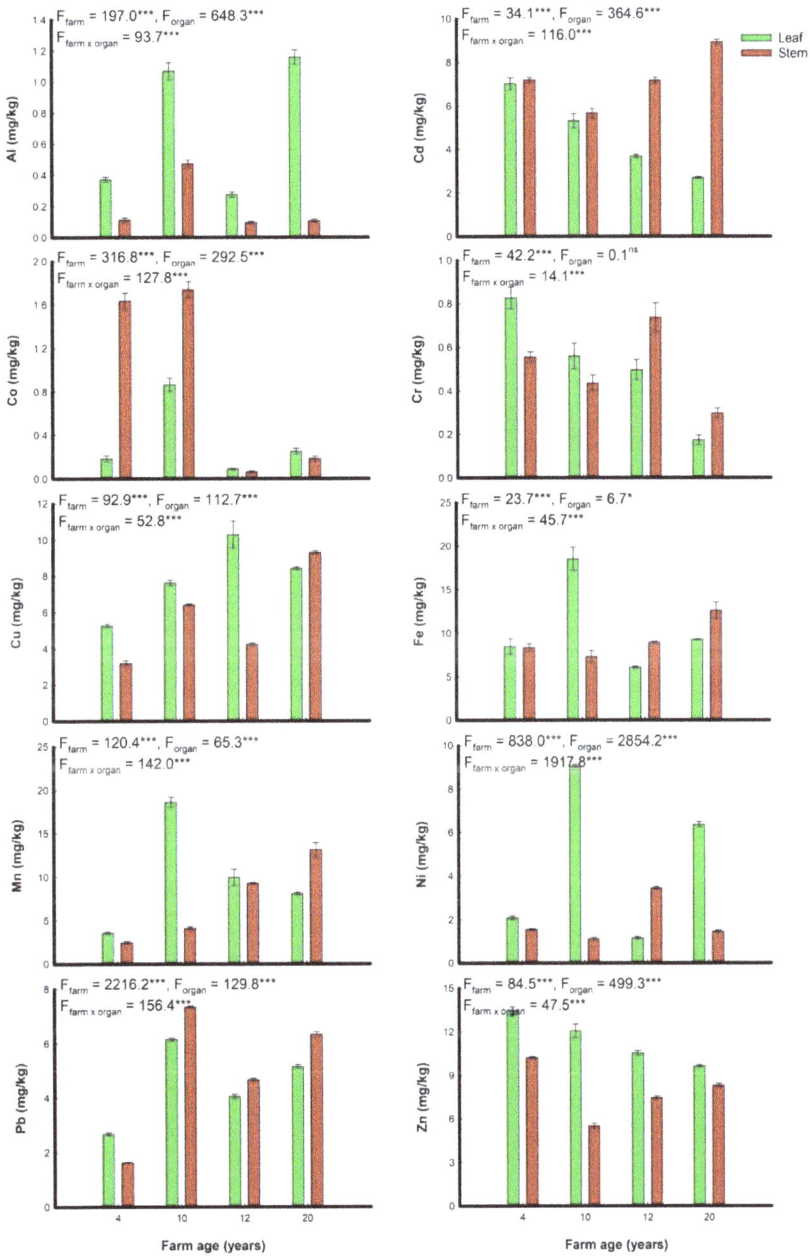

Figure 3. Mean heavy metal concentrations (mg kg^{-1}) in the leaves and stems of Taif's rose populations grown on four farms in the mountainous area of Taif City in Saudi Arabia. The standard errors of the means ($n = 6$) are indicated by vertical bars. F-values demonstrate the two-way analysis of variance (ANOVA). *: $p < 0.05$; ***: $p < 0.001$; ns: not significant (i.e., $p > 0.05$).

3.4. Heavy Metals Removal Efficiency

The statistical study (ANOVA I) demonstrated that Taif's rose shrubs of different ages had significantly differing heavy metal contents (Figure 4). The current findings demonstrated that plant stems could remove higher quantities of all heavy metals than leaves, except for Al. The highest concentrations of Cd, Cr, Cu, Fe, Mn, Pb, and Zn were found in the stems of the oldest Taif's rose shrubs (34.94, 1.16, 36.29, 49.32, 51.22, 24.76, and 32.51 g ha^{-1}, respectively), while the highest concentrations of Co and Ni were found in the stems of plants that were 10 and 12 years old (3.21 and 9.54 g ha^{-1}, respectively). On the other hand, the leaves of the youngest plants had the lowest levels of the majority of heavy metals (apart from Al and Cr). Additionally, the oldest shrubs' leaves had the largest quantity of Al (1.51 g ha^{-1}), whilst the youngest plant stem had the lowest (0.11 g ha^{-1}).

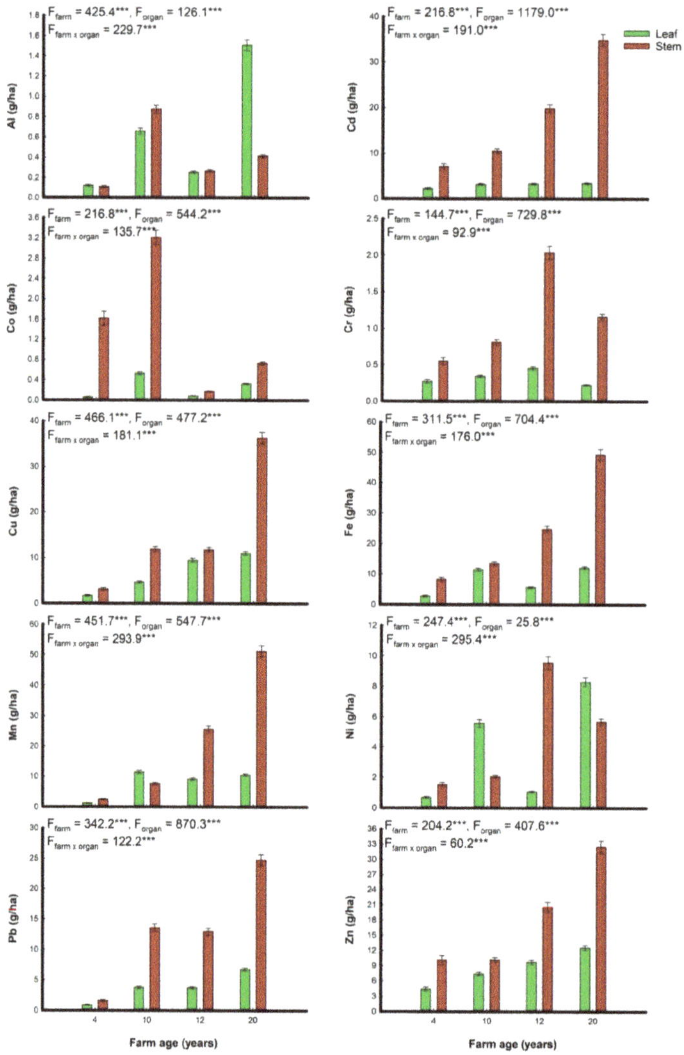

Figure 4. Mean heavy metal contents (g ha^{-1}) in the leaves and stems of Taif's rose populations grown on four farms in the mountainous area of Taif City in Saudi Arabia. The standard errors of the means (n = 6) are indicated by vertical bars. F-values demonstrate the two-way analysis of variance (ANOVA). ***: $p < 0.001$.

3.5. Bioaccumulation Factor (BAF)

The bioaccumulation factors (BAF) from soil to the stems and leaves of Taif roses were statistically evaluated (ANOVA), and they significantly varied among the measured heavy metals and the different plant ages (Table 1). The current research demonstrated that the

Taif rose stem and leaves of various ages may accumulate the majority of heavy metals under investigation with BAF greater than one. The youngest Taif's rose plants showed the highest BAF of Cu, Pb, and Zn (respectively 48.59, 68.99, and 279.50) in the leaves and the highest BAF of Cd and Co (31.47 and 7.24) in the stem. In addition, the stem and leaves of the 12-year-old plant exhibited the highest BAF of Cr and Mn values (6.88 and 63.88, respectively). Additionally, the BAF of Fe (175.64) in the stem and Al and Ni (30.40 and 27.52) in the leaves were also highest in the oldest plants.

Table 1. Bioaccumulation factors (BAFs), from soil to stems, and leaves, of 10 heavy metals in Taif's rose populations grown on four farms in the mountainous area of Taif City in Saudi Arabia (means ± standard error, n = 6).

Metal	Factor	Farm Age (Years)				*F*-Value
		4	10	12	20	
Al	BAF_{stem}	1.16 ± 0.06a	2.27 ± 0.47a	1.11 ± 0.30a	2.82 ± 0.88a	2.6 ^{ns}
	BAF_{leaf}	3.98 ± 0.18a	5.13 ± 1.05a	3.12 ± 0.79a	30.40 ± 9.57b	7.5 **
Cd	BAF_{stem}	31.47 ± 1.06b	13.41 ± 1.94a	23.02 ± 4.30b	11.33 ± 0.97a	14.2 ***
	BAF_{leaf}	30.58 ± 0.37c	12.74 ± 2.17b	11.45 ± 1.69b	3.42 ± 0.29a	67.4 ***
Co	BAF_{stem}	7.24 ± 1.28c	6.00 ± 1.40bc	3.23 ± 0.41ab	1.39 ± 0.14a	7.4 **
	BAF_{leaf}	0.73 ± 0.05a	3.06 ± 0.76bc	4.93 ± 0.92c	1.89 ± 0.19ab	8.7 **
Cr	BAF_{stem}	0.86 ± 0.03a	0.58 ± 0.02a	6.88 ± 1.12b	0.47 ± 0.04a	31.0 ***
	BAF_{leaf}	1.32 ± 0.17a	0.74 ± 0.01a	4.65 ± 0.77b	0.27 ± 0.03a	25.4 ***
Cu	BAF_{stem}	27.49 ± 4.22bc	11.69 ± 1.44a	19.25 ± 2.16ab	36.93 ± 4.53c	10.5 ***
	BAF_{leaf}	48.59 ± 10.39b	13.97 ± 1.89a	44.94 ± 2.35b	33.27 ± 3.63ab	7.5 **
Fe	BAF_{stem}	18.93 ± 1.43a	7.26 ± 0.30a	90.46 ± 6.77b	175.64 ± 15.37c	84.8 ***
	BAF_{leaf}	21.15 ± 4.76a	19.39 ± 2.48a	61.34 ± 3.52b	129.71 ± 7.70c	106.2 ***
Mn	BAF_{stem}	6.61 ± 0.43a	6.18 ± 0.20a	60.82 ± 2.62b	60.60 ± 3.30b	218.2 ***
	BAF_{leaf}	9.54 ± 0.46a	28.11 ± 1.59b	63.81 ± 3.12d	37.51 ± 1.58c	136.5 ***
Ni	BAF_{stem}	3.57 ± 0.14ab	1.91 ± 0.37a	20.14 ± 2.13c	6.41 ± 0.86b	50.9 ***
	BAF_{leaf}	4.81 ± 0.25a	14.92 ± 2.14b	6.82 ± 0.91a	27.52 ± 2.54c	35.6 ***
Pb	BAF_{stem}	43.36 ± 12.61b	6.46 ± 0.25a	11.22 ± 1.52a	6.78 ± 0.69a	7.8 **
	BAF_{leaf}	68.99 ± 18.43b	5.43 ± 0.29a	9.67 ± 1.20a	5.52 ± 0.55a	11.3 ***
Zn	BAF_{stem}	206.81 ± 41.54b	4.17 ± 0.40a	94.38 ± 34.32a	7.83 ± 0.40a	12.5 ***
	BAF_{leaf}	279.50 ± 63.27b	9.16 ± 0.93a	126.40 ± 44.41a	9.07 ± 0.41a	11.0 ***
F-value$_{BAFstem}$		20.3 ***	19.9 ***	10.0 ***	108.9 ***	
F-value$_{BAFleaf}$		16.1 ***	28.9 ***	8.2 ***	85.7 ***	

F-values represent one-way ANOVA; degrees of freedom (df) = 3; means in the same row followed by different letters are significantly different at $p < 0.05$, according to Tukey's HSD test; **: $p < 0.01$; ***: $p < 0.001$; ns: not significant (i.e., $p > 0.05$).

3.6. Plant Stem-Soil Correlations

According to a simple linear correlation coefficient between heavy metal concentrations in Taif's rose stems and the soil chemical properties of the four study farms, the soil pH was negatively correlated (r = −0.88 and −0.82) with stem Al and Pb but positively correlated (r = 0.53 and 0.80) with stem Ni and Zn (Figure 5). Additionally, the electrical conductivity (EC) had positive correlations with Al and Pb (r = 0.56 and −0.59) and negative correlations with Zn (r = −0.60). Al, Co, and Pb had higher concentrations in plant stems (r = 0.78, 0.76, and 0.94), while Cr, Fe, Mn, Ni, and Zn had lower concentrations (r = −0.54, −0.51, −0.63, −0.67, and −0.75) as their soil contents increased. Additionally, there was a strong positive correlation between several heavy metals in the plant stem and some soil metals. For example, the correlation between the metals Al, Cu, and Co in the stem and the metals Fe, Mn, and Ni in the soil was 0.92, 0.92, and 0.85, respectively. However, some stem heavy metal concentrations had significant negative correlations with some soil metals. For example, stem Cr and soil Cd, Pb, and Zn had r values of −0.76, −0.56, and −0.72, and stem Cd and soil Fe, Mn, and Ni had r values of −0.75, −0.67, and −0.57, and stem Zn had r values of −0.51, −0.85, and −0.74 respectively.

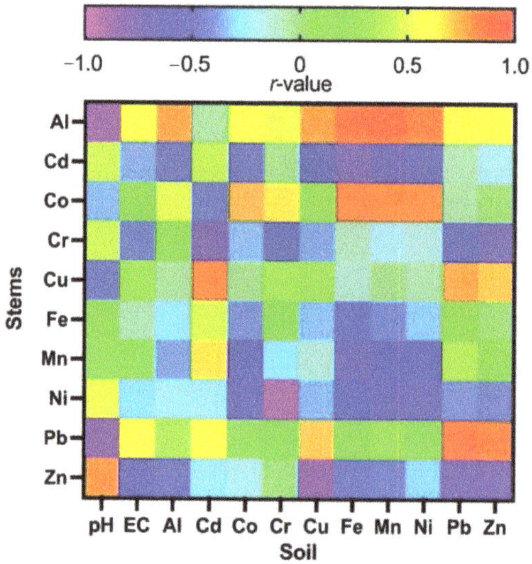

Figure 5. Pearson correlation coefficient (*r*-values, *n* = 24) between heavy metal concentrations in Taif's rose stems and the chemical characteristics of the soil from four farms in the mountainous area of Taif City in Saudi Arabia; EC: electrical conductivity.

3.7. Plant Leaves-Soil Correlations

Regarding the simple linear correlation coefficient between heavy metal concentrations in Taif's rose leaves and the soil chemical characteristics, the soil pH was negatively correlated with leaves Al, Co, Fe, Mn, Ni, and Pb (r = −0.75, −0.90, −0.83, −0.90, and −0.89), while EC was positively correlated with leaves (r = 0.52, 0.55, 0.63, 0.67, 0.59, and 0.65) (Figure 6). Co, Fe, Mn, Ni, and Pb concentrations in plant leaves increased as the soil content increased (r = 0.56, 0.74, 0.65, 0.57, and 0.94); however, Cd concentrations fell (r = −0.71) as soil Cd concentrations increased. Additionally, there was a strong positive correlation between some heavy metals found in plant leaves and certain soil metals, including leaf Al with soil Cr, Pb, and Zn (r = 0.68, 0.88, and 0.77), leaf Co with soil Fe, Mn, and Ni (r = 0.90, 0.92, and 0.86), and leaf Ni with soil Cu, Pb, and Zn (r = 0.75, 0.85, and 0.83). However, the concentration of some heavy metals in the leaves was negatively correlated with some soil metals. For example, the correlations between Cd, Pb, and Zn in the leaves and Cd in the soil were −0.91, −0.57, and −0.62, respectively.

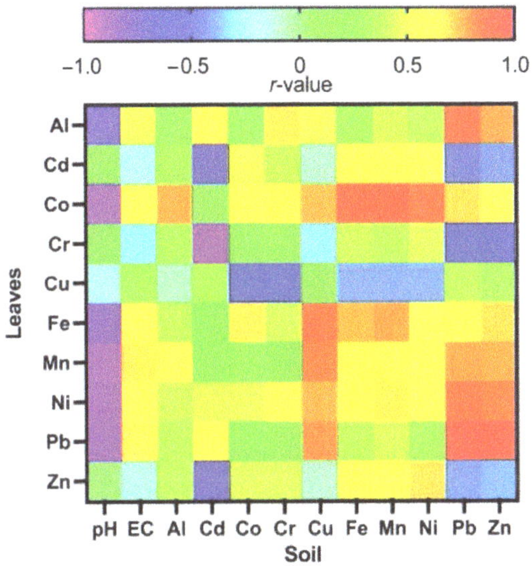

Figure 6. Pearson correlation coefficient (*r*-values, $n = 24$) between heavy metal concentrations in Taif's rose leaves and the chemical characteristics of the soil from four farms in the mountainous area of Taif City in Saudi Arabia; EC: electrical conductivity.

4. Discussion

Taif's rose may grow in a variety of environmental conditions, although the availability of nutrients and the soil's capability to hold water are dependent on the soil's physico-chemical characteristics [34]. The current findings showed a significant difference in all soil parameters between the sample farms. Farm 2 (10 years old) had the lowest soil pH (6.3) along with the greatest EC (0.87 dS m^{-1}), highest Cr, Cu, Fe, Mn, Ni, Pb, Zn, Al, and Co contents, and lowest EC. The soil on the youngest farm (F1: 4 years) had the greatest pH value (8.4) together with the lowest EC (0.61 dS m^{-1}) and the least amount of Cu, Pb, Zn, and Cd. According to Pal and Singh [34], roses thrive in soil reactions with pH ranges between 6.0 and 7.5 that are slightly acidic to slightly alkaline, whereas acidic soil limits plant growth and diminishes blossom output most likely due to an imbalance of micronutrients. Additionally, according to Brichet [35], rose plants are susceptible to both salty and alkaline soil. Additionally, lowering soil pH raises the amount of heavy metals that are available in the soil [19,36]. However, in Indian conditions, alkali–saline soil with a pH range of 8 to 9 is relatively appropriate [37]. The recommended soil micronutrient values for roses, according to Karlik et al. [38], are (in mg kg^{-1}) Fe (0.3–3.0), Mn (0.2–3.0), Zn (0.03–3.0), and Cu (0.001–0.5). Therefore, the soil content of these metals is suitable for the growth of Taif's rose, except Cu in farm 2, which exceeded the required limit.

Pruning flowering plants is a successful agricultural technique for improving development and blossoming [39]. The biomass of the pruning wastes and the plant height and diameter both considerably increased as the age of the plant rose. As a result, the average biomass of the stem, leaves, and aboveground biomass pruning wastes was 2.4, 0.8, and 3.2 t ha^{-1}, respectively. Pruning wastes are disposed of by drying, burning, or storing them, which pollutes the environment [40]. Taif's rose is grown on over 860 farms spread across Taif governorate and its suburbs [11]; as a result, roughly 2752 tons of pruning debris, similar to the 2730 tons noted by Galal et al. [41], may be created and result in a serious environmental issue. Therefore, it is of great importance and urgent need for recycling these agricultural wastes and reusing them for various economic purposes.

Al, Cr, Cu, Fe, Mn, Ni, and Zn were more abundant in Taif rose shrubs' leaves than in their stems, which were more abundant in Cd, Co, and Pb. The concentration of the analyzed metals (except for Co and Ni) in the stem and leaves of Taif roses did not surpass the safe level for typical plants based on the normal level of heavy metals in plants [31,42–44]. The WHO's recommended limit for Ni in plants is 10 mg kg^{-1} [45]. Plant growth and biomass production are hampered by heavy metals such as Ni and Co, which also have a deleterious effect on their chlorophyll content [46,47]. The largest concentrations of Cr and Zn were collected by the youngest plants (4 years), whereas Co, Pb, Fe, Mn, and Ni were deposited by shrubs 10 years old, Cu by shrubs 12 years old, and Al by shrubs 20 years old. *Phragmites australis* has a high capacity for absorbing heavy metals for up to 10 years, according to reports from Cicero-Fernández et al. [48] and Eid et al. [49].

Information on a plant's chemical makeup and biomass is needed to calculate how well it removes heavy metals from the environment, which is useful for determining a species' specific heavy metal amounts per unit area [41]. By multiplying the element concentration in tissue by its biomass, the amount of heavy metals was determined. Plant biomass is regarded as the primary component for calculating the standing stock of heavy metals, according to Vymazal [50]. The current findings show that plant stems can remove larger quantities of all heavy metals than leaves, except for Al. This may be due to stems' higher biomass than leaves. The highest concentrations of Cd, Cr, Cu, Fe, Mn, Pb, and Zn were found in the stems of the oldest Taif's rose shrubs, whereas the highest concentrations of Co and Ni were found in the stems of plants between 10 and 12 years old. It is important to note that Taif's rose shrubs are supported for prospective application in heavy metals clean-up from contaminated soils by the heavy metal contents eliminated by their pruning wastes.

The bioaccumulation factor (BAF) was used to determine the relationship between the concentration of heavy metals in the soil and plant tissues [51]. This indicator of the ability of available metals to be taken up by a plant from its environment provides insight into whether the plant is an excluder, accumulator, or indicator [52]. The BAF from soil to the stem and leaves of the Taif's rose demonstrated that plants of various ages might acquire the majority of the heavy metals under investigation with a BAF greater than one. The majority of heavy metals found in Taif's rose plants had high BAF levels, indicating a high likelihood that this plant will concentrate those metals in its tissues [53]. Younger plants have a greater capacity to acquire heavy metals. The stem and leaves of the youngest Taif's rose plants had the highest BAFs of Cd and Co, as well as the highest BAFs of Cu, Pb, and Zn. Additionally, the stem and leaves of the 12-year-old plant showed the highest levels of BAF for Cr and Mn, respectively. Furthermore, the BAF of Fe in the stem and Al and Ni in the leaves were both highest in the oldest plants. High metal content and high BAF, as reported by Fawzy et al. [29], suggest that this species may be a good carrier for absorbing heavy metals from nutrient-rich soils. According to Pandey et al. [5], aromatic plants, which operate as possible phytostabilizers, hyper-accumulators for particular heavy metals, bio-monitors, and facultative metallophytes, have a significant potential for phytoremediation of heavy metal-contaminated soils.

The simple linear correlation coefficient revealed a strong negative link with soil pH and a substantial positive correlation between the heavy metal contents in Taif's rose stems, leaves, and the soil EC. The pH of the soil plays a crucial role in regulating how readily available trace metals are to plants [54]. According to Jung [55] and Eid et al. [56], soil pH, which has a negative correlation with the presence of heavy metals in plants, is crucial for controlling the uptake of heavy metals by plants. Significant relationships between various heavy metals in the soil and the same in the stems (Al, Co, and Pb) and leaves of Taif roses have been observed (Co, Fe, Mn, Ni, and Pb). This finding suggests that heavy metal concentrations in plant tissues were dependent on their levels in the soils [19]. Eid et al. [56], who observed that heavy-metal concentrations in plants rise with an increase in their levels in the soil, made similar predictions. Additionally, the use of this plant as a bioindicator and biomonitor of these heavy metals in contaminated soils is supported by the significantly positive correlations between the majority of heavy metal concentrations in the soil and rose tissues [53,57]. Additionally, these relationships imply that Taif's rose plants might represent the cumulative impact of environmental pollution as a result of soil contamination, with heavy metal concentrations rising in plant tissues as they do so in the soil [19,58].

Many studies had shown that growing aromatic plants such as lavender [59], basil [60,61], rosemary [62], and Mentha [46,63] in contaminated soils did not significantly increase the risk of metal contamination in essential oils. According to Zheljazkov and Nielsen [64], after being harvested, several aromatic plants considerably acquire heavy metals from heavy metal-polluted locations. In addition, Cr, Pb, Zn, and Cu were discovered by Onursal and Ekinci [65] in rose oil processing wastes at quantities that were significantly lower than those allowed by law. By absorbing certain hazardous substances through their roots and storing them in less harmful forms throughout the plant, plants are believed to help the environment become less polluted [66]. Most plants, when grown in contaminated environments, ingest and translocate harmful components to the harvestable sections, which is a process known as phytoremediation [67]. The right plant must be chosen for phytoremediation projects for them to be successful. It must be a high-value economic crop with no or little risk of contamination when used to make final products and unusual characteristics (such as high biomass and high heavy-metal extraction efficiency) [5]. Taif rose is a viable and secure crop for the phytoremediation of soils affected by heavy metals. The pruning wastes of Taif's rose could be converted to ash and packed in a safe place, or the absorbed heavy metals could also be recovered for economic purposes.

5. Conclusions

Taif rose may grow in a variety of soil types with different physicochemical characteristics and environmental conditions. The biomass of the pruning wastes and the plant height and diameter both considerably increased as the age of the plant rose. Al, Cr, Cu, Fe, Mn, Ni, and Zn were more abundant in Taif rose shrubs' leaves than in their stems, which were more abundant in Cd, Co, and Pb. The current findings show that plant stems can remove larger quantities of all heavy metals than leaves, except for Al. This may be due to stems' higher biomass than leaves. The BAF from soil to the stem and leaves of the Taif's rose demonstrated that plants of various ages might acquire the majority of the heavy metals under investigation with a BAF greater than one. The use of this plant as a bioindicator and biomonitor of these heavy metals in contaminated soils is supported by the significantly positive correlations between the majority of heavy metal concentrations in the soil and rose tissues. The current findings revealed that Taif's rose is a promising viable and safe crop for heavy metals phytoremediation if it is grown in polluted soil because there is little to no risk of contamination in the use of its end products, high biomass of pruning wastes, and high efficiency of heavy metal removal.

Author Contributions: Conceptualization, E.A.F., E.F.A. and T.M.G.; methodology, E.A.F., E.F.A., T.M.G., E.M.E., A.M. and H.M.A.-Y., software, T.M.G. and E.M.E., validation, E.A.F., E.F.A., T.M.G., E.M.E., A.M. and H.M.A.-Y.; formal analysis, E.A.F., E.F.A., T.M.G., E.M.E. and H.M.A.-Y.; investigation, E.A.F., E.F.A. and A.M.; resources, E.F.A., T.M.G., A.M. and H.M.A.-Y.; data curation, E.F.A., E.M.E., A.M. and H.M.A.-Y.; writing—original draft preparation, E.A.F., E.F.A., T.M.G., E.M.E., A.M. and H.M.A.-Y.; writing—review and editing, E.F.A. and T.M.G.; visualization, E.M.E., A.M. and H.M.A.-Y.; supervision, E.F.A. and T.M.G.; project administration, E.F.A. and T.M.G.; funding acquisition, E.A.F., E.F.A. and T.M.G.; methodology, E.A.F., E.F.A., T.M.G., E.M.E. and A.M. All authors have read and agreed to the published version of the manuscript.

Funding: This research was funded by the High-Altitude Research Centre at Taif University, Taif, Saudi Arabia, under project number 1-442-42.

Institutional Review Board Statement: Not applicable.

Informed Consent Statement: Not applicable.

Data Availability Statement: Not applicable.

Acknowledgments: The authors acknowledge the Dean of Scientific Research at Taif University, Taif, Saudi Arabia, for funding this work through the High-Altitude Research Centre, under project number 1-442-42.

Conflicts of Interest: The authors declare no conflict of interest.

References

1. Deniz, F. Green purification of heavy metal pollution from aquatic environment by biorefinery waste biomass of *Nigella sativa* L.: A novel and effective treatment agent. *Environ. Technol. Innov.* **2022**, *25*, 102118. [CrossRef]
2. Chai, W.S.; Cheun, J.Y.; Kumar, P.S.; Mubashir, M.; Majeed, Z.; Banat, F.; Ho, S.-H.; Show, P.L. A review on conventional and novel materials towards heavy metal adsorption in wastewater treatment application. *J. Clean. Prod.* **2021**, *296*, 126589. [CrossRef]
3. Kolluru, S.S.; Agarwal, S.; Sireesha, S.; Sreedhar, I.; Kale, S.R. Heavy metal removal from wastewater using nanomaterials-process and engineering aspects. *Process Saf. Environ. Prot.* **2021**, *150*, 323–355. [CrossRef]
4. Lubbe, A.; Verpoorte, R. Cultivation of medicinal and aromatic plants for specialty industrial materials. *Ind. Crops Prod.* **2011**, *34*, 785–801. [CrossRef]
5. Pandey, J.; Verma, R.K.; Singh, S. Suitability of aromatic plants for phytoremediation of heavy metal contaminated areas: A review. *Int. J. Phytoremediation* **2019**, *21*, 405–418. [CrossRef] [PubMed]
6. Verma, S.K.; Singh, K.; Gupta, A.K.; Pandey, V.C.; Trivedi, P.; Verma, R.K.; Patra, D.D. Aromatic grasses for phytomanagement of coal fly ash hazards. *Ecol. Eng.* **2014**, *73*, 425–428. [CrossRef]
7. Galal, T.M.; Al-Yasi, H.M.; Fawzy, M.A.; Abdelkader, T.G.; Hamza, R.Z.; Eid, E.M.; Ali, E.F. Evaluation of the Phytochemical and Pharmacological Potential of Taif's Rose (*Rosa damascena* Mill var. *trigintipetala*) for Possible Recycling of Pruning Wastes. *Life* **2022**, *12*, 273. [CrossRef]
8. Nasir, M.H.; Nadeem, R.; Akhtar, K.; Hanif, M.A.; Khalid, A.M. Efficacy of modified distillation sludge of rose (*Rosa centifolia*) petals for lead(II) and zinc(II) removal from aqueous solutions. *J. Hazard. Mater.* **2007**, *147*, 1006–1014. [CrossRef]
9. Nunes, H.S.; Miguel, M.G. *Rosa damascena* essential oils: A brief review about chemical composition and biological properties. Trends in Phytochemical Research (TPR). *Trends Phytochem. Res.* **2017**, *1*, 111–128.
10. Najem, W.; Beyrouthy, M.E.; Wakim, L.H.; Neema, C.; Ouaini, N. Essential oil composition of *Rosa damascene* Mill. from different localities in Lebanon. *Acta Bot. Gall.* **2011**, *158*, 365–373. [CrossRef]
11. Al-Yasi, H.; Attia, H.; Alamer, K.; Hassan, F.; Esmat, F.; Elshazly, S.; Siddiqued, K.H.M.; Hessiniae, K. Impact of drought on growth, photosynthesis, osmotic adjustment, and cell wall elasticity in Damask rose. *Plant Physiol. Biochem.* **2020**, *150*, 133–139. [CrossRef] [PubMed]
12. Shawl, A.S.; Adams, R. Rose Oil in Kashmiri India, an emerging cash crop benefiting industry and local agribusiness. *Perfum. Flavorist* **2009**, *34*, 22–25.
13. Pal, P.K. Evaluation, Genetic Diversity, Recent Development of Distillation Method, Challenges and Opportunities of *Rosa damascena*: A Review. *J. Essent. Oil Bear. Plants* **2013**, *16*, 1–10. [CrossRef]
14. Rabbani, D.; Mahmoudkashi, N.; Mehdizad, F.; Shaterian, M. Green approach to wastewater treatment by application of *Rosa damascena* waste as nano-biosorbent. *J. Environ. Sci. Technol.* **2016**, *9*, 121–130. [CrossRef]
15. Yadav, S. Heavy metals toxicity in plants: An overview on the role of glutathione and phytochelatins in heavy metal stress tolerance of plants. *S. Afr. J. Bot.* **2010**, *76*, 167–179. [CrossRef]
16. Maham, S.G.; Rahimi, A.; Smith, D.L. Environmental assessment of the essential oils produced from dragonhead (*Dracocephalum moldavica* L.) in conventional and organic farms with different irrigation rates. *J. Clean. Prod.* **2018**, *204*, 1070–1086. [CrossRef]
17. Chipomho, J.; Rugare, J.T.; Mabasa, S.; Zingore, S. Short-term impacts of soil nutrient management on maize (*Zea mays* L.) productivity and weed dynamics along a toposequence in Eastern Zimbabwe. *Heliyon* **2020**, *6*, e05223. [CrossRef]
18. Wu, C.; Wu, G.; Wang, Z.; Zhang, Z.; Qian, Y.; Ju, L. Soil mercury speciation and accumulation in rice (*Oryza sativa* L.) grown in wastewater-irrigated farms. *Appl. Geochem.* **2018**, *89*, 202–209. [CrossRef]
19. Eid, E.M.; Khedher, K.M.; Ayed, H.; Arshad, M.; Mouldi, A.; Shaltout, K.H.; Sewelam, N.A.; Galal, T.M.; El-Bebany, A.F.; Alshehri, A.M.A. Prediction models based on soil properties for evaluating the heavy metal uptake into *Hordeum vulgare* L. grown in agricultural soils amended with different rates of sewage sludge. *Int. J. Environ. Health Res.* **2022**, *32*, 106–120. [CrossRef]
20. Wortman, S.E.; Drijber, R.A.; Francis, C.A.; Lindquist, J.L. Arable weeds, cover crops, and tillage drive soil microbial community composition in organic cropping systems. *Appl. Soil Ecol.* **2013**, *72*, 232–241. [CrossRef]
21. Abd Elsalam, H.E.; El-Sharnouby, M.E.; Mohamed, A.E.; Raafat, B.M. Effect of sewage sludge compost usage on corn and faba bean growth, carbon and nitrogen forms in plants and soil. *Agronomy* **2021**, *11*, 628. [CrossRef]
22. Mansoora, N.; Kausar, S.; Amjad, S.F.; Yaseen, S.; Id, M.E.; Mustafa, G.; Ali, S.A.; Id, S.D. Application of sewage sludge combined with thiourea improves the growth and yield attributes of wheat (*Triticum aestivum* L.) genotypes under arsenic-contaminated soil. *PLoS ONE* **2021**, *16*, e0259289. [CrossRef] [PubMed]
23. Zhang, X.; Zhao, B.; Liu, H.; Zhao, Y.; Li, L. Effects of pyrolysis temperature on biochar's characteristics and speciation and environmental risks of heavy metals in sewage sludge biochars. *Environ. Technol. Innov.* **2022**, *26*, 102288. [CrossRef]
24. Liu, T.; Liu, Z.; Zheng, Q.; Lang, Q.; Xia, Y.; Peng, N.; Gai, C. Effect of hydrothermal carbonization on migration and environmental risk of heavy metals in sewage sludge during pyrolysis. *Bioresour. Technol.* **2018**, *247*, 282–290. [CrossRef]

25. Kumar, R.; Sharma, S.; Kaundal, M.; Sharma, S.; Thakur, M. Response of damask rose (*Rosa damascena* Mill.) to foliar application of magnesium (Mg), copper (Cu) and zinc (Zn) sulphate under western Himalayas. *Ind. Crops Prod.* **2016**, *83*, 596–602. [CrossRef]
26. Anser, M.K.; Hanif, I.; Vo, X.V.; Alharthi, M. The long-run and short-run influence of environmental pollution, energy consumption, and economic activities on health quality in emerging countries. *Environ. Sci. Pollut. Res.* **2020**, *27*, 32518–32532. [CrossRef] [PubMed]
27. Tokay, B.; Akpınar, I. A comparative study of heavy metals removal using agricultural waste biosorbents. *Bioresour. Technol. Rep.* **2021**, *15*, 100719. [CrossRef]
28. Peng, Y.; Huang, H.; Zhang, Y.; Kang, C.; Chen, S.; Song, L.; Liu, D.; Zhong, C. A versatile MOF-based trap for heavy metal ion capture and dispersion. *Nat. Commun.* **2018**, *9*, 187. [CrossRef]
29. Fawzy, M.A.; Al-Yasi, H.M.; Galal, T.M.; Hamza, R.Z.; Abdelkader, T.G.; Ali, E.F.; Hassan, S.H.A. Statistical optimization, kinetic, equilibrium isotherm and thermodynamic studies of copper biosorption onto *Rosa damascena* leaves as a low-cost biosorbent. *Sci. Rep.* **2022**, *12*, 8583. [CrossRef]
30. Razzak, S.A.; Faruque, M.O.; Alsheikh, Z.; Alsheikhmohamad, L.; Alkuroud, D.; Alfayez, A.; Hossain, S.M.Z.; Hossain, M.M. A comprehensive review on conventional and biological-driven heavy metals removal from industrial wastewater. *Environ. Adv.* **2022**, *7*, 100168. [CrossRef]
31. Allen, S.E. *Chemical Analysis of Ecological Materials*; Blackwell Scientific Publications: London, UK, 1989.
32. Wade, T.L.; Brooks, J.M.; Kennicutt, M.C.; McDonald, T.J.; Sericano, J.L.; Jackson, T.L. GERG Trace Metals and Organic Contaminants Analytical Techniques. In *Sampling and Analytical Methods of the National Status and Trend Program*; National Benthic Surveillance and Mussel Watch Projects 1984–1992, 121–139; NOAA Technical Memorandum NOS ORCA 71; Lauenstein, G.G., Cantillo, A.Y., Eds.; NOAA: Silver Spring, MD, USA, 1993.
33. SPSS. *IBM SPSS Statistics Version 21.0*; IBM: New York, NY, USA, 2012.
34. Pal, P.K.; Singh, R.D. Understanding crop- ecology and agronomy of *Rosa damascena* Mill for higher productivity. *Aust. J. Crop Sci.* **2013**, *7*, 196–205.
35. Brichet, H. Distribution and Ecology: Continental Asia and Japan. In *Encyclopedia of Rose Science*; Roberts, A.V., Debener, T., Gudin, S., Eds.; Elsevier: Amsterdam, The Netherlands, 2003.
36. Eid, E.M.; Shaltout, K.H. Bioaccumulation and translocation of heavy metals by nine native plant species grown at a sewage sludge dump site. *Int. J. Phytoremed.* **2016**, *18*, 1075–1085. [CrossRef] [PubMed]
37. Sharma, B.M. Plant biomass in the semi-arid zone of India. *J. Arid Environ.* **1982**, *5*, 29–33. [CrossRef]
38. Karlik, J.F.; Becker, J.O.; Pemberton, H.B.; Schuch, U.K. Production and Marketing: Field Rose production. In *Encyclopedia of Rose Science*; Roberts, A.V., Debener, T., Gudin, S., Eds.; Elsevier: Amsterdam, The Netherlands, 2003.
39. Ali, E.F.; Al-Yasi, H.M.; Hassan, F.A.S.; Alamer, K.H.; Hessini, K.; Attia, H.; El-Shazly, S. Effect of the pruning system and P-Fertilizer on growth and productivity of *Rosa damascena* mill. var. *trigentipetala plant. Egypt. J. Bot.* **2021**, *61*, 565–578. [CrossRef]
40. Atay, O.A.; Ekinci, K. Characterization of pellets made from rose oil processing solid wastes/coal powder/pine bark. *Renew. Energy* **2020**, *149*, 933–939. [CrossRef]
41. Galal, T.M.; Gharib, F.A.; Al-Yasi, H.M.; Al-Mutairi, K.A.; Mansour, K.H.; Eid, E.M. Nutrient Remediation Efficiency of the Sedge Plant (*Cyperus alopecuroides* Rottb.) to Restore Eutrophic Freshwater Ecosystems. *Sustainability* **2022**, *14*, 2823. [CrossRef]
42. Misra, S.G.; Mani, D. *Soil Pollution*; Ashish Publishing House: Punjabi Bagh, India, 1991.
43. Nagajyoti, P.C.; Lee, K.D.; Sreekanth, T.V.M. Heavy metals, occurrence and toxicity for plants: A review. *Environ. Chem. Lett.* **2010**, *8*, 199–216. [CrossRef]
44. Chiroma, T.M.; Ebewele, R.O.; Hymore, F.K. Comparative assessment of heavy metal levels in soil, vegetables and urban grey waste water used for irrigation in Yola and Kano. *Int. Refereed J. Eng. Sci.* **2014**, *3*, 1–9.
45. Hassan, F.A.S.; Ali, E.F.; Mahfouz, S.A. Comparison between different fertilization sources, irrigation frequency and their combinations on the growth and yield of coriander plant. *Aust. J. Basic Appl. Sci.* **2012**, *6*, 600–615.
46. Gharib, F.A.; Mansour, K.H.; Ahmed, E.Z.; Galal, T.M. Heavy metals concentration, and antioxidant activity of the essential oil of the wild mint (*Mentha longifolia* L.) in the Egyptian watercourses. *Int. J. Phytoremediation* **2021**, *23*, 641–651. [CrossRef]
47. Raklami, A.; Oufdou, K.; Tahiri, A.; Mateos-Naranjo, E.; Navarro-Torre, S.; Rodríguez-Llorente, I.D.; Meddich, A.; Redondo-Gómez, S.; Pajuelom, E. Safe cultivation of Medicago sativa in metal-polluted soils from semi-arid regions assisted by heat- and metallo-resistant PGPR. *Microorganisms* **2019**, *7*, 212. [CrossRef] [PubMed]
48. Cicero-Fernández, D.; Peña-Fernández, M.; Expósito-Camargo, J.A.; Antizar-Ladislao, B. Long-term (two annual cycles) phytoremediation of heavy metal-contaminated estuarine sediments by *Phragmites australis*. *New Biotechnol.* **2017**, *38*, 56–64. [CrossRef] [PubMed]
49. Eid, E.M.; Shaltout, K.H.; Al-Sodany, Y.M.; Haroun, S.A.; Galal, T.M.; Ayed, H.; Khedher, K.M.; Jensen, K. Seasonal potential of *Phragmites australis* in nutrient removal to eliminate the eutrophication in Lake Burullus, Egypt. *J. Freshw. Ecol.* **2020**, *35*, 135–155. [CrossRef]
50. Vymazal, J. Removal of nutrients in constructed wetlands for wastewater treatment through plant harvesting—Biomass and load matter the most. *Ecol. Eng.* **2020**, *155*, 105962. [CrossRef]
51. Galal, T.M.; Hassan, L.M.; Ahmed, D.A.; Alamri, S.A.M.; Alrumman, S.A.; Eid, E.M. Heavy metals uptake by the global economic crop (*Pisum sativum* L.) grown in contaminated soils and its associated health risks. *PLoS ONE* **2021**, *16*, e0252229. [CrossRef]

52. Eid, E.M.; Alrumman, S.A.; Galal, T.M.; El-Bebany, A.F. Regression models for monitoring trace metal accumulations by *Faba sativa* Bernh. plants grown in soils amended with different rates of sewage sludge. *Sci. Rep.* **2019**, *9*, 5443. [CrossRef]
53. Galal, T.M.; Shehata, H.S. Evaluation of the invasive macrophyte *Myriophyllum spicatum* L. as a bioaccumulator for heavy metals in some watercourses of Egypt. *Ecol. Indic.* **2014**, *41*, 209–214. [CrossRef]
54. Eid, E.M.; Alrumman, S.A.; El-Bebany, A.F.; Fawy, K.F.; Taher, M.A.; Hesham, A.; El-Shaboury, G.A.; Ahmed, M.T. Evaluation of the potential of sewage slude as a valuable fertilizer for wheat (*Triticum aestivum* L.) crops. *Environ. Sci. Pollut. Res.* **2019**, *26*, 392–401. [CrossRef]
55. Jung, M.C. Heavy metal concentrations in soils and factors affecting metal uptake by plants in the vicinity of a Korean Cu-W mine. *Sensors* **2008**, *8*, 2413–2423. [CrossRef]
56. Eid, E.M.; Galal, T.M.; El-Bebany, A.F. Prediction models for monitoring heavy metals accumulation by wheat (*Triticum aestivum* L.) plants grown in soil amended with different rates of sewage sludge. *Int. J. Phytoremediation* **2020**, *22*, 1000–1008. [CrossRef]
57. Galal, T.M.; Shehata, H.S. Bioaccumulation and translocation of heavy metals by *Plantago major* L. grown in contaminated soils under the effect of traffic pollution. *Ecol. Indic.* **2015**, *48*, 244–251. [CrossRef]
58. Galal, T.M.; Gharib, F.A.; Ghazi, S.M.; Mansour, K.H. Metal uptake capability of *Cyperus articulatus* L. and its role in mitigating heavy metals from contaminated wetlands. *Environ. Sci. Pollut. Res.* **2017**, *24*, 21636–21648. [CrossRef] [PubMed]
59. Zheljazkov, V.D.; Astatkie, T. Effect of residual distillation water of 15 plants and three plant hormones on Scotch spearmint (*Mentha* × *gracilis* Sole). *Ind. Crop Prod.* **2011**, *33*, 704–709. [CrossRef]
60. Siddiqui, F.; Krishna, S.K.; Tandon, P.K.; Srivastava, S. Arsenic accumulation in *Ocimum* spp. and its effect on growth and oil constituents. *Acta Physiol. Plant.* **2013**, *35*, 1071–1079. [CrossRef]
61. Kunwar, G.; Pande, C.; Tewari, G.; Singh, C.; Kharkwal, G.C. Effect of heavy metals on terpenoid composition of *Ocimum basilicum* L. and *Mentha spicata* L. *J. Essent. Oil Bear Plants* **2015**, *18*, 818–825. [CrossRef]
62. Parra, A.; Zornoza, R.; Conesa, E.; Gómez-López, M.D.; Faz, A. Seedling emergence, growth and trace elements tolerance and accumulation by Lamiaceae species in a mine soil. *Chemosphere* **2014**, *113*, 132–140. [CrossRef]
63. Chand, S.; Pandey, A.; Patra, D.D. Influence of nickel and lead applied in combination with vermicompost on growth and accumulation of heavy metals by *Mentha arvensis* Linn. cv.'Kosi'. *Indian J. Nat. Prod. Resour.* **2012**, *3*, 256–261.
64. Zheljazkov, V.D.; Nielsen, N.E. Effect of heavy metals on peppermint and cornmint. *Plant Soil.* **1996**, *178*, 59–66. [CrossRef]
65. Onursal, E.; Ekinci, K. Co-composting of rose oil processing waste with caged layer manure and straw or sawdust: Effects of carbon source and C/N ratio on decomposition. *Waste Manag. Res.* **2015**, *33*, 332–338. [CrossRef]
66. Schnoor, J.L.; Licht, L.A.; McCutcheon, S.C.; Wolfe, N.L.; Carreira, L.H. Phytoremediation of organic and nutrient contaminants. *Environ. Sci. Technol.* **1995**, *29*, 318A–323A. [CrossRef]
67. Meagher, R.B. Phytoremediation of toxic elemental and organic pollutants. *Curr. Opin. Plant Biol.* **2000**, *3*, 153–162. [CrossRef]

agriculture

MDPI

Article

Evaluating the Nutrient Contents and Nutritive Value of Taif's Rose (*Rosa damascena* Mill var. *trigintipetala*) Waste to Be Used as Animal Forage or Soil Organic Fertilizers

Tarek M. Galal [1,2], Esmat F. Ali [1,2,*], Ebrahem M. Eid [3,4], Hatim M. Al-Yasi [1,2], Ali Magrashi [1,2], Fayez Althobaiti [2,5] and Emad A. Farahat [6,*]

1 Department of Biology, College of Sciences, Taif University, P.O. Box 11099, Taif 21944, Saudi Arabia
2 High Altitude Research Center, Taif University, P.O. Box 11099, Taif 21944, Saudi Arabia
3 Biology Department, College of Science, King Khalid University, P.O. Box 9004, Abha 61321, Saudi Arabia
4 Botany Department, Faculty of Science, Kafrelsheikh University, Kafr El-Sheikh 33516, Egypt
5 Department of Biotechnology, College of Science, Taif University, P.O. Box 11099, Taif 21944, Saudi Arabia
6 Botany and Microbiology Department, Faculty of Science, Helwan University, Cairo 11790, Egypt
* Correspondence: a.esmat@tu.edu.sa (E.F.A.); emad.farahat@science.helwan.edu.eg (E.A.F.)

Abstract: This study aimed to evaluate the nutrient content and nutritive value of pruning waste of the Taif Damask rose for its potential use as an organic fertilizer or animal forage in the Taif region, Saudi Arabia. For this purpose, the pruning waste of Taif's rose and soil samples supporting the plant growth at different ages were collected from four farms (F1: 4, F2: 10, F3: 12, and F4: 20 years old). The total aboveground biomass (AGB) of the plants, stems, and leaves were measured in addition to the stem height and crown diameter. The results showed that the maximum stem height and crown diameter (184.20 and 243.5 cm) were recorded in the oldest farm (F4). Moreover, the stem, leaves, and aboveground biomass (AGB) of the waste were maximal (3.91, 1.30 and 5.21 t ha^{-1}) at F4. F1 had the highest N content (154.30 mg kg^{-1}) in the plant leaves, while F2 had the highest stem N and P (172.33 and P 9.40 mg kg^{-1}). Moreover, F3 had the highest concentrations of leaf P (7.17 mg kg^{-1}), leaf and stem K (112.47 and 277.30 mg kg^{-1}), stem Ca^{2+} (251.93 mg kg^{-1}), and leaf and stem Mg^{2+} (122.27 and 123.57 mg kg^{-1}). The stems had higher percentages of total proteins, fibers, ash, and NFE (total carbohydrates) than the leaves in F1 and F2, while the opposite was observed in F3 and F4. The leaves of F2 rose plants had the highest percentage of neutral detergent fibers (NDF), and their stems had the highest percentages of total proteins (10.71%). The leaves of F3 plants had the highest percentage of acid detergent lignin (ADL) and the lowest crude fibers (7.63 and 13.27%), while the stems had the highest NFE (72.71%). The plant–soil relationship expressed by the CCA biplot showed that all the measured plant parameters were at higher positions on the Mg axis, except for the plant height and crown diameter, which were at low positions on the N and NO_3 axes, respectively. In contrast, Cl^-, NO_3^-, HCO_3^-, and SO_4^{2-} had high positive correlations with axis 1 and negative values with axis 2, while EC, the total P, and Ca^{2+} had high positive correlations with, and pH had high negative values in relation to, axis 2. Due to its considerable high inorganic and organic nutrient contents, Taif's rose could be used in the manufacturing of organic fertilizer. Additionally, the analysis of the nutritive value of the pruning waste supports its use as animal forage. We strongly recommend that further studies be conducted on the application of plant waste as a soil amendment and animal forage in the field.

Keywords: damask rose; soil amendments; forage quality; inorganic elements; organic nutrients

Citation: M. Galal, T.; Ali, E.F.; Eid, E.M.; Al-Yasi, H.M.; Magrashi, A.; Althobaiti, F.; Farahat, E.A. Evaluating the Nutrient Contents and Nutritive Value of Taif's Rose (*Rosa damascena* Mill var. *trigintipetala*) Waste to Be Used as Animal Forage or Soil Organic Fertilizers. *Agriculture* **2022**, *12*, 1481. https://doi.org/10.3390/agriculture12091481

Academic Editor: Dongyang Liu

Received: 24 July 2022
Accepted: 13 September 2022
Published: 16 September 2022

1. Introduction

Agricultural waste is improperly disposed of, which not only pollutes the environment but also wastes a large amount of precious biomass resources. To control pollution, an effective transformation of the process of agricultural waste recycling and utilization is regarded as a critical step in environmental conservation [1]. Recently, an enormous quantity of agricultural waste has been created annually all over the world. This waste has increased at a rate of 5–10% each year, on average. Air pollution, soil degradation, and other issues would from its haphazard abandonment and inappropriate use. Burning agricultural waste produces many hazardous gases, smoke, and dust, which significantly damage the air quality [2–4].

Agricultural waste recycling can take several forms, including the gasification of crops, the use of crop stalks as feed, fertilizers, or new building materials, and the production of manure from livestock dung [5,6]. Fruit, vegetable, and crop wastes are valuable sources of natural compounds and chemicals that are dependent on their fundamental properties and composition [7]. For instance, Torkashvand et al. [8] reported that peanut shell compost could be mixed with low-porosity substrates and applied as an appropriate growth medium to ornamental plants, as an alternative to peat. Great amounts of apple orchard waste (e.g., pruning branches, fruit thinning, and trunks) are produced, which might be considered as a promising alternative energy source for fuel and material manufacturing [9]. According to Sharma et al. [10], the agricultural recycling of organic waste is an eco-friendly and long-term waste management strategy. They also stated that organic waste is a rich source of beneficial plant macro- and micro-nutrients and organic matter. Organic waste additions boost plant production by improving the soil's physicochemical and biological qualities.

In Saudi Arabia, Taif's rose (*Rosa damascena* Mill. var. *trigintipetala*) is a conventional and interesting agricultural plant utilized for essential oil production [11–13]. Taif's rose cultivation is a long-standing tradition in the Taif region, and it has helped to transform this city in Saudi Arabia into a popular tourist destination. Saudi Arabia's rose oil output currently accounts for less than 5% of the total global production [14]. Essential oil and rose water are produced from Taif's rose [15] and are applied in medicine, the food sector, and perfumery. Water distillation is used to process roses in this part of the world. Many studies have been conducted on Taif's rose populations, including the chemical analysis of its oil and antimicrobial and antioxidant activities [14]; the phytochemical and pharmacological potential of its pruning wastes [11]; the effects of pruning systems and P-fertilizers on its growth and productivity [16]; and the effects of salinity on its oil and flower production [17–20]. Dragoev et al. [21] studied the chemical composition and quantity of the polyphenol content in dry rose petals, dry-pressed distilled rose petals, and wastewater, as well as the possibility for the re-utilization of dry-pressed distilled rose petals as feed material in animal production.

The waste generated by rose pruning in Taif province amounts to about 2700 tons, which can cause environmental concerns [11]. This pruning waste can be recycled so as to be used for various purposes such, as industrial and medicinal applications, forage, or fertilizers [11,12]. To our knowledge, the chemical composition of the major macronutrients, as well as the potential nutritive value, of the Taif rose plants have not yet been investigated. Accordingly, the objective of this study is to investigate the inorganic and organic macronutrients, as well as the potential nutritive value, of Taif's rose pruning waste destined to be used as soil organic fertilizer or animal forage. Such a study can be applied in the context of the recycling of waste materials and prevention of environmental pollution resulting from burning this waste. Additionally, it may aid in decision-making regarding the safe use of soil amendments such as sewage sludge or treated wastewater for improving soil quality.

2. Materials and Methods

For this study, we chose four farms in Al-Shafa highland, Taif Province, Saudi Arabia, to investigate the main morphological and biomass characteristics of Taif's Rose in addition to its nutrient contents and the potential nutritive value of its pruning waste. The selected farms had the same soil and climatic conditions, as well as the same agricultural practices. The four sampled farms (F1: 4 years; F2: 10 years; F3: 12 years; F4: 20 years) comprised an area of one hectare each. Taif's rose produces flowers once a year, during March and

April, for a period of 40–45 days. Annual pruning, which is performed once per year at the end of December and the beginning of January, is considered one of the most important agricultural practices of Taif's rose aiming to maximize its flower production. To increase the flower yield significantly, irrigation should be prevented for two months before and after pruning. After that time, irrigation is allowed and is usually achieved by natural dripping according to the needs of the plant during the rest of the year. Organic fertilizers, at a rate of 7.5 t/ha, are added immediately after the annual pruning.

2.1. Morphological and Biomass Characteristics

The stem maximum height (cm) and crown diameter (cm) were measured on about 10 individual plants of Taif's rose in each studied farm. Then, these individuals were pruned to maintain their height at about 80–90 cm. The fresh biomass of the stems and leaves of pruning waste was determined. The average biomass of each pruning waste was expressed as kg ha^{-1}, which was then multiplied by the number of individuals per hectare in order to calculate the total aboveground biomass (AGB) as t ha^{-1}.

2.2. Plant Chemical Analyses

From each of the ten individual rose plants, a sample of the stems and leaves were taken and combined to form three composite samples from each farm (N = 24). The oven-dried leaves and stems of Taif rose vegetative waste were ground separately in a metal-free plastic mill, passed through 2 mm-sized mesh, and finally stored in labelled plastic containers. For the chemical analyses, 1 g of either the milled leaf or stem samples was digested using a sulphury and perchloric acid mixture [22]. The determination of N, P, K, Ca, Mg, and Na in the extracts was carried out using Agilent 4210 MP-AES (Microwave Plasma-Atomic Emission Spectrometer, Agilent Inc., Santa Clara, CA, USA) at the Ecology Laboratory of the Faculty of Science, Helwan University. The instrumental settings and operational procedures were adjusted according to the manufacturer's user manual. The final concentrations were expressed as mg kg^{-1} of biomass dry matter for each element.

The ash content was estimated by igniting the ground leaves and stems in a muffle furnace for 3 h at 550 °C. The total N was measured by the Kjeldahl method, according to Chapman and Pratt [22]. The crude protein (CP) was calculated by multiplying the total N% by 6.25. The total lipids or fats were obtained by ether extract (EE), and they were determined by extracting the leaf and stem materials with diethyl ether using a Soxhlet extractor, according to Allen [23]. The crude fiber (CF) was gravimetrically determined after the chemical digestion and solubilization of the other present materials, according to Allen [23]. The nitrogen-free extract (NFE, i.e., total carbohydrates) was calculated in the leaves and stem samples according to Le Houérou [24], as follows:

NFE (% DM) = 100 − (CP + CF + Fat + Ash). The digestible crude protein (DCP) was determined according to NRC [25], as follows: % digestible crude protein (%DCP) = 0.85 X − 2.5, where X = crude protein % on the DM basis.

The fiber fractions of the cell walls consist of neutral detergent fiber (NDF), acid detergent fiber (ADF), and acid detergent lignin (ADL), which were determined according to Goering and Van Soest [26] and Van Soest et al. [27]. Additionally, the hemicellulose and cellulose were calculated by difference, as follows: hemicellulose = NDF − ADF; cellulose = ADF − ADL.

The gross energy (kcal kg^{-1} DM) was calculated according to Blaxter [28], where each gram of crude protein = 5.65 kcal, each gram of fat = 9.40 kcal, and each gram of crude fiber and carbohydrate = 4.15 kcal. The animal obtains its energy through feed and loses energy through heat, feces, urine, and gases [29]. The digestible energy (kcal kg^{-1} DM) was calculated according to NRC [25], as follows: digestible energy (DE) = gross energy × 0.76. The metabolizable energy (kcal kg^{-1} DM) was calculated according to NRC [30], as follows: metabolizable energy (ME) = digestible energy × 0.82. The net energy (kcal kg^{-1} DM) was calculated according to NRC [30], as follows: net energy

(NE) = metabolizable energy × 0.56. The caloric values were expressed in Mcal kg^{-1} DM. The total amount of digestible nutrients (%) was calculated according to NRC [30]:

$$\%\text{Total digestible nutrients (TND)} = \text{digestible energy}/44.3$$

2.3. Soil Characteristics

Five composite soil samples were collected from the profiles, including the topsoil at a depth of 0–50 cm from each studied farm ($N = 24$). In the laboratory, the soil samples were air-dried, sieved through a 2 mm sieve, and then packed in paper bags for further chemical analysis. Soil water extracts (1:5, w:v) were prepared and used for the pH, EC, Cl^-, NO_3^-, SO_4^{2-}, HCO_3^-, total N, total P, Ca^{2+}, Mg^{2+}, Na^+, and K^+ analysis. The pH of the soil samples was measured using a glass electrode pH meter (model 9107 BN, ORION type). The electrical conductivity (EC) of the soil water extract was measured with a multi-range Cryson-HI8734 electrical conductivity meter (Crison Instruments, S.A., Barcelona, Spain). The total N was measured by the Kjeldahl method. Bicarbonates were determined by titration against 0.1 N HCl, using methyl orange as an indicator, while chlorides were determined using silver nitrate solution [25,31]. Soluble soil cations (total P, Ca^{2+}, Mg^{2+}, K^+, and Na^+) were determined using Agilent 4210 MP-AES, as mentioned above in the case of the plant analysis. The concentrations of soil elements were expressed as mg kg^{-1} of dry weight.

2.4. Multivariate Analysis

The relationships between the morphological and growth measurements of the Taif rose plants and the soil factors were analyzed by canonical correspondence analysis [32], using Canoco for Windows version 4.0 [33]. The soil variables in the CCA biplots were represented by arrows pointing in the direction of maximum variation, and their length was proportional to the rate of change [34]. Each arrow determined an axis upon which the morphological and growth variable points could be projected. Intra-set correlations determined from the CCA were used to evaluate the importance of the soil variables.

2.5. Statistical Analysis

A one-way analysis of variance (ANOVA I) test followed by Tukey's HSD test was used to assess the significant differences between the analyzed morphological and growth variables of Taif's rose at the studied farms. A two-way analysis of variance (ANOVA II) test was used to examine the significant differences in the nutrient contents between the studied farms and the interaction between the farm age and organ type on the nutrient concentrations and nutritive value of the Taif's rose populations grown in the studied four farms. When there were significant variations between the farms, a post hoc test (Tukey's test) was used. The statistical analysis was conducted using (SPSS software version 21.0 Armonk, NY, USA: IBM Corp) [35].

3. Results

3.1. Morphological and Growth Characteristics

The results of the current study showed that all the measured morphological and biomass parameters of Taif's rose were significantly different between the sampled farms, at $p \leq 0.001$ (Table 1). Additionally, Tukey's test showed significant intra-specific variations in the morphological and growth parameters. The maximum stem height (184.20 ± 3.10 cm) was recorded in the oldest farm (F4), while the minimum (110.50 ± 3.70 cm) was recorded in the youngest (F1). The crown diameter of the trees followed the same trend as the stem height, with the maximum (243.50 ± 3.10 cm) at F4 and the minimum (112.7 ± 3.9 cm) at F1. The highest stem and leaf waste biomasses were 3.52 ± 0.13 and 1.17 ± 0.04 kg $ind.^{-1}$, respectively, at F4, while the lowest were 0.89 ± 0.08 and 0.30 ± 0.03 kg $ind.^{-1}$ at F1. The maximum stem and leaf waste production levels were at F4 (3.91 ± 0.1 and 1.30 ± 0.05 t ha^{-1}, respectively). Likewise, the aboveground biomass (AGB) of the plants showed the same trend.

Table 1. Mean (±standard error, n = 10) morphological and biomass variables of the Taif rose populations grown in four farms in the mountainous area of Taif City in Saudi Arabia.

Variable	Farm No. (Age, Years)				*F*-Value
	1 (4 Years)	2 (10 Years)	3 (12 Years)	4 (20 Years)	
Height (cm)	110.50 ± 3.70 a	148.70 ± 3.70 b	155.30 ± 4.00 b	184.20 ± 3.10 c	69.8 ***
Diameter (cm)	112.70 ± 3.90 a	182.90 ± 2.60 b	204.50 ± 5.70 c	243.50 ± 3.10 d	186.4 ***
Stem biomass (kg ind.$^{-1}$)	0.89 ± 0.08 a	1.67 ± 0.07 b	2.49 ± 0.11 c	3.52 ± 0.13 d	128.5 ***
Leaf biomass (kg ind.$^{-1}$)	0.30 ± 0.03 a	0.56 ± 0.02 b	0.83 ± 0.04 c	1.17 ± 0.04 d	128.5 ***
AGB (kg ind.$^{-1}$)	1.19 ± 0.10 a	2.22 ± 0.09 b	3.32 ± 0.15 c	4.69 ± 0.17 d	128.5 ***
Stem biomass (t ha^{-1})	0.99 ± 0.08 a	1.85 ± 0.08 b	2.77 ± 0.12 c	3.91 ± 0.14 d	128.5 ***
Leaf biomass (t ha^{-1})	0.33 ± 0.03 a	0.62 ± 0.03 b	0.92 ± 0.04 c	1.30 ± 0.05 d	128.5 ***
AGB (t ha^{-1})	1.32 ± 0.11 a	2.47 ± 0.11 b	3.69 ± 0.16 c	5.21 ± 0.19 d	128.5 ***

F-values demonstrate the one-way analysis of variance (ANOVA), with degrees of freedom (*df*) = 3. Means in the same row followed by different letters are significantly different at $p < 0.05$, according to Tukey's HSD test. AGB: above-ground biomass; *** $p < 0.001$.

3.2. Soil Characteristics and Their Relationships with the Plant Variables

The chemical characteristics of the farms' soils showed that all the measured variables significantly differed between the four farms, with no influence of the farm age on the soil chemistry (Figure 1). The youngest farm (F1) had the highest values of soil pH (8.4), Cl^- (15.4 mg kg^{-1}), NO_3^- (4.4 mg kg^{-1}), SO_4^{2-} (1.0 mg kg^{-1}), and HCO_3^- (3.8 mg kg^{-1}), while F2 (10 years old) had the highest values of soil EC (0.9 mS cm^{-1}), total N (55.2 mg kg^{-1}), K^+ (104.5 mg kg^{-1}), Ca^{2+} (155.3 mg kg^{-1}), and Na^+ (170.5 mg kg^{-1}). On the other hand, the highest soil P content (8.9 mg kg^{-1}) was recorded in F3 (12 years old), while the highest soil Mg^{2+} (35.6 mg kg^{-1}) was recorded in the oldest farm (F4).

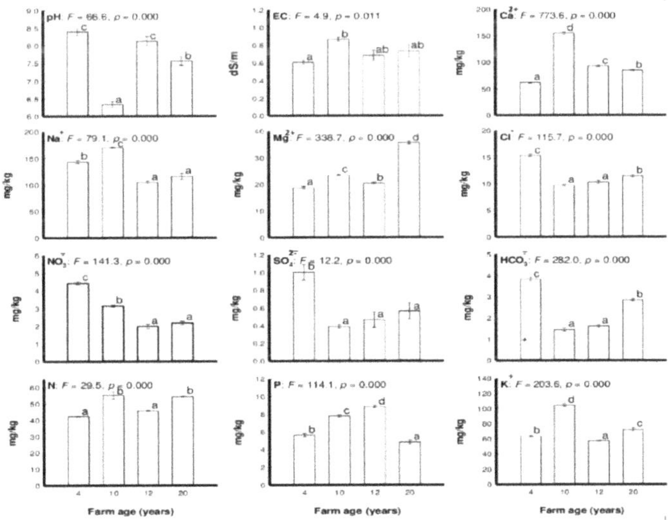

Figure 1. Mean soil characteristics of the four sampling farms in the mountainous area of Taif City in Saudi Arabia, supporting the growth of the Taif rose populations. The standard errors of the means (n = 6) are indicated by vertical bars. *F*-values demonstrate the one-way analysis of variance (ANOVA), with degrees of freedom (*df*) = 3. Means followed by different letters are significantly different at $p < 0.05$, according to Tukey's HSD test.

The plant–soil relationship was expressed using a CCA biplot, which showed that the plant height and crown diameter were at low positions on the N and NO_3 axes, respectively (Figure 2). In contrast, all the other measured plant parameters were at a higher position on the Mg axis. The interest correlations of the soil parameters (Table 2) showed that Cl^-,

NO_3^-, HCO_3^-, and SO_4^{2-} had high positive correlations with axis 1 and negative values with axis 2. In contrast, EC, total P, and Ca^{2+} had high positive correlations, while pH had high negative values, with axis 2. Moreover, Mg^{2+} was negatively correlated with axes 1 and 2.

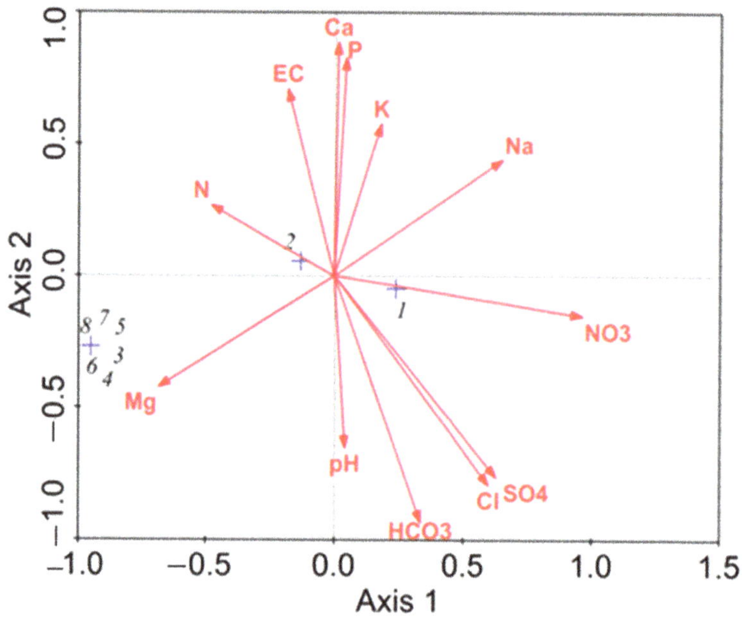

Figure 2. CCA biplot with the soil characteristics (→) and parameters (+) of Taif rose. 1: height (cm); 2: diameter (cm); 3: leaf biomass (kg ind.$^{-1}$); 4: stem biomass (kg ind.$^{-1}$); 5: above-ground biomass (kg ind.$^{-1}$); 6: leaf biomass (t ha^{-1}); 7: stem biomass (t ha^{-1}); 8: above-ground biomass (t ha^{-1}).

Table 2. Inter-set correlations of the soil characteristics with CCA axes.

Soil Characteristic	Axis 1	Axis 2
pH	0.0424	−0.6525
EC (dS m^{-1})	−0.1797	0.7102
Cl^- (mg kg^{-1})	0.6106	−0.7924
NO_3^- (mg kg^{-1})	0.9681	−0.1516
SO_4^{2-} (mg kg^{-1})	0.6303	−0.7626
HCO_3^- (mg kg^{-1})	0.3367	−0.9351
Total N (mg kg^{-1})	−0.4778	0.2682
Total P (mg kg^{-1})	0.0451	0.8280
K^+ (mg kg^{-1})	0.1805	0.5779
Ca^{2+} (mg kg^{-1})	0.0140	0.8868
Mg^{2+} (mg kg^{-1})	−0.6800	−0.4213
Na^+ (mg kg^{-1})	0.6225	0.4416

EC: electrical conductivity.

3.3. Inorganic and Organic Nutrient Contents

The results of the statistical analysis (ANOVA II) showed that the concentrations of the analyzed inorganic macronutrients in the leaves and stems of Taif's rose significantly differed between the studied farms and the estimated organs, with a significant positive intercept between the farm age and plant organ (Table 3). F1 had the highest N content (154.30 ± 7.99 mg kg^{-1}) in the plant leaves but the second-highest concentrations of stem N, P, Ca^{2+}, and Mg^{2+} and leaf Mg^{2+}. Additionally, F2 had the highest concentrations of stem N (172.33 ± 3.70 mg kg^{-1}) and P (9.40 ± 0.29 mg kg^{-1}). Moreover, F3 had the highest concentrations of leaf P (7.17 ± 0.09 mg kg^{-1}), leaf and stem K (112.47 ± 3.90 and 277.30 ± 8.52 mg kg^{-1}, respectively), stem Ca^{2+} (251.93 ± 7.99 mg kg^{-1}), and leaf and stem Mg^{2+} (122.27 ± 2.31 and 123.57 ± 1.86 mg kg^{-1}, respectively), while F4 had the highest concentration of leaf Ca^{2+} (206.30 ± 7.99 mg kg^{-1}).

Table 3. Mean (±standard error, n = 6) inorganic nutrient contents of Taif Rose populations grown in four farms in the mountainous area of Taif City in Saudi Arabia.

Element	Organ	Farm No. (Age, Years)				F_{Farm}	F_{Organ}	$F_{Farm \times Organ}$
		1 (4 Years)	2 (10 Years)	3 (12 Years)	4 (20 Years)			
N (mg kg^{-1})	Leaf	154.30 ± 7.99	42.20 ± 1.48	92.00 ± 2.65	103.50 ± 4.15	84.6 ***	163.8 ***	94.4 ***
	Stem	161.30 ± 6.61	172.33 ± 3.70	85.07 ± 2.72	136.20 ± 2.62			
P (mg kg^{-1})	Leaf	6.10 ± 0.11	6.70 ± 0.24	7.17 ± 0.09	6.33 ± 0.09	68.4 ***	0.2 ns	65.7 ***
	Stem	6.33 ± 0.16	9.40 ± 0.29	5.47 ± 0.15	5.33 ± 0.09			
K (mg kg^{-1})	Leaf	94.70 ± 1.66	103.43 ± 4.25	112.47 ± 3.90	95.57 ± 3.14	139.0 ***	412.9 ***	92.5 ***
	Stem	91.87 ± 1.57	174.67 ± 4.71	277.30 ± 8.52	154.87 ± 8.01			
Ca (mg kg^{-1})	Leaf	152.30 ± 5.58	171.47 ± 1.64	165.63 ± 8.99	206.30 ± 7.99	44.6 ***	4.1 *	70.1 ***
	Stem	150.83 ± 0.72	125.03 ± 4.68	251.93 ± 7.99	134.03 ± 3.41			
Mg (mg kg^{-1})	Leaf	114.10 ± 1.94	20.57 ± 0.26	122.27 ± 2.31	84.77 ± 1.05	974.9 ***	78.2 ***	50.3 ***
	Stem	111.87 ± 1.88	56.17 ± 1.81	123.57 ± 1.86	92.87 ± 1.69			
Na (mg kg^{-1})	Leaf	170.70 ± 2.98	186.23 ± 3.76	173.03 ± 3.23	140.60 ± 3.29	77.7 ***	216.7 ***	307.3 ***
	Stem	142.53 ± 4.04	68.67 ± 1.28	128.00 ± 2.14	205.40 ± 2.54			

F-values demonstrate the two-way analysis of variance (ANOVA); * $p < 0.05$; *** $p < 0.001$; *ns*: not significant (i.e., $p > 0.05$).

The statistical analysis (ANOVA II) of the organic nutrients in the leaves and stems of Taif's rose plants revealed significant variations in all the investigated nutrients between farms and organs and the intercept between farms and plant organs (Table 4). The leaves of F1 rose plants contributed the highest percentage of fats and total ash (0.79 and 14.07%, respectively), while their stems had the lowest ADL and percentage of fats (1.87 and 0.11%). Additionally, the leaves of the F2 rose plants had the highest percentage of NDF and the lowest total proteins (45.01 and 2.63%), while their stems had the highest percentages of total proteins and the lowest ash content (10.71 and 5.17%). Moreover, the leaves of the F3 plants had the highest percentage of ADL and the lowest of crude fibers (7.63 and 13.27%), while the stems had the highest NFE (i.e., total carbohydrates, 72.71%). The stems of the oldest rose plants in F4 had the highest ADF and crude fibers (32.60 and 45.90%) but the lowest NFE (38.37%), while the leaves had the lowest ADF and NDF (15.30 and 27.50%). It is noticeable that the stems had higher percentages of total proteins, fibers, ash, and NFE than the leaves in F1 and F2, while the opposite was observed in F3 and F4 (Table 4).

Table 4. Mean (± standard error, n = 6) organic nutrient contents of Taif Rose populations grown in four farms in the mountainous area of Taif City in Saudi Arabia.

Element	Organ	Farm Age (Years)				F_{Farm}	F_{Organ}	$F_{Farm \times Organ}$
		1 (4 Years)	2 (10 Years)	3 (12 Years)	4 (20 Years)			
ADF (%)	Leaf	18.20 ± 0.53	26.40 ± 1.24	31.60 ± 1.02	15.30 ± 0.83	40.1 ***	4.2 *	82.8 ***
	Stem	17.33 ± 0.40	25.13 ± 0.59	21.60 ± 0.61	32.60 ± 1.39			
ADL (%)	Leaf	2.67 ± 0.02	2.83 ± 0.09	7.63 ± 0.09	2.57 ± 0.06	690.9 ***	697.7 ***	391.1 ***
	Stem	1.87 ± 0.09	2.37 ± 0.09	3.03 ± 0.09	2.70 ± 0.04			
NDF (%)	Leaf	36.13 ± 1.04	45.10 ± 1.68	44.10 ± 0.79	27.50 ± 0.90	24.3 ***	5.8*	52.5 ***
	Stem	33.27 ± 0.38	36.10 ± 0.68	36.10 ± 1.04	40.57 ± 0.95			
Fat (%)	Leaf	0.79 ± 0.02	0.57 ± 0.01	0.24 ± 0.01	0.41 ± 0.01	198.5 ***	1627.5 ***	166.8 ***
	Stem	0.11 ± 0.01	0.23 ± 0.01	0.11 ± 0.01	0.11 ± 0.01			
Crude fiber (%)	Leaf	28.83 ± 1.52	28.27 ± 0.71	13.27 ± 0.71	17.07 ± 0.63	27.4 ***	161.3 ***	230.1 ***
	Stem	19.43 ± 1.08	17.43 ± 0.61	42.57 ± 1.02	45.90 ± 1.61			
Ash (%)	Leaf	14.07 ± 0.42	11.63 ± 0.82	8.03 ± 0.22	12.93 ± 0.58	4.7 **	147.8 ***	49.2 ***
	Stem	6.03 ± 0.28	5.17 ± 0.15	11.07 ± 0.79	7.10 ± 0.28			

Table 4. *Cont.*

Element	Organ	Farm Age (Years)				F_{Farm}	F_{Organ}	$F_{Farm \times Organ}$
		1 (4 Years)	2 (10 Years)	3 (12 Years)	4 (20 Years)			
Total Protein (%)	Leaf	9.65 ± 0.49	2.63 ± 0.09	5.75 ± 0.16	6.48 ± 0.25	87.6 ***	166.4 ***	95.8 ***
	Stem	10.08 ± 0.40	10.71 ± 0.24	5.31 ± 0.17	8.52 ± 0.17			
NFE (%)	Leaf	46.67 ± 0.61	56.91 ± 0.34	72.71 ± 0.66	63.11 ± 0.29	44.4 ***	117.6 ***	331.1 ***
	Stem	64.34 ± 0.97	66.47 ± 0.25	40.94 ± 1.97	38.37 ± 1.18			

ADF: acid detergent fibers, ADL: acid detergent lignin, NDF: neutral detergent fibers. *F*-values demonstrate the two-way analysis of variance (ANOVA); * $p < 0.05$; ** $p < 0.01$; *** $p < 0.001$.

3.4. Nutritive Value

The different calculated parameters of the nutritive values (except for the GE between farms) had significant variations between the studied farms and plant organs and the intercept between the farms and plant organs (Table 5). The range of DCP (%) in the leaves and stems of Taif's rose of this study ranged from 1.08–5.44% DM in F2 and F1 and from 1.42–6.43% DM in F3 and F2, respectively. The TDN (%) ranged from 55.97 to 60.85% in the leaves of F1 and F2 plants and from 54.77 to 58.56% in the stems of F2 and F3. Additionally, the ranges of DE and ME were 2.31–3.22 Mcal kg^{-1} in F1 and F3 and 1.89–2.64 Mcal kg^{-1} in the F2 and F3 plant leaves, respectively, while they were 2.07–3.15 Mcal kg^{-1} and 1.69–2.58 Mcal kg^{-1} in the F3 and F2 stems, respectively. Moreover, NE and GE varied significantly between the leaves (0.99–1.32 and 377.02–391.72 Mcal kg^{-1}, respectively) and stems (0.85–1.29 and 400.35–424.25 Mcal kg^{-1}, respectively) of the studied farms.

Table 5. Mean (± standard error, n = 6) nutritive values of Taif's Rose populations grown in four farms in the mountainous area of Taif City in Saudi Arabia.

Element	Organ	Farm No. (Age, Years)				F_{Farm}	F_{Organ}	$F_{Farm \times Organ}$
		4	10	12	20			
DCP (%)	Leaf	5.44 ± 0.46	1.08 ± 0.08	1.82 ± 0.15	2.50 ± 0.23	81.8 ***	98.8 ***	48.7 ***
	Stem	5.85 ± 0.37	6.43 ± 0.22	1.42 ± 0.16	4.40 ± 0.16			
TDN (%)	Leaf	55.97 ± 0.34	60.85 ± 0.06	58.34 ± 0.11	57.95 ± 0.17	78.5 ***	222.5 ***	96.5 ***
	Stem	55.13 ± 0.30	54.77 ± 0.18	58.56 ± 0.13	56.25 ± 0.13			
DE (Mcal kg^{-1})	Leaf	2.31 ± 0.02	2.44 ± 0.02	3.22 ± 0.02	2.74 ± 0.02	35.1 ***	9.2 **	418.9 ***
	Stem	3.02 ± 0.06	3.15 ± 0.01	2.07 ± 0.05	2.18 ± 0.02			
ME (Mcal kg^{-1})	Leaf	1.89 ± 0.02	1.99 ± 0.02	2.64 ± 0.02	2.25 ± 0.01	35.1 ***	9.2 **	418.9 ***
	Stem	2.48 ± 0.05	2.58 ± 0.01	1.69 ± 0.04	1.79 ± 0.02			
NE (Mcal kg^{-1})	Leaf	0.95 ± 0.01	0.99 ± 0.01	1.32 ± 0.01	1.12 ± 0.01	35.1 ***	9.2 **	418.9 ***
	Stem	1.24 ± 0.02	1.29 ± 0.01	0.85 ± 0.02	0.90 ± 0.01			
GE (Mcal kg^{-1})	Leaf	388.83 ± 1.94	385.15 ± 3.66	391.72 ± 1.75	377.02 ± 2.29	2.0 ns	324.4 ***	28.9 ***
	Stem	411.09 ± 0.91	414.76 ± 0.71	400.35 ± 2.15	424.25 ± 2.09			

DCP: digestible crude protein, TDN: total digestible nutrients, DE: digestible energy, ME: metabolized energy, NE: net energy and GE: gross energy. *F*-values demonstrate the two-way analysis of variance (ANOVA); ** $p < 0.01$; *** $p < 0.001$; *ns*: not significant (i.e., $p > 0.05$).

4. Discussion

One of the global goals is to ensure the long-term sustainability of agricultural systems. Organic fertilizers can help to improve sustainability by lowering the usage of chemical fertilizers. Keeping this in mind, the addition of vegetative waste to the soil during rose cultivation aims to the return of nutrients removed by the plants, which can help to reduce the contamination of the environment and play a role in the operation of this industry as a component of a bioeconomy [36]. As shown by the current results, the plant height and crown diameter increased with the increase in the farm age. The effect of age on the biomass of the plant stems and leaves was higher. The AGB value of 1.32–5.21 t/ha, due to the yearly pruning of rose plants in Taif, should be exploited in soil fertilization or animal feeding. The application of organic fertilizers can reduce the cost of chemical fertilizers, as well as the energy consumption during their production [37]. Composting rose waste is, in theory, an excellent way of producing stable organic fertilizers. This compost is usually rich in numerous microbial communities, including bacteria, fungi, and actinomycetes, which decompose organic matter in the presence of oxygen, resulting in a sustainable humic substance with a high nutritive quality [38]. Recently, following an unregulated biodegradation process, several industries have begun to add rose waste to the soil as compost.

The application of organic amendments, such as sewage sludge is, required in order to maintain the high organic matter content of rose-growing soils, but these amendments must be sustainable [36]. Since the same fertilization practices were applied to the four farms, the old farms accumulated more mineral nutrients than the youngest ones over the same time and, thus, the N contents in the old soils were higher than those in the young soils. The macronutrient contents of rose pruning wastes are promising for their use as compost and for increasing soil quality. Our results revealed that the age of the farms had a significant effect on the nutrient content of Taif rose plants, but without an exponential increase with age. The nutrient content of the Taif rose was lower than that reported for herbal plant residues [39] and olive mill waste [40]. However, in general, composting agriculture residue and converting it into organic fertilizer can help to reduce the need for chemical fertilizers and nutrient requirements [41]. Despite the importance of organic fertilizers to soil quality, they may not be an absolute substitute for chemical fertilizers, since their nutrient release rate is too slow to meet the crop requirements within a short time [42].

Forage quality, expressed mainly by the levels of crude protein (CP), crude fiber (CF), digestibility, and other associated characteristics, are essential for animal productivity [43]. CP and CF are classically viewed as indicators of the nutritional value of food for grazing animals [44]. CP is used as energy and helps to build tissue. Its calculation is based on a laboratory nitrogen analysis, from which the total protein content in foodstuff can be calculated by multiplying the total nitrogen by 100/16 or 6.25. This works on the assumption that nitrogen is derived from protein containing 16% nitrogen [45]. However, a certain portion of the N in most feeds is non-protein nitrogen and, therefore, the value calculated by multiplying N $\times$ 6.25 is referred to as the crude rather than true protein. According to MAFF [46], the minimum protein level in animal diets ranges between 6 to 12% DM, and accordingly, most of the recorded values of the CP content in the leaves and stems of the Taif rose lie within this range. Additionally, Taif rose has higher CP than that reported for certain grasses, such as *Hyperrhenia hirta* and *Chloris pycnothrix* [47]. Taif rose lies within the required range of CP for gestating cows (7–9%), as reported by NRC [25]. The crude fiber (CF) in plants represents all the cell wall fractions that are resistant to the action of digestive enzymes and includes the insoluble residue of acid hydrolysis, as well as alkaline hydrolysis [48]. The CF content is considered as a major indicator of the chemical composition when determining the energy feeding value of the forage [49]. In this study, the crude fiber was higher in the leaves than in the stems of young plants (F1 and F2), while the opposite was true for old plants (F3 and F4). The range of the CF in Taif's rose (13.27–45.90% DM) was either within or slightly higher than that which has been reported for some known wild forage plants, such as *Phragmites australis* (29.9% DM), *Panicum repens* (27.3% DM), and *Cynodon dactylon* (20.5% DM) [50,51]. Moreover, the range of the CF content in the young leaves and old stems of Taif's rose was higher than the mean content of temperate legumes (25.3%) and grasses (20.0%) [52].

Based on the plant age, the CP, ash, and CF contents were significantly affected, whereby they increased according to the N fertilization practices. Similar results were reported by Čop et al. [53] on the CF, Mohammed et al. [54] on the ash content, and Dindová, et al. [55] on the CP, CF, ash, and lipids. The lipid, or fat, percentage in Taif's rose in this study was very low compared to the dry leaves and shoots of *Vossia cuspidaa* [56,57] and corn stover (2.2%) [58]. The NFE was considerably high in Taif's rose leaves and stems (38.37–72.71%), similarly to other studied plants, such as *Echinochloa. stagnina* and *Eichhornia crassipes* (≈54%), and higher than *C. demersum* (33.4%) [59]. This high carbohydrate content is effective in providing the rumen microbes with enough energy, which finally benefits lactating cows [60].

The nutritive value of forage is mainly the result of the chemical composition, including the CP content and fiber fractions such as NDF, ADF, and ADL [46]. Long-term fertilization showed consistent effects on the NDF, ADF, hemicellulose, cellulose, and ADL, as did the whole plant ash content [61]. After the ignition and/or oxidation of plant organic matter at a high temperature, the inorganic residue of the remaining chemical elements is called ash. The ash content in Taif's rose pruning waste was comparable with that of many plants, such as *V. cuspidata* (10.44%, [56]), *Panicum turgidum* (9.1%) [62], and *E. stagnina* (12.9%) [59]. In this study, the ADF was lower than that of wheat straw (46.5–50.8%, [63]), but the NDF was comparable to that reported for *Trifolium alexandrinum* [64] and *Leucaena lanceolata* (32.1%) [65]. Additionally, it was found that the ADF, NDF, and ADL increased, to some extent, with the farm age, receiving more long-term fertilization, in line with the study of Coblentz et al. [61] on oat plants. Furthermore, the percentages of CP, NDF, and ADF of Taif's rose were lower than 4.9%, 74.4%, and 44.4%, respectively, which were the values recorded in central Oklahoma [66], while values of 10.0%, 57.5 %, and 31.6 % were reported in western Washington [67] for forage intermediate wheatgrass. In the same context, Favre et al. [68] reported values of 11.4% CP, 59.7% NDF, and 33.7% ADF for Kernza intermediate wheatgrass in Wisconsin.

The DCP is an important component of proteins that is ingested and absorbed by the animal and not excreted in feces [69]. In this study, Taif's rose had a low DCP content (1.08–6.43%), which is comparable to that reported for *E. stagnina* (2.8%) and *E. crassipes* (3.7%) [60]. Additionally, it was lower than that reported for the shoots of *T. alexandrinum* (9 %) [70] and the leaves of *V. cuspidata* (13.82%, [56]). The total digestible nutrients (TDN) are defined as the energy content of the feeds available to animals after the digestion-induced losses [49]. The calculated TDN values of Taif's rose in this study (54.7–60.8%) were comparable to or higher than those of certain investigated plants, such as the leaves of *P. australis* (41.58%) [71], the shoots of *E. crassipes* and *L. stolonifera* (54.2% and 51.5%, respectively) [72], and the shoots of *C. demersum* (48.6%) [60]. This study revealed that Taif's rose has suitable contents of TDN for mature dry gestating beef cows, which require 55–60% [73].

The estimated values of the DE and ME of Taif's rose pruning waste were comparable to the values of 2.65, and 2.17 Mcal kg^{-1} DM, respectively, recorded for the hay of alfalfa (*Medicago sativa*), and 2.43 and 1.99 Mcal kg^{-1} DM for red clover (*Trifolium pratense*) [25]. The values of GE of Taif's rose pruning waste (377.02–424.25 Mcal kg^{-1} DM) were remarkably higher than those reported for *Cynodon dactylon* (389 Mcal kg^{-1} DM) and *Panicum repens* (398 Mcal kg^{-1} DM) [51] and the shoots of many grazable plants on the western Mediterranean coast of Egypt (399 Mcal kg^{-1} DM) [74].

5. Conclusions

The AGB value of 1.32–5.21 t/ha, due to the yearly pruning of rose plants in Taif, should be exploited in soil fertilization or animal feeding. The recycling of Taif rose pruning waste could be used in the manufacturing of organic fertilizer as a co-compost cover material. Taif rose pruning waste can also be considered as an interesting raw material due to its high content of digestible proteins and high energy compared with other sources of protein frequently used in animal feed. Moreover, the positive response, in terms of the growth performance of the rose plants, to the soil nutrient contents may allow us to consider the safe use of soil amendments, such as sewage sludge and treated wastewater, for improving production.

Author Contributions: Conceptualization, E.A.F., E.F.A. and T.M.G.; methodology, E.A.F., E.F.A., T.M.G., E.M.E., A.M. and H.M.A.-Y.; software, T.M.G. and E.M.E.; validation, E.A.F.; E.F.A., T.M.G., E.M.E., A.M. and H.M.A.-Y.; formal analysis, E.A.F., E.F.A., T.M.G., E.M.E. and H.M.A.-Y.; investigation, E.A.F., E.F.A. and A.M.; resources, E.F.A., T.M.G., A.M. and H.M.A.-Y.; data curation, E.F.A., E.M.E., A.M., H.M.A.-Y. and F.A.; writing—original draft preparation, E.A.F., E.F.A., T.M.G., E.M.E., A.M., H.M.A.-Y. and F.A.; writing—review and editing, E.F.A. and T.M.G., visualization, E.M.E., A.M., H.M.A.-Y. and F.A.; supervision, E.F.A. and T.M.G.; project administration, E.F.A. and T.M.G.; funding acquisition, E.A.F., E.F.A. and T.M.G.; methodology, E.A.F., E.F.A., T.M.G., E.M.E. and A.M. All authors have read and agreed to the published version of the manuscript.

Funding: This research was funded by the High-Altitude Research Centre at Taif University, Taif, Saudi Arabia, under project number 1-442-42.

Institutional Review Board Statement: Not applicable.

Data Availability Statement: Not applicable.

Acknowledgments: The authors acknowledge the Dean of Scientific Research at Taif University, Taif, Saudi Arabia, for funding this work through the High-Altitude Research Centre, under project number 1-442-42.

Conflicts of Interest: The authors declare no conflict of interest.

References

1. Wang, B.; Dong, F.; Chen, M.; Zhu, J.; Tan, J.; Fu, X.; Wang, Y.; Chen, S. Advances in recycling and utilization of agricultural wastes in China: Based on environmental risk, crucial pathways, influencing factors, policy mechanism. *Procedia Environ. Sci.* **2016**, *31*, 12–17. [CrossRef]
2. Varma, V.S.; Yadav, J.; Das, S.; Kalamdhad, A.S. Potential of waste carbide sludge addition on earthworm growth and organic matter degradation during vermicomposting of agricultural wastes. *Ecol. Eng.* **2015**, *83*, 90–95. [CrossRef]
3. Wang, H.Y.; Liu, S.; Zhai, L.M.; Zhang, J.Z.; Ren, T.Z.; Fan, B.Q.; Liu, H.B. Preparation and utilization of phosphate biofertilizers using agricultural waste. *J. Integr. Agric.* **2015**, *14*, 158–167. [CrossRef]
4. Mengqi, Z.; Shi, A.; Ajmal, M.; Ye, L.; Awais, M. Comprehensive review on agricultural waste utilization and high-temperature fermentation and composting. *Biomass Convers. Biorefinery* **2021**, 1–24. [CrossRef]
5. Yaqoob, H.; Teoh, Y.H.; Din, Z.U.; Sabah, N.U.; Jamil, M.A.; Mujtaba, M.A.; Abid, A. The potential of sustainable biogas production from biomass waste for power generation in Pakistan. *J. Clean. Prod.* **2021**, *307*, 127250. [CrossRef]
6. Koul, B.; Yakoob, M.; Shah, M.P. Agricultural waste management strategies for environmental sustainability. *Environ. Res.* **2022**, *206*, 112285. [CrossRef] [PubMed]
7. De Corato, U. Agricultural waste recycling in horticultural intensive farming systems by on-farm composting and compost-based tea application improves soil quality and plant health: A review under the perspective of a circular economy. *Sci. Total Environ.* **2020**, *738*, 139840. [CrossRef]
8. Torkashvand, M.A.; Alidoust, M.; Mahboub Khomami, A. The reuse of peanut organic wastes as a growth medium for ornamental plants. *Int. J. Recycl. Org. Waste Agric.* **2015**, *4*, 85–94. [CrossRef]
9. Duan, Y.; Mehariya, S.; Kumar, A.; Singh, E.; Yang, J.; Kumar, S.; Li, H.; Kumar Awasthi, M. Apple orchard waste recycling and valorization of valuable product—A review. *Bioengineered* **2021**, *12*, 476–495. [CrossRef]
10. Sharma, B.; Vaish, B.; Singh, U.K.; Singh, P.; Singh, R.P. Recycling of organic wastes in agriculture: An environmental perspective. *Int. J. Environ. Res.* **2019**, *13*, 409–429. [CrossRef]
11. Galal, T.M.; Al-Yasi, H.M.; Fawzy, M.A.; Abdelkader, T.G.; Hamza, R.Z.; Eid, E.M.; Ali, E.F. Evaluation of the Phytochemical and Pharmacological Potential of Taif's Rose (*Rosa damascena* Mill var. *trigintipetala*) for Possible Recycling of Pruning Wastes. *Life* **2022**, *12*, 273. [CrossRef] [PubMed]
12. Hamza, R.Z.; Al-Yasi, H.M.; Ali, E.F.; Fawzy, M.A.; Abdelkader, T.G.; Galal, T.M. Chemical Characterization of Taif Rose (*Rosa damascena* Mill var. *trigintipetala*) Waste Methanolic Extract and Its Hepatoprotective and Antioxidant Effects against Cadmium Chloride ($CdCl_2$)-Induced Hepatotoxicity and Potential Anticancer Activities against Liver Cancer Cells (HepG2). *Crystals* **2022**, *12*, 460. [CrossRef]
13. Ali, E.F.; Al-Yasi, H.M.; Issa, A.A.; Hessini, K.; Hassan, F.A.S. Ginger extract and fulvic acid foliar applications as novel practical approaches to improve the growth and productivity of damask rose. *Plants* **2022**, *11*, 412. [CrossRef]
14. Kurkcuoglu, M.; Baser, K.H.C.; Akterian, S.G.; Fidan, H.N.; Stoyanova, A.S. Chemical composition, sensory evaluation and antimicrobial activity of Taif rose (*Rosa damascena* Mill.) essential oils. *Bulg. Chem. Commun.* **2020**, *460*.
15. Ali, E.F.; Issa, A.A.; Al-Yasi, H.M.; Hessini, K.; Hassan, F.A.S. The efficacies of 1-methylcyclopropene and chitosan nanoparticles in preserving the postharvest quality of damask rose and their underlying biochemical and physiological mechanisms. *Biology* **2022**, *11*, 242. [CrossRef] [PubMed]

16. Ali, E.F.; Alyasi, H.M.; Hassan, F.A.; Alamer, K.H.; Hessini, K.; Attia, H.; Elshazly, S. Effect of the Pruning System and P-Fertilizer on Growth and Productivity of *Rosa damascena* mill. *Egypt. J. Bot.* **2021**, *61*, 565–578. [CrossRef]
17. Al-Yasi, H.; Attia, H.; Alamer, K.; Hassan, F.; Ali, E.; Elshazly, S. Impact of drought on growth, photosynthesis, osmotic adjustment, and cell wall elasticity in damask rose. *Plant Physiol. Biochem.* **2020**, *150*, 133–139. [CrossRef]
18. Hassan, F.; Al-Yasi, H.; Ali, E.; Alamer, K.; Hessini, K.; Attia, H.; El-Shazly, S. Mitigation of salt-stress effects by moringa leaf extract or salicylic acid through motivating antioxidant machinery in damask rose. *Can. J. Plant Sci.* **2020**, *101*, 157–165. [CrossRef]
19. El-Sharnouby, M.E.; Montaser, M.M.; Abdallah, S.M. Oil and Flower Production in Rosa *damascena trigintipetala* Dieck under Salinity Stress in Taif Region, Saudi Arabia. *Sustainability* **2021**, *13*, 4547. [CrossRef]
20. Hessini, K.; Wasli, H.; Al-Yasi, H.M.; Ali, E.F.; Issa, A.A.; Hassan, F.A.S.; Siddique, K.H.M. Graded moisture deficit effect on secondary metabolites, antioxidant, and inhibitory enzyme activities in leaf extracts of *Rosa damascena* Mill. var. *trigintipetala*. *Horticulturae* **2022**, *8*, 177. [CrossRef]
21. Dragoev, S.; Vlahova-Vangelova, D.; Balev, D.; Bozhilov, D.; Dagnon, S. Valorization of waste by-products of rose oil production as feedstuff phytonutrients. *Bulg. J. Agric. Sci.* **2020**, *27*, 209–219. [CrossRef]
22. Chapman, H.D.; Pratt, P.F. *Methods of Analysis for Soils, Plants, and Water*; Department of Soil, Plant Nutrition, University of California Press: Berkeley, CA, USA, 1961.
23. Allen, S.E. *Chemical Analysis of Ecological Materials*; Blackwell Scientific Publications: London, UK, 1989.
24. Le Houérou, H.N. Chemical composition and nutritive value of browse in tropical West Africa. In *Browse in Africa, the Current State of Knowledge*; Le Houérou, H.N., Ed.; ILCA: Addis Ababa, Ethiopia, 1980; pp. 261–289.
25. NRC (National Research Council). *Nutrient Requirements of Dairy Cattle*, 8th ed.; NRC (National Research Council): Ottawa, ON, Canada, 2001.
26. Goering, H.K.; Van Soest, P.J. *Forage Fiber Analyses (Apparatus, Reagents, Procedures, and Some Applications)*; Agriculture Handbook No. 379; ARSUSDA: Washington, DC, USA, 1970.
27. Van Soest, P.J.; Robertson, J.B.; Lewis, B.A. Methods for dietary fiber, neutral detergent fiber and non-starch polysaccharides in relation to animal performance. *J. Dairy Sci.* **1991**, *74*, 3583–3597. [CrossRef]
28. Blaxter, K.L. *The Energy Metabolism of Ruminants*, 2nd ed.; Charles Thomas Publisher: Spring Field, IL, USA, 1968.
29. McDonald, P.; Edwards, R.A.; Greenhalgh, J.F.D.; Morgan, C.A.; Sinclair, L.A.; Wilkinson, R.G. *Animal Nutrition*, 7th ed.; Harlow, Pearson: London, UK, 2011.
30. NRC (National Research Council). *Nutrient Requirements of Beef Cattle*, 7th ed.; updated 2000; The National Academies Press: Washington, DC, USA, 2000.
31. Maff, R. *The Analysis of Agricultural Materials*, 3rd ed.; HMSO: London, UK, 1986; p. 427.
32. Leps, J.; Šmilauer, P. *Multivariate Analysis of Ecological Data using CANOCO*; Cambridge University Press: Cambridge, UK, 2003; p. 269.
33. Ter Braak, C.J.F.; Smilauer, P. *CANOCO Reference Manual for User's Guide to Canoco for Windows: Software for Canonical Community Ordination (Version 4.0)*; Microcomputer Power: Ithaca, NY, USA, 1998; p. 352.
34. Ter Braak, C.J. Canonical correspondence analysis: A new eigenvector technique for multivariate direct gradient analysis. *Ecology* **1986**, *67*, 69–77. [CrossRef]
35. SPSS. IBM SPSS statistics version 21.0. Copyright of IBM and other(s); IBM Corp.: Armonk, NY, USA, 2012.
36. Idrovo-Novillo, J.; Gavilanes-Terán, I.; Veloz-Mayorga, N.; Erazo-Arrieta, R.; Paredes, C. Closing the cycle for the cut rose industry by the reuse of its organic wastes: A case study in Ecuador. *J. Clean. Prod.* **2019**, *220*, 910–918. [CrossRef]
37. Galal, T.M.; Gharib, F.A.; Al-Yasi, H.M.; Al-Mutairi, K.A.; Mansour, K.H.; Eid, E.M. Nutrient Remediation Efficiency of the Sedge Plant (*Cyperus alopecuroides* Rottb.) to Restore Eutrophic Freshwater Ecosystems. *Sustainability* **2022**, *14*, 2823. [CrossRef]
38. Bernal, M.P.; Sommer, S.G.; Chadwick, D.; Qing, C.; Guoxue, L.; Michel, F.C., Jr. Current approaches and future trends in compost quality criteria for agronomic, environmental, and human health benefits. In *Advances in Agronomy*; Sparks, D.L., Ed.; Academic Press: Burlington, NY, USA, 2017; Volume 144, pp. 143–233.
39. Khater, E.S.G. Some physical and chemical properties of compost. *Int. J. Waste Resour.* **2015**, *5*, 172. [CrossRef]
40. Roig, A.; Cayuela, M.L.; Sánchez-Monedero, M.A. An overview on olive mill wastes and their valorisation methods. *Waste Manag.* **2006**, *26*, 960–969. [CrossRef]
41. Chang, K.H.; Wu, R.Y.; Chuang, K.C.; Hsieh, T.F.; Chung, R.S. Effects of chemical and organic fertilizers on the growth, flower quality and nutrient uptake of Anthurium andreanum, cultivated for cut flower production. *Sci. Hortic.* **2010**, *125*, 434–441. [CrossRef]
42. Naeem, M.; Ansari, A.; Gill, S. (Eds.) V Role of Organic Fertilizers in Improving Soil Fertility. In *Contaminants in Agriculture*; Springer: Cham, Switzerland, 2020. [CrossRef]
43. Vasileva, V.; Naydenova, Y.; Stoycheva, I. Nutritive value of forage biomass from sainfoin mixtures. *Saudi J. Biol. Sci.* **2019**, *26*, 942–949. [CrossRef]
44. Heneidy, S.Z.; Halmy, M.W. The nutritive value and role of *Panicum turgidum* Forssk. in the arid ecosystems of the Egyptian desert. *Acta Bot. Croat.* **2009**, *68*, 127–146.
45. A.O.A.C. *Official Methods of Analysis. Association of Official Analytical Chemists*, 14th ed.; A.O.A.C.: Arlington, TX, USA, 1984.
46. MAFF (Ministry of Agriculture, Fisheries, and Food). *Energy Allowances and Feeding Systems for Ruminants*; Technical Bulletin 33; Her Majesty's Stationary Office: London, UK, 1975.

47. Lee, M.A. A global comparison of the nutritive values of forage plants grown in contrasting environments. *J. Plant Res.* **2018**, *131*, 641–654. [CrossRef] [PubMed]
48. Wu, G. *Principles of Animal Nutrition*; CRC Press: Boca Raton, FL, USA; London, UK; New York, NY, USA, 2018.
49. Krachunov, I. Estimation of energy feeding value of forages for ruminants II. Energy prediction through crude fiber content. *J. Mt. Agric. Balk.* **2007**, *10*, 122–134.
50. El-Kady, H.F. Seasonal variation in phytomass and nutrient status of *Phragmites australis* along the watercourses in the Middle Delta region. *Taekholmia* **2000**, *20*, 123–138. [CrossRef]
51. Shaltout, K.H.; Galal, T.M.; Komi, T.M. Nutrients and heavy metals accumulation in the aboveground biomass of two perennial grasses along the watercourses of Nile Delta, Egypt. In *Journal of Botany, 3rd International Congress 17–18 April 2013*; Helwan University: Cairo, Egypt, 2013; pp. 201–218.
52. Norton, B.W. Differences between species in forage quality. In *Nutritional Limits to Animal Production from Pastures*; Hacker, J.B., Ed.; CAB: Farnham Royal, UK, 1982; pp. 98–110.
53. Čop, J.; Vidrih, M.; Hacin, J. Influence of cutting regime and fertilizer application on the botanical composition, yield and nutritive value of herbage of wet grasslands in Central Europe. *Grass Forage Sci.* **2009**, *64*, 454–465. [CrossRef]
54. Mohammed, G.; Trolard, F.; Bourrié, G.; Gillon, M.; Tronc, D.; Charron, F. A long- term data sequence (1960–2013) to analyse the sustainability of hay quality in irrigated permanent grasslands under climate change. *Am. J. Agric. For.* **2016**, *4*, 140–151.
55. Dindová, A.; Hakl, J.; Hrevušová, Z.; Nerušil, P. Relationships between long-term fertilization management and forage nutritive value in grasslands. *Agric. Ecosyst. Environ.* **2019**, *279*, 139–148. [CrossRef]
56. Farahat, E.; Fahmy, G.; Farrag, H.; Mahmoud, W.; Awad, H. Potential Nutritional Value of the Macrophyte *Vossia cuspidata* (Roxb.) Griff. (Poaceae) in a River Nile System, Egypt. *Egypt. J. Bot.* **2021**, *6*, 867–877. [CrossRef]
57. Galal, T.M.; Gharib, F.A.; Al-Yasi, H.M.; Mansour, K.H.; Hassan, M.M. Evaluation of the nutrient status and forage quality of the hippo grass (*Vossia cuspidata* (Roxb.) Griff.) along Ismailia canal, Egypt. *J. Freshw. Ecol.* **2021**, *36*, 63–76. [CrossRef]
58. Wei, Z.; Zhang, B.; Liu, J. Effects of the dietary non-fiber carbohydrates content on lactation performance, rumen fermentation, and nitrogen utilization in mid-lactation dairy cows receiving corn stover. *J. Anim. Sci. Biotechnol.* **2018**, *9*, 1–7. [CrossRef]
59. Shaltout, K.H.; Galal, T.M.; El-Komi, T. Evaluation of the nutrient status of some hydrophytes in the watercourses of Nile Delta, Egypt. *J. Bot.* **2009**, *2009*, 1–11. [CrossRef]
60. Batajoo, K.K.; Shaver, R.D. Impact of nonfiber carbohydrate on intake, digestion, and milk production by dairy cows. *J. Dairy Sci.* **1994**, *77*, 1580–1587. [CrossRef]
61. Coblentz, W.K.; Akins, M.S.; Cavadini, J.S.; Jokela, W.E. Net effects of nitrogen fertilization on the nutritive value and digestibility of oat forages. *J. Dairy Sci.* **2017**, *100*, 1739–1750. [CrossRef] [PubMed]
62. Heneidy, S.Z. Palatability and nutritive value of some common plant species from the Aqaba Gulf area of Sinai, Egypt. *J. Arid. Environ.* **1996**, *34*, 115–123. [CrossRef]
63. Blümmel, M.; Updahyay, S.R.; Gautam, N.; Barma, N.C.D.; Abdul Hakim, M.; Hussain, M.; Joshi, A.K. Comparative assessment of food-fodder traits in a wide range of wheat germplasm for diverse biophysical target domains in South Asia. *Field Crops Res.* **2019**, *236*, 68–74. [CrossRef]
64. Kholif, A.E.; Gouda, G.A.; Morsy, T.A.; Salem, A.Z.M.; Lopez, S.; Kholif, A.M. *Moringa oleifera* leaf meal as a protein source in lactating goat's diets: Feed intake, digestibility, ruminal fermentation, milk yield and composition, and its fatty acids profile. *Small Rumin. Res.* **2015**, *129*, 29–137. [CrossRef]
65. Martínez-Martínez, R.; López-Ortiz, S.; Ortega-Cerrilla, M.E.; Soriano-Robles, R.; Herrera-Haro, J.G.; López-Collado, J.; Ortega-Jiménez, E. Preference, consumption and weight gain of sheep supplemented with multinutritional blocks made with fodder tree leaves. *Livest. Sci.* **2012**, *149*, 185–189. [CrossRef]
66. Nelson, M.L.; Finley, J.W.; Scarnecchia, D.L.; Parish, S.M. Diet and forage quality of intermediate wheatgrass managed under continuous and short-duration grazing. *J. Range Manag.* **1989**, *42*, 474–477. [CrossRef]
67. Coleman, S.W.; Rao, S.C.; Volesky, J.D.; Phillips, W.A. Growth and nutritive value of perennial C3 grasses in the southern *Great Plains*. *Crop Sci.* **2010**, *50*, 1070–1078. [CrossRef]
68. Favre, J.R.; Castiblanco, T.M.; Combs, D.K.; Wattiaux, M.A.; Picasso, V.D. Forage nutritive value and predicted fiber digestibility of Kernza intermediate wheatgrass in monoculture and in mixture with red clover during the first production year. *Anim. Feed. Sci. Technol.* **2019**, *258*, 114298. [CrossRef]
69. Cherian, G. *A Guide to the Principles of Animal Nutrition*; Open textbook library, Oregon State Open Educational Resources: Corvallis, IL, USA, 2020; p. 163.
70. Shoukry, M.M. *An Actual Vision about the Availability of the Utilization of Water Hyacinth in Feeding Ruminants*; National Symposium on Water Hyacinth, Assiut University: Assiut, Egypt, 1992; pp. 75–92.
71. Al-Sodany, Y.M.; El-Sheikh, M.A.; Baraka, D.M.; Shaltout, K.H. Element's accumulation and nutritive value of *Phragmites australis* (Cav.) Trin, ex Steudel in Lake Burullus: A Ramsar site, Egypt. *Catrina* **2012**, *8*, 51–63.
72. Galal, T.M.; Abu Alhmad, M.F.; Al-Yasi, H.M. Nutrient sequestration potential of water primrose *Ludwigia stolinefera* (Guill. & Perr.) P.H. Raven: A strategy for restoring wetland eutrophication. *Saudi J. Biol. Sci.* **2021**, *28*, 2438–2446. [PubMed]

73. Gill, K.S.; Omokanye, A. Spring triticale varieties forage yield, nutrients composition, and suitability for beef cattle production. *J. Agric. Sci.* **2016**, *8*, 1–14. [CrossRef]
74. Heneidy, S.Z. Browsing and nutritive value of the most common range species in Matruh area, a coastal Mediterranean region, Egypt. *Ecol. Mediterr.* **2002**, *28*, 39–49. [CrossRef]

agriculture

Review

Advances in Applications of Cereal Crop Residues in Green Concrete Technology for Environmental Sustainability: A Review

Md Amir Suhail [1], Sandeep Shrivastava [2], Kunwar Paritosh [1], Nidhi Pareek [3], Andrey A. Kovalev [4,*], Dmitriy A. Kovalev [4], Yuri V. Litti [5], Vladimir Panchenko [6], Vadim Bolshev [4] and Vivekanand Vivekanand [1,*]

1 Centre for Energy and Environment, Malaviya National Institute of Technology Jaipur, Jaipur 302017, Rajasthan, India
2 Department of Civil Engineering, Malaviya National Institute of Technology Jaipur, Jaipur 302017, Rajasthan, India
3 Department of Microbiology, School of Life Sciences, Central University of Rajasthan Bandarsindri, Kishangarh, Ajmer 305801, Rajasthan, India
4 Federal State Budgetary Scientific Institution "Federal Scientific Agroengineering Center VIM", 1st Institutskiy Proezd, 5, 109428 Moscow, Russia
5 Federal Research Center "Fundamentals of Biotechnology" of the Russian Academy of Sciences, Leninsky Prospekt, 33, 2, 119071 Moscow, Russia
6 Department of Theoretical and Applied Mechanics, Russian University of Transport, 127994 Moscow, Russia
* Correspondence: kovalev_ana@mail.ru (A.A.K.); vivekanand.cee@mnit.ac.in (V.V.); Tel.: +7-926-347-79-55 (A.A.K.); +91-141-2713573 (V.V.)

Abstract: Concrete is mainly employed as a construction material. Due to the manufacturing of cement and the extent of concrete usage, numerous environmental issues and water suction have presented challenges. There is an immediate need to overcome these problematic issues by substituting natural resources with wastes and by-products of different biological processes in the production of concrete in order to make green concrete. Green concrete provides a relatively low-impact material to satisfy potential concrete demand and offers a cheaper, robust and highly reliable alternative that could fulfil future construction requirements in an environmentally safer way. The present review highlights the possible use of waste residues of agricultural origin from cereal farming in concrete as alternative materials to cement, fine aggregate and fiber reinforcement. The review also considers appropriate methods of treatment, the selection of residual resources and the blending ratios that may allow the development of next-generation green concrete with better physicochemical and mechanical properties. It also explores in-depth studies and the wider range of innovations in cereal farming residues for appropriate use in green construction for environmental sustainability. Green concrete could be an alternative material that could replace those used in conventional methods of construction and help make a further step towards environmental sustainability and a circular bioeconomy.

Keywords: green concrete; wheat straw; corn cob; rice husk; barley; crop residues

Citation: Suhail, M.A.; Shrivastava, S.; Paritosh, K.; Pareek, N.; Kovalev, A.A.; Kovalev, D.A.; Litti, Y.V.; Panchenko, V.; Bolshev, V.; Vivekanand, V. Advances in Applications of Cereal Crop Residues in Green Concrete Technology for Environmental Sustainability: A Review. *Agriculture* **2022**, *12*, 1266. https://doi.org/10.3390/agriculture12081266

Academic Editor: Maria Roulia

Received: 22 July 2022
Accepted: 18 August 2022
Published: 19 August 2022

1. Introduction

Due to the rapid rate of urbanization, the demand for construction and infrastructure development is increasing around the world. To fulfill this demand, large amounts of concrete are being employed; however, there are numerous demerits to this usage, such as higher water utilization and CO_2 emissions. Continuous urban development and negative environmental issues have led researchers to look for a better eco-friendly option of green concrete for construction purpose to reduce the environmental burden on nature due to current practices. One of the main motives is the mitigation of the undesirable environmental impact of the use of building materials, such as sand, aggregates and cement. Furthermore, the production of cement is totally dependent upon natural resources, the exhaustion of which harms flora and fauna and leads to negative environmental impacts in the long run. In addition, excessive and illegal mining, as well as processing in the extraction of these resources, have led to the mutilation of eco-systems and to air, water and soil pollution, with consequences for human health [1–3].

The conventional methods of cement production are energy-intensive processes and have the drawbacks of emitting particulate matter and greenhouse gases (GHGs), mainly CO_2. Cement manufacturing is a key contributor, which is reported to contribute around 1.8 gigatons (GT) of CO_2 per annum, accounting for roughly 5–7% of the CO_2 generated all across the world [4]. Life cycle reports indicate that approximately 0.8 tons of CO_2 is released in the manufacture of 1 ton of cement [5]. Presently, eco-friendly initiatives are being implemented all over the world to bring forward the sustainable and efficient reuse of waste in order to reduce overall energy consumption in the conventional brick industry. Recent research and development efforts have supported the concept of green concrete and taken a pragmatic approach in addressing the effective and efficient use of industrial waste, agricultural waste and municipal waste resources [6–8]. Massive quantities of industrial waste, such as residual ash, have been predominantly used globally [9–12], while the use of industrial waste in concrete making can easily be approved and developed. However, the assimilation of waste residues in concrete manufacture is still under-developed and in the investigation phase. In this regard, there has been specific interest in utilizing waste residues from the cereal farming industries.

India is an agricultural country and has larger farms devoted to agriculture where farming and cultivation are the prime sources of occupation in rural areas; however, no effort has been made to dispose of the bio-degradable waste generated by agricultural processes [13]. The major proportion of such waste is either fed to livestock as food or is dried and burnt. These harmful processes create environmental issues, such as smog, air pollution and contamination, which is evident from the pollution levels during winter in the northern states of India, including Haryana, Punjab, Delhi and Rajasthan. Therefore, waste utilization is an issue of high priority that needs to be addressed in the wake of alarming environmental conditions. Several studies have been undertaken over the past few years regarding the utilization of agricultural waste residues in concrete [2,3,6–8]. Cereal-farming waste residues may represent the best alternatives to cement, sand, aggregate, fiber reinforcement and supplementary cementitious material (SCM) as binding elements. Nowadays, there is an emerging trend of utilization of farming waste residues as substitutes in concrete making, such as the straw of wheat, rice, corn and barley [13,14]. Agriculture residues from cultivation have widely been used by researchers as partial cement substitutes in concrete. Vegetation growing in soil receives certain amounts of inorganic compounds, including silica (SiO_2) and minerals, and silica, specifically, is obtained at high levels per annum in full-grown vegetation [15]. Farming residues can be a good potential source of cement replacement materials with good binding properties as well as pozzolanic reactivity [15]. Another benefit is that they can also be used for fiber reinforcement, which may strengthen the eventual concrete structures. The reasons for the employment of natural fibers in cement making are due to various beneficial factors, such as the ones mentioned below [16]:

(i) Low cost and ease of availability;
(ii) Does not require high degrees of industrialization;
(iii) Eco-friendly waste utilization;
(iv) Few natural fibers have greater strength than synthetic fibers.

Some cereal waste resources have already been utilized as partial sand and aggregate substitutes in building materials to help the environment as well as minimize reliance on traditional aggregates (i.e., natural sand, gravel and granite) [17–19].

Whether a substance such as cereal waste is pozzolanic or not depends on the reactive quality of silicon and alumina, as well as the percentage of amorphous structure. Other common parameters, such as fineness modulus, maximum grain size and reaction between pozzolana and calcium hydroxide, are considered [20–22]. Danchenko et al. (2020)

has observed that a blend of buckwheat and oat husk forms a uniform filled-composite structure with minimum internal stresses [23]. Mixtures of sorghum husk ash and recycled concrete aggregate were used in a concrete mixture which showed better strength and other mechanical properties. A similar kind of research was performed with kenaf fiber and sorghum husk ash in concrete by Ogunbode et al. (2021); the resultant material was found to be technically and environmentally sustainable [24,25].

Bheel et al. [26,27] utilized millet ash with other materials in concrete, and it was noted that the compressive and split tensile strength were significantly enhanced, which shows the importance of blending the agricultural residual materials with conventional concrete to produce green concrete.

There has been research on the compressive strength of different Portland cement combinations, which may constitute a step towards the future development of an environmentally friendly concrete that integrates and optimizes cereal-farming residues [28]. This review brings forward the holistic approach of utilizing cereal waste by effective and efficient methods for the production of green concrete. The research and development in this domain have to be examined with regard to the common behavior of different waste resources, including their advantages and disadvantages, in concrete making, as well as their characterization, such as physical and chemical composition.

2. Manufacturing of Green Concrete

The production procedure for green concrete varies and mainly depends on the ingredients used and the projected applications. The required properties of concrete and cement quality must be described in the first phase. The properties of the raw materials must be experimentally analyzed, characterized and calculated [29]. There are five steps to be followed, as listed below [29,30]:

I. Estimation of optimal relevant properties of chosen concrete ingredients;
II. Obtainment of a suitable water–cement ratio on the basis of the strength of the concrete and cement contents;
III. Determination of the close interconnections within the particle grain distribution;
IV. Determination of the packing density appropriate to the decreased water content;
V. Production of fresh concrete and estimation of its properties related to compactness and predicted compressive strength.

The methods of optimization that can be used in green concrete production require the development of atom packaging by granular optimization of all concrete ingredients [30,31]. After microanalysis data have been obtained, mathematical modeling and calcium silicate hydrate (CSH) gel evaluation are to be conducted [32–34]; other forms of optimization require the micro-optimization of ranking fine gradient arrangements [35]. The allocation of grain size for all granular elements by the mixture of grain size templates should be optimized. The packing density of grain structures is closely intertwined with grain size distribution. Therefore, it is necessary to choose the percentages of materials available according to an optimal distribution for grain size compaction [30]. The benefit of optimization in green concrete lies in the relation of air pressure to maximum strength and optimizing the properties of end-product materials. It was suggested to separately grind each supplementary cementitious material (SCM) to obtain higher compressive strengths for ternary mixed cement concrete [36]. Few properties, such as physical, thermochemical, mechanical, workmanship and fire resistance properties, of concrete are altered in the manufacturing of green concrete, as is shown in Figure 1. These properties define the capability and life of the concrete in all aspects. Mechanical properties of materials include strength (tensile and compressive), stiffness, elasticity, plasticity, hardness, brittleness and resilience, which describes the behavior under the action of external loads. Thermal and fire resistance are also considered as important parameters in the manufacture of green concrete of high quality.

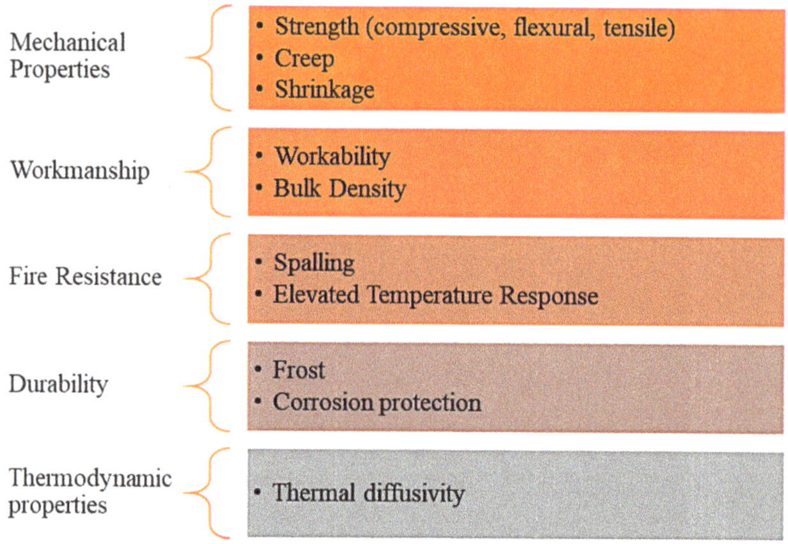

Figure 1. Properties of concrete.

2.1. Cereal-Farming Residues

Cereal farming is highly prevalent in India, with most of the cereal crops harvested being used as sources of food grains for Indian people. States such as Haryana, Punjab, Uttar Pradesh, Bihar and West Bengal are India's largest producers of cereal-based agricultural products. However, after the crops are reaped and gathered in the fields and they are effectively and efficiently segregated, there is a lot of waste left over, such as straw, husks, stalks and leaves, depending on the cereals in question. These residues also reduce the fertility and biodiversity of agricultural soil. Therefore, efforts to promote efficient collection and transportation must be prioritized to increase soil fertility and the amounts of valuable products that can be obtained from it [30]. Scientists have recently started to use the aforementioned waste materials as partial substitutes for traditional concrete products and have come up with interesting results [3]. The following section reviews the major crop residues used in concrete production and their characterization.

2.1.1. Waste from Wheat Cereal Harvesting

Wheat is one of the main cereal sources of food production and consumption in India. According to the Ministry of Agriculture (Government of India), wheat production increased to a record 101.20 million tons for the 2018–2019 (July–June) crop year and by 1.3 percent year-on-year in the third annual crop output survey. Wheat straw (WS) residue is one of the primary by-products of wheat production and is usually burnt by farmers in open areas, with a major impact on environmental pollution and human health [37]. Wheat straw ash (WSA) is obtained when it is burnt at a high temperature. Ash has an excess silica content and higher fineness compared to cement [17]. Therefore, wheat straw ash may be a source of alternative material that can be used in concrete production processes [38]. However, depending on the process of ignition, WSA has different chemical properties. The burning temperature of WS was found to be in the range of 570–670 °C for 5 h (Figure 2); hence, the color of white and grey ash denoted the absolute burning of ash [15].

Figure 2. Wheat straw (**a**). Wheat straw ash (WSA) (**b**).

Further, it has been observed that if WS is burnt at 900 °C for 6 h it gives a black-colored ash [15,18,38,39]. It has also been reported that cement hydration was accelerated when WSA was pretreated in a comparison with untreated WSA, which retards cement hydration. Results also demonstrated a significant (32%) surge in compressive strength when WSA was treated and compared with normal concrete mortar made with untreated WSA [39]. The positive outcomes for WSA as an accompanying concrete material in mortar were reflected in a superior compressive strength of around 25% [39]. Moreover, when 8% of WSA was mixed in cement, the achieved compressive strength was quite satisfactory as compared to concrete made without WSA after the 180th day [40]. The primary reason was a slow chemical reaction between pozzolana and calcium hydroxide. When WSA was mixed in cement up to 16%, the flexural strength of the concrete was enhanced on the 28th day [40].

The most important criterion of concrete quality is its durability, and many researchers have been conducting studies on the durability of concrete material combined with WSA as a partial cement substitute. It was found that when WSA was mixed in concrete with magnesium sulfate solution, the concrete showed up to 8% higher performance [40]. A further positive result for the freezing and thawing resistance of WSA mixed concrete as compared to normal concrete was also shown [41]. The WSA replacement level was increased from 5–15%, which helped to increase the freezing and thawing resistance of the concrete. Furthermore, the resistance of WSA mixed concrete to deterioration by alkali and silica was a bit higher compared to equivalent normal concrete made without WSA. If the WSA content is increased to 15%, it has a higher resistance to the reactions that occur between alkali and silica [42].

Most important is the influence of WSA on the oxidation of alkali and silica in concrete mixed with a lower water–cement ratio [42]. In due course, some researchers have also explored the prospect of the efficient usage of WSA as a partial substitute for sand in concrete. Promising results were achieved when WSA was partly mixed with cement up to 10.9%, due to the higher fineness of WSA as compared to cement, which has reduced workability, though water demand was found to be increased [17]. Along with this, another important factor, namely, setting time of new concrete, was enhanced by up to 92% given the presence of WSA as 10.9% of a sand substitute, the effect on setting time being one of the major reasons for the particular binding characteristics of WSA [17]. In terms of strength features or properties, up to 10.9% mixing of WSA with fine aggregate showed increments in tensile, flexural and compressive strengths of concrete of up to 67, 71 and 87%, respectively [17]. Similarly, as reported by Binici et al. [18], with partial replacement with WSA up to 6%, WSA concrete's compressive strength was enhanced in comparison with normal concrete after the 28th day, and it was observed that the compressive strength on the 7th day was similar [17]. It was found that mixing of WSA as a partial fine aggregate replacement (6%) resulted in the concrete with the best durability. According to reports in the literature, sulphate resistance is the main area of concern in concrete production and in which WSA has shown promising results, with water penetration and abrasion resistance having been found to be increased in concrete made with WSA due to denser pore formation [18]. Concrete exposed to a thermal cycling process shows a decrease in compressive strength and it was found that this decrease was less sharp for WSA concrete in comparison with normal concrete. It has been well reflected that WSA concrete has better resistance when exposed to thermal cycling, even when WSA as a fine aggregate partial substitute was used up to 15% [43]. Moreover, the cracks in WSA concrete were observed to develop at a much later stage due to the presence of fibers in WSA, which helps in binding the other ingredients of concrete as well. Electrical resistivity further enhances the resistance of WSA concrete at higher temperatures [42]. With regard to the use of WS for fiber reinforcement in concrete, it was concluded that the tensile strength of wheat straw fiber (WSF) was around 40 megapascals (MPa) (Table 1) [44]. This was ascribed to the irregular surface attributes of WSF promoting bond formation in the cement and fiber matrix, which was composed of a mix of low-strength fibers, and the breakdown of the concrete was considered to involve the pulling out of fibers rather than their rupture [44].

Table 1. Natural fiber properties of WS [44–46].

Physical Properties	Bulk Density (kg/m^3)	Moisture Content (%)	Tensile Strength (MPa)	Modulus of Elasticity (GPa)
Wheat straw	160.75	6.4	40	3.1–3.7

MPa: Megapascal, GPa: Gigapascal.

Effect of the mixing ratio of WSA with Cement, Sand and Fiber

The mixing of WSA with cement and sand may alter the chemical and physical as well as the mechanical composition of cement because WSA has its own oxide composition and physical, chemical and natural fiber properties. Varying the percentage of WSA mixed in cement affects the final properties of concrete to different extents. Hence, the study of the different ratios is essential to establish the appropriate proportions of WSA and adequate cement mixing is required to achieve desired properties in concrete. Previous studies have revealed the blending ratios for the desirable properties of concrete, which are mentioned below.

Effect of partial replacement of WSA with Cement

When cement was replaced with 8, 16 and 24% WSA [40], the compressive strength was reduced on 28th day; the equivalent strength at 180 days confirmed the superiority of the replacement level of 8%. This provides higher flexural strength and also enhanced the sulfate tolerance along with the compressive strength. When cement was replaced with 5, 10 and 15% WSA [42], the reliability of alkali–silica reactions was improved, and when cement was replaced with 20% pretreated WSA [39], the compressive strength was raised.

Effect of partial replacement of WSA with Sand

When sand was replaced with 3.6, 7.3 and 11% WSA [17], the flow properties were reduced, initial setting time improved and the compressive, flexural and splitting tensile strengths increased. When sand was replaced with 5, 10 and 15% WSA [43], resistance to thermal cycling was strengthened. When sand was replaced with 2, 4 and 6% WSA [18], compressive strength, sulfate resistance and abrasion resistance were improved and the rate of water penetration was reduced.

Effect of partial replacement of WSF addition in Cement

When wheat straw fiber was added to cement (0.19%) [44], a nominal rise in fracture strength was observed.

The oxide composition and physical and chemical properties of wheat straw materials are detailed in Tables 1–4.

Table 2. Oxide composition of WSA [15–18,40–42].

Oxide Composition	SiO_2	CaO	Al_2O_3	Fe_2O_3	Na_2O	MgO	K_2O	LOI
WSA (%)	4.9–7.8	9.4–24.4	0.1–4.6	0.1–1.3	0.1–5.4	0.6–4.6	0.7–24.7	1.1–29

Table 3. Physical properties of WSA [17,18,39,41,42].

Physical Properties	Specific Gravity	Blaine's Specific Surface Area (m^2/Kg)	BET Specific Surface Area (m^2/g)
WSA	1.97–2.89	430–552	8.3–168

Table 4. Natural fiber chemical and morphological characteristics of WS [44,47–55].

Chemical Properties	Lignin (%)	Cellulose (%)	Hemicellulose (%)	Length (mm)	Width (%)
Wheat straw	6.5–22	30.1–49.1	22.2–34	40	-

2.1.2. Waste from Rice Cereal Harvesting

As per the Ministry of Agriculture, India, rice production was projected at 115.6 million tons in 2018–2019, which was higher than in 2017–2018. There is a lot of waste generated in rice production, such as the straw and husks left after harvesting. Rice husks (RH)—a cereal waste obtained from the crushing of rice—constitute another good substitute material for cement manufacturing [56]. They are considered one of the major globally accessible waste residues. RH from rice fields is one of the best alternative materials that can be used as a partial replacement for ash to enhance the properties of concrete in terms of strength, workability and durability and can also reduce the quantity of cement by weight.

RH is an exterior paddy grain cover and accounts for 20–30% of the weight of paddy rice collected [56]. RH is usually used as a fuel in boilers due to heat generation during combustion (13–16 MJ/kg) [57]. Rice husk ash (RHA) is obtained after the burning of RH, which accounts for 18–25% of RH's preliminary weight. RHA presents a land-filling problem, as it is not utilized. Generated by rice mills, it is a source of environmental contamination and there has been debate over the issue of dumping [58]. If this accessible waste is not treated or handled properly, it may contribute to pollution. The color of RHA depends on the raw material resources and can vary from black to white gray depending on the incineration process, the duration of burning and burning temperature. RH is exposed to a self-controlled high temperature between 600–800 °C for 1 h in a closed environment in a furnace/incinerator. The ash produced is cooled after the firing process, either slowly or quickly. The quick cooling process is carried out by continuously spreading the ash in dishes at an ambient lab temperature of 21°C after achieving the appropriate temperature of 800 °C for 1 h. In this way, RHA is obtained when the furnace/incinerator is cooled down at a lower rate [59–61].

In RHA, silica content is a major factor in producing C-S-H gel through reaction with calcium hydroxide and lime. C-S-H is mostly essential for the strength and other micro-structural properties of concrete [62]. RHA has a similar character to WSA in that it has excess silica content. It consists of 80–95% silica, which is an important factor in concrete; however, when compared to conventional concrete, ordinary Portland cement only contains 21% silica [63–65]. Due to the higher silica content of RHA, it can be used to enhance the transition zone and surface area between the cement paste and the microscopic framework [39]. Thermo-chemicals, such as dilute acid, increase the rate of pozzolanic reactions by increasing the surface area and volume of amorphous silica and reducing RHA carbon content [66]. Researchers are committed to finding economic routes to extract nano-silica from RHA due to an increasingly extensive market for amorphous silica.

Alkali extraction is a cheap method that can be used to prepare high-purity silica sol or nano-silica powder for various mullite ceramics [67–69]. When RHA is treated well, a C-S-H gel is formed which can fill the gaps between cracks in concrete and protect it from leaching degradation and corrosion. The silica present in RHA reacts with $Ca(OH)_2$ to form a resistive layer that acts as a protective layer for the materials when subjected to acidic conditions. When RHA was used as a partial replacement concrete-making material, it was found that it enhanced the tensile strength, compressive strength and modulus of elasticity of the mixed concrete as compared to normal concrete [70]. RHA consists of minerals that are used as SCMs in mortar and concrete making. The water absorption and porosity of the two samples decreased with an improvement of up to 15% in the RHA content and then began to increase. For the normal concrete, water absorption was 6.4, 4.8 and 4.3% at the 7th, 28th and 91st day and declined to 5.5, 3.7 and 3.5%, respectively, after mixing with 15% RHA. On the other hand, the porosity analysis revealed a decrease of 26, 17 and 14% for concrete mixed with 15% RHA as compared to the control combination at the 7th, 28th and 91st day. Such changes are due to the creation of additional C-S-H gel by the pozzolanic reaction between calcium hydroxide and silica found in RHA, with decreased pores and improved concrete capacity [71,72]. When RHA was at a partial replacement level of 15%, water demand was increased because of the lower fineness modulus and higher surface area of RHA particles. Nonetheless, decreased concrete workability likely resulted in pores and voids being created, which in effect improved the concrete's susceptibility to water absorption [71]. In fact, there is a lack of $Ca(OH)_2$ to react with the higher amount of SiO_2 present at such high doses. Consequently, a significant amount of silica remains unreacted, which, through growing porosity, breaks down the consistency of the concrete's micro-structure. The maximum water permeability of concrete mixed with 15% RHA was 72, 64.7 and 87.5% lower than that of normal concrete at the 7th, 28th and 91st day. Concrete water permeability is a feature of porosity and mixing of RHA up to 15% as partial cement replacement decreased the voids in concrete due to pozzolanic action and micro-filling, thereby decreasing water permeability [72].

The amount of nano-silica present in RHA changes the character of concrete made with RHA and is responsible for the development of the rough surface texture as well as the layered polymerized structure which enhances the cohesion between cement mortar and aggregates. Using RHA to make green concrete would help to reduce global carbon emissions from the manufacturing of concrete. RHA-modified concrete is currently in the development stage at laboratory scale. RHA-mixed strong castable concrete has led to economic and ecological advantages in the refractory industries [66]. However, the manufacture of green concrete needs to be promoted to the upper level and be adopted in mass manufacturing at commercial scale [29].

Effect of partial replacement of RHA in Cement

When partial replacement of cement with RHA is performed, the chemical and physical properties of the cement will be changed, different proportions of RHA influencing to different degrees the final properties of the concrete. Previous reports in the literature have revealed the outcomes of different blending percentages on the properties of concrete, as described below. When cement replacement at 25% was performed [73]:

- ➢ The lowest water absorption rates on the 7th day were about 4.8% and on the 28th day of curing about 3%;
    - There was a change of 6.9% in compressive strength at the 7th day and 6.8% at the 28th day;
    - The most improved tensile strength was achieved with up to 6.8% RHA, then it tended to fall;
    - This contributed to a dramatic improvement in the permeability of the mixed concrete relative to normal concrete;
    - Water permeability was reduced to 26%;
    - This led to a reduction in chloride permeation of 78%.

➢ When cement replacement at 15% was performed [74–76]:
- The compressive strength rose by 15%; at the 7th, 28th, 56th and 91st day, by 9, 12, 13 and 16% above that of the control combination, respectively;
- The average elasticity element was 14% higher than the 0% combination;
- The 28th-day flexural strength of the maximum flexural strength was 21% higher than that of the 0% combination.

The oxide composition and physical and chemical properties of RHA have been summarized in Tables 5–7.

Table 5. Oxide composition of RHA [62–65,77].

Oxide Compo-sition	SiO_2	CaO	Al_2O_3	Fe_2O_3	Na_2O	MgO	K_2O	LOI
RHA (%)	85–95	0.31–1.5	0.05–0.3	0.06–0.2	0.06–1.36	0.35–0.6	1.4–2.3	0.8–1.4

Table 6. Physical properties of RHA [3,78–80].

Physical Properties	Specific gravity	Blaine's Specific Surface Area (m^2/Kg)	BET Specific Surface Area (m^2/g)
RHA	2.14	350–376.8	117.6

Table 7. Natural fiber chemical and morphological characteristics of RHA [41,81–85].

Chemical Properties	Lignin (%)	Cellulose (%)	Hemicellulose (%)	Length (mm)	Width (%)
Rice husk	6.8–20.4	33.4–50.1	3.8–21.7	-	-

2.1.3. Waste from Corn Cereal Harvesting

Corn is produced as a major cereal product in India, along with rice and wheat [86]. Corn cob is a cereal waste that is obtained from corn or maize. It is another substitute for utilization in concrete making in place of cement, sand and coarse aggregate to reduce the quantity of raw materials required for concrete production. Corn cob ash (CCA) is obtained by burning maize cob waste at a temperature of 600 °C for 3 h and 650 °C for a period of 8 h [18,87]. It is pozzolanic in nature and has a significant quantity of loss on ignition (LOI) and reactive amorphous silica content [18]. According to previous studies, the percentage of silica contained in CCA is around 37–66% [18,86] (Table 8). When CCA was mixed with cement up to a level of 25%, as the percentage of CCA increased, the LOI of the mixed cement was enhanced due to an increase in the organic percentage present in CCA [88]. This has an undesirable effect on cement properties because cement works as a binding material in concrete, providing strength [88]. Due to this phenomenon, the reliability of the mixed cement will be compromised, while the soundness as well as the setting times of the mixed cement will be enhanced [88]. The increase in the setting time and soundness of cement was attributed to the CCA, which decreased the surface area of the cement. Since CCA is very light-weight in comparison to cement, it delayed the hydration process [88]. Similarly, the workability decreased as the CCA percentage increased due to this. Water demand also increased.

Table 8. Oxide composition of CCA [18,87,88].

Oxide Composition	SiO_2	CaO	Al_2O_3	Fe_2O_3	Na_2O	MgO	K_2O	LOI
CCA (%)	37–66.4	11.6–13	2.4–7.5	1.2–4.4	0.3–0.4	2.1–7.4	4.9–15	22.5

Regarding compressive strength, the production of mixed concrete with CCA worked equally in comparison to normal cementitious materials. It was found that the early strength of CCA concrete was lower, but after some time it exhibited higher strength due to the effective pozzolanic reactions between SiO_2 and $Ca(OH)_2$ in the CCA [87]. When CCA and WSA were used up to 6% as partial fine aggregate replacements, it was found that the compressive strength of CCA mixed concrete was enhanced by up to 40% in comparison with WSA mixed concrete at the age of 365 days [33]. If we look CCA used as a partial fine aggregate replacement, then the overall development factors and performance of CCA concrete show it to be superior to WSA concrete. Similarly, it was also observed that the durability performance, e.g., abrasion resistance, sulphate resistance, and water penetration resistance, of CCA concrete was enhanced in comparison to WSA concrete [33]. It was found that when CCA was used as a coarse aggregate replacement, better light-weight concrete resulted as compared with other light-weight concretes made with long-drawn-out clay aggregates [89]. Finally, it was observed that CCA concrete has a compressive strength (0.12 MPa) that is lower in comparison with that of long-drawn-out clay concrete (1.36 MPa). The overall thermal performance and density were consistent, suggesting that CCA concrete might be used in light-weight structures or single-storey buildings.

Effect of Mixing of CCA with Cement, Sand and Aggregate

When partial replacement of CCA with cement, sand, and aggregate is performed, then the chemical composition and physical properties of cement change because CCA and cement have different chemical and physical properties. It is therefore important to find different ratios to establish the idol composition of CCA and cement mixing necessary to achieve appropriate concrete properties. These properties were shown above, in Tables 8–10.

Table 9. Physical properties of CCA [18,88].

Physical Properties	Specific Gravity	Blaine's Specific Surface Area (m^2/Kg)	BET Specific Surface Area (m^2/g)
CCA	2.97	270–385	-

Table 10. Natural fiber chemical and morphological characteristics of Cornstalk [85,90–95].

Chemical Properties	Lignin (%)	Cellulose (%)	Hemicellulose (%)	Length (mm)	Width (%)
Cornstalk	3.8–21.7	37.5–49.3	22.5–30	0.7–0.9	0.023–0.029

Effect of partial replacement of Cement with CCA

➢ Cement replaced with 2 to 25% CCA [87,88]:
- Reduced workability;
- Decreased early strength;
- Improved strength gain;
- Enhanced initial and final setting time.

Effect of partial replacement of Sand with CCA

➢ Sand replaced with 2, 4 and 6% CCA [18,88]:
- Enhanced compressive strength;
- Improved sulfate resistance;
- Higher aberration resistance;
- Reduced rate of water penetration.

Effect of replacement of coarse aggregate with 100% CCA

➢ Coarse aggregate replaced with 100% CCA [89]:
- Similar thermal properties to lightweight concrete.

2.1.4. Waste from Barley Cereal Harvesting

Barley is also produced in India as a key cereal crop, along with rice, corn and wheat. Consequently, straw of barley is also accumulated excessively and burnt due to a lack of efficient utilization methods [95]. Similar to wheat straw ash, barley straw could also be utilized for ash generation and barley straw ash (BSA) may be used as a pozzolanic substance for concrete production, though there has been insufficient investigative work carried out on using BSA as SCM. In general, BSA has a high amount of potassium and a slightly lower amount of silica (21.2%) in comparison to WSA, as shown in Table 11 [96]. Furthermore, BSA structure and crystallography can rely on pozzolanic reactivity. Similar molecules undergo different reactions during the solution process in interaction with saturated $Ca(OH)_2$, impacting the speed and length of the reaction. Owing to the occurrence of KCl, the pozzolanic action of BSA may be lesser in comparison to traditional pozzolans, viz., fly ash, and it led to low divergence in compressive strength at the 7th day and the 28th day [96]. Barley straw with a tensile strength of approximately 115 MPa and a modulus of elasticity of 9.92 GPa was used as an alternative to wood shavings to produce sand concrete of lower weight [97]. The findings of the study described that the addition of barley straw fiber led to a 35% increase in thermal diffusivity, along with an improvement in the compressive strength and toughness of the produced concrete. In this study, since the pozzolanic substance had the same particle size and fineness, there was improved pozzolanic reactivity throughout the BSA and fiber addition as compared to the other samples [96].

Table 11. Oxide composition of BSA [96].

Oxide Composition	SiO_2	CaO	Al_2O_3	Fe_2O_3	Na_2O	MgO	K_2O	LOI
BSA (%)	21.2	10.0	2.8	3.5	4.1	-	38.0	-

Effect of Fiber addition in Concrete

When partial replacement of fiber addition in cement concrete is performed, the chemical composition and physical properties of cement would be changed because BSA and cement have different chemical and physical properties (Tables 11–13).

Table 12. Natural fiber chemical and morphological characteristics of WSA [97,98].

Chemical Properties	Lignin (%)	Cellulose (%)	Hemicellulose (%)	Length (mm)	Width (%)
Barley straw	13.72–15.8	37.6	34.9	35	-

Table 13. Natural fiber properties of barley straw [97,99].

Physical Properties	Bulk Density (kg/m^3)	Moisture Content (%)	Tensile Strength (MPa)	Modulus of Elasticity (GPa)
Barley straw	30–130	-	115	9.92

Fiber addition in concrete [97] leads to:

- Lower shrinkage;
- Increased porosity;
- Reduced thermal diffusivity;
- Improved ductility;
- Higher compressive strength.

3. Inferences from the Literature Related to the Usage of Agricultural Waste Residue Ash in Concrete

From the data reported in the literature, it can be inferred that natural agricultural wastes obtained from cereals, such as wheat, rice, barley and corn, can serve as alternative materials in the production of green concrete. The addition of residual ash to concrete affects the mechanical and physicochemical properties of the end product. The compressive strength of green concrete is reduced at first but becomes equivalent to conventional concrete after 180 days due to the stability of the residue ash. Nevertheless, with appropriate pretreatment of residues (mainly thermal treatment) the durability of concrete may be improved. Similarly, the higher content of silica in waste residue ash promotes pozzolanic reactivity, which leads to the higher strength of the final concrete. The foremost significance of the utilization of waste residues lies in the mitigation of CO_2, CH_4 and NO_x emissions, reduced dependence on fossil fuels and increase in the recycling of materials, which indirectly reduces the degradation of the environment. Additionally, it is reported that the water absorption of green concrete (made with waste residue ash) is lower after the 7th, 15th and 28th days of curing as compared to conventional concrete. Moreover, the early strength of green concrete is reduced, though it increases later on due to effective pozzolanic reactions between SiO_2 and $Ca(OH)_2$.

4. Conclusions

Cereal waste handling and environmental degradation are major concerns for all nations, especially agriculturally rich and developing countries, such as India. Cereal waste residues have potential applications as partial substitutes for cement, sand, coarse aggregate and fiber reinforcement as they have significant mechanical properties. Residues from agriculture can be used to make green concrete, increasing the strength and properties of the final concrete depending on the type and amount of residue used in its production. In addition, if appropriate pretreatment of waste materials is carried out, the waste may be incorporated into concrete to enhance its mechanical and durability properties. In addition, optimization of the packing density of grain, the grain size distribution of blends, and water–cement ratios may further improve the performance of cereal residue-based concrete. Green concrete consisting of one-quarter customary cement may show adequate robustness with respect to composition and conditions during corrosive exposures. Consequently, a more productive design for green concrete made with cereal residues may be a promising area to explore further in the future that will result in the recycling of waste and the reduction of undesirable effects on the environment. Eventually, it will contribute to innovation in the sustainable building sector and the construction industry and at the same time promote a cleaner environment.

Green concrete with desirable properties and high sustainability may increase the possibilities and the feasibility of using cereal crop residues in future applications. The limitations include starting up, the economics of the process and the sustainable supply of agricultural residues due to seasonal variations, geographical factors and climatic conditions.

Author Contributions: Conceptualization, M.A.S. and V.V.; methodology, D.A.K.; software, K.P.; validation, V.V. and Y.V.L.; formal analysis, S.S. and N.P.; investigation, M.A.S.; resources, V.P. and D.A.K.; data curation, V.P. and V.B.; writing—original draft preparation, M.A.S.; writing—review and editing, S.S., K.P., N.P., V.V., A.A.K. and Y.V.L.; visualization, K.P. and A.A.K.; supervision, Y.V.L.; project administration, V.V. and A.A.K.; funding acquisition, V.V., A.A.K., Y.V.L., V.B. and V.P. All authors have read and agreed to the published version of the manuscript.

Funding: The authors would like to thank the Department of Science and Technology, Ministry of Science and Technology, Government of India, for an Indo-Russian grant i.e. Indo Russia Project DST/INT/RUS/RSF/P-62/2021 and for supporting this work financially (V.V.). The research was supported by the Russian Science Foundation (grant no. 22-49-02002, https://rscf.ru/en/project/22-49-02002/ (A.A.K., V.P. and Y.V.L.)). D.A.K. and V.B. were supported by the Ministry of Science and Higher Education of the Russian Federation.

Institutional Review Board Statement: Not applicable.

Informed Consent Statement: Not applicable.

Conflicts of Interest: The authors declare no conflict of interest.

References

1. Sakthieswaran, N.; Sophia, M. Prosopis juliflora fiberreinforced green building plaster materials—An eco-friendly weed control technique by effective utilization. *Environ. Technol. Innov.* **2020**, *20*, 101158.
2. Shafigh, P.; Mahmud, H.B.; Jumaat, M.Z.; Zargar, M. Agricultural wastes as aggregate in concrete mixtures—A review. *Constr. Build. Mater.* **2014**, *53*, 110–117. [CrossRef]
3. Aprianti, E.; Shafigh, P.; Bahri, S.; Farahani, J.N. Supplementary cementitious materials origin from agricultural wastes—A review. *Constr. Build. Mater.* **2015**, *74*, 176–187. [CrossRef]
4. Gao, T.; Shen, L.; Shen, M.; Chen, F.; Liu, L.; Gao, L. Analysis on differences of carbon dioxide emission from cement production and their major determinants. *J. Clean. Prod.* **2015**, *103*, 160–170. [CrossRef]
5. Flower, D.J.; Sanjayan, J.G. Green-house gas emissions due to concrete manufacture. *Int. J. Life Cycle Assess.* **2007**, *12*, 282. [CrossRef]
6. Luhar, S.; Cheng, T.W.; Luhar, I. Incorporation of natural waste from agricultural and aquacultural farming as supplementary materials with green concrete: A review. *Compos. Part B Eng.* **2019**, *175*, 107076. [CrossRef]
7. He, J.; Kawasaki, S.; Achal, V. The utilization of agricultural waste as agro-cement in concrete: A review. *Sustainability* **2020**, *12*, 6971. [CrossRef]
8. Chinnu, S.N.; Minnu, S.N.; Bahurudeen, A.; Senthilkumar, R. Recycling of industrial and agricultural wastes as alternative coarse aggregates: A step towards cleaner production of concrete. *Constr. Build. Mater.* **2021**, *287*, 123056. [CrossRef]
9. Zhang, B.; Poon, C.S. Use of furnace bottom ash for producing lightweight aggregate concrete with thermal insulation properties. *J. Clean. Prod.* **2015**, *99*, 94–100. [CrossRef]
10. Zhao, H.; Sun, W.; Wu, X.; Gao, B. The properties of the self-compacting concrete with fly ash and ground granulated blast furnace slag mineral admixtures. *J. Clean. Prod.* **2015**, *95*, 66–74. [CrossRef]
11. Thiruvenkitam, M.; Pandian, S.; Santra, M.; Subramanian, D. Use of waste foundry sand as a partial replacement to produce green concrete: Mechanical properties, durability attributes and its economical assessment. *Environ. Technol. Innov.* **2020**, *19*, 101022. [CrossRef]
12. Wu, Z.; Jiang, Y.; Guo, W.; Jin, J.; Wu, M.; Shen, D.; Long, Y. The long-term performance of concrete amended with municipal sewage sludge incineration ash. *Environ. Technol. Innov.* **2021**, *23*, 101574. [CrossRef]
13. Karade, S.R. Cement-bonded composites from lignocellulosic wastes. *Constr. Build. Mater.* **2010**, *24*, 1323–1330. [CrossRef]
14. Pappu, A.; Saxena, M.; Asolekar, S.R. Solid wastes generation in India and their recycling potential in building materials. *Build. Environ.* **2007**, *42*, 2311–2320. [CrossRef]
15. Biricik, H.; Aköz, F.; Lhan Berktay, I.; Tulgar, A.N. Study of pozzolanic properties of wheat straw ash. *Cem. Concr. Res.* **1999**, *29*, 637–643. [CrossRef]
16. Pacheco-Torgal, F.; Jalali, S. Cementitious building materials reinforced with vegetable fibres: A review. *Constr. Build. Mater.* **2011**, *25*, 575–581. [CrossRef]
17. Al-Akhras, N.M.; Abu-Alfoul, B.A. Effect of wheat straw ash on mechanical properties of autoclaved mortar. *Cem. Concr. Res.* **2002**, *32*, 859–863. [CrossRef]
18. Binici, H.; Yucegok, F.; Aksogan, O.; Kaplan, H. Effect of corncob, wheat straw, and plane leaf ashes as mineral admixtures on concrete durability. *J. Mater. Civ. Eng.* **2008**, *20*, 478–483. [CrossRef]
19. Al-Akhras, N.M.; Abdulwahid, M.Y. Utilisation of olive waste ash in mortar mixes. *Struct. Concr.* **2010**, *11*, 221–228. [CrossRef]
20. Malhotra, S.K.; Dave, N.G. Investigations into the effect of addition of flyash and burnt clay pozzolana on certain engineering properties of cement composites. *Cem. Concr. Compos.* **1999**, *21*, 285–291. [CrossRef]
21. Larbi, J.A.; Fraay, A.L.A.; Bijen, J.M.J.M. The chemistry of the pore fluid of silica fume-blended cement systems. *Cem. Concr. Res.* **1990**, *20*, 506–516. [CrossRef]
22. Taylor, H.F.W. *Cement Chemistry*; Tomas Helford: London, UK, 1997; p. 480.
23. Danchenko, Y.; Kariev, A.; Lebedev, V.; Barabash, E.; Obizhenko, T. Physic-mechanical properties of composites based on secondary polypropylene and dispersed of plant waste. *Mater. Sci. Forum* **2020**, *1006*, 227–232. [CrossRef]
24. Tijani, M.A.; Ajagbe, W.O.; Agbede, O.A. Combined reusing of sorghum husk ash and recycled concrete aggregate for sustainable pervious concrete production. *J. Clean. Prod.* **2022**, *343*, 131015. [CrossRef]
25. Ogunbode, E.B.; Dodo, Y.A.; Moveh, S. Mechanical and Fresh Properties of Sustainable Kenaf Fibrous Concrete Incorporating Sorghum Husk Ash. *Int. J. Sustain. Constr. Eng. Technol.* **2021**, *12*, 97–107.
26. Bheel, N.; Ali, M.O.A.; Kirgiz, M.S.; de Sousa Galdino, A.G.; Kumar, A. Fresh and mechanical properties of concrete made of binary substitution of millet husk ash and wheat straw ash for cement and fine aggregate. *J. Mater. Res. Technol.* **2021**, *13*, 872–893. [CrossRef]

27. Bheel, N.; Ali, M.O.A.; Khahro, S.H.; Keerio, M.A. Experimental study on fresh, mechanical properties and embodied carbon of concrete blended with sugarcane bagasse ash, metakaolin, and millet husk ash as ternary cementitious material. *Environ. Sci. Pollut. Res.* **2022**, *29*, 5224–5239. [CrossRef]
28. Couvidat, J.; Diliberto, C.; Meux, E.; Cotelle, S.; Bojic, C.; Izoret, L.; Lecomte, A. Greening effect of slag cement-based concrete: Environmental and ecotoxicological impact. *Environ. Technol. Innov.* **2021**, *22*, 101467. [CrossRef]
29. Müller, H.S.; Breiner, R.; Moffatt, J.S.; Haist, M. Design and properties of sustainable concrete. *Procedia Eng.* **2014**, *95*, 290–304. [CrossRef]
30. Kapoor, R.; Ghosh, P.; Kumar, M.; Sengupta, S.; Gupta, A.; Kumar, S.S.; Vijay, V.; Kumar, V.; Vijay, V.K.; Pant, D. Valorization of agricultural waste for biogas based circular economy in India: A research outlook. *Bioresour. Technol.* **2020**, *304*, 123036. [CrossRef]
31. Elrahman, M.A.; Hillemeier, B. Combined effect of fine fly ash and packing density on the properties of high-performance concrete: An experimental approach. *Constr. Build. Mater.* **2014**, *58*, 225–233. [CrossRef]
32. Kar, A.; Ray, I.; Unnikrishnan, A.; Davalos, J.F. Microanalysis and optimization-based estimation of C–S–H contents of cementitious systems containing fly ash and silica fume. *Cem. Concr. Compos.* **2012**, *34*, 419–429. [CrossRef]
33. Hua, S.; Wang, K.; Yao, X. Developing high performance phosphogypsum-based cementitious materials for oil-well cementing through a step-by-step optimization method. *Cem. Concr. Compos.* **2016**, *72*, 299–308. [CrossRef]
34. Proske, T.; Hainer, S.; Rezvani, M.; Graubner, C.A. Eco-friendly concretes with reduced water and cement contents—Mix design principles and laboratory tests. *Cem. Concr. Res.* **2013**, *51*, 38–46. [CrossRef]
35. Cepuritis, R.; Jacobsen, S.; Onnela, T. Sand production with VSI crushing and air classification: Optimising fines grading for concrete production with micro-proportioning. *Miner. Eng.* **2015**, *78*, 1–14. [CrossRef]
36. Binici, H.; Aksogan, O.; Cagatay, I.H.; Tokyay, M.; Emsen, E. The effect of particle size distribution on the properties of blended cements incorporating GGBFS and natural pozzolan (NP). *Powder Technol.* **2007**, *177*, 140–147. [CrossRef]
37. Binici, H.; Aksogan, O. The use of ground blast furnace slag, chrome slag and corn stem ash mixture as a coating against corrosion. *Constr. Build. Mater.* **2011**, *25*, 4197–4201. [CrossRef]
38. Al-Akhras, N.M.; Al-Akhras, K.M.; Attom, M.F. Performance of olive waste ash concrete exposed to elevated temperatures. *Fire Saf. J.* **2009**, *44*, 370–375. [CrossRef]
39. Ataie, F.F.; Riding, K.A. Thermochemical pretreatments for agricultural residue ash production for concrete. *J. Mater. Civ. Eng.* **2012**, *25*, 1703–1711. [CrossRef]
40. Biricik, H.; Aköz, F.; Türker, F.; Berktay, I. Resistance to magnesium sulfate and sodium sulfate attack of mortars containing wheat straw ash. *Cem. Concr. Res.* **2000**, *30*, 1189–1197. [CrossRef]
41. Al-Akhras, N.M. Durability of wheat straw ash concrete exposed to freeze–thaw damage. *Proc. Inst. Civ. Eng. Constr. Mater.* **2011**, *164*, 79–86. [CrossRef]
42. Al-Akhras, N.M. Durability of wheat straw ash concrete to alkali-silica reaction. *Proc. Inst. Civ. Eng. Constr. Mater.* **2012**, *166*, 65–70. [CrossRef]
43. Al-Akhras, N.M.; Al-Akhras, K.M.; Attom, M.F. Thermal cycling of wheat straw ash concrete. *Proc. Inst. Civ. Eng. Constr. Mater.* **2008**, *161*, 9–15. [CrossRef]
44. Merta, I.; Tschegg, E.K. Fracture energy of natural fiberreinforced concrete. *Constr. Build. Mater.* **2013**, *40*, 991–997. [CrossRef]
45. Zhang, Y.; Ghaly, A.E.; Li, B. Physical properties of wheat straw varieties cultivated under different climatic and soil conditions in three continents. *Am. J. Eng. Appl. Sci.* **2012**, *5*, 98–106.
46. Kumar, S.; Gandhi, P.; Yadav, M.; Paritosh, K.; Pareek, N.; Vivekanand, V. Weak alkaline treatment of wheat and pearl millet straw for enhanced biogas production and its economic analysis. *Renew. Energy* **2019**, *139*, 753–764. [CrossRef]
47. Chandra, R.; Takeuchi, H.; Hasegawa, T. Methane production from lignocellulosic agricultural crop wastes: A review in context to second generation of biofuel production. *Renew. Sustain. Energy Rev.* **2012**, *16*, 1462–1476. [CrossRef]
48. Kaparaju, P.; Serrano, M.; Thomsen, A.B.; Kongjan, P.; Angelidaki, I. Bioethanol, biohydrogen and biogas production from wheat straw in a biorefinery concept. *Bioresour. Technol.* **2009**, *100*, 2562–2568. [CrossRef]
49. Chandra, R.; Takeuchi, H.; Hasegawa, T.; Kumar, R. Improving biodegradability and biogas production of wheat straw substrates using sodium hydroxide and hydrothermal pretreatments. *Energy* **2012**, *43*, 273–282. [CrossRef]
50. Jaffar, M.; Pang, Y.; Yuan, H.; Zou, D.; Liu, Y.; Zhu, B.; Li, X. Wheat straw pretreatment with KOH for enhancing biomethane production and fertilizer value in anaerobic digestion. *Chin. J. Chem. Eng.* **2016**, *24*, 404–409. [CrossRef]
51. Sambusiti, C.; Monlau, F.; Ficara, E.; Carrère, H.; Malpei, F. A comparison of different pre-treatments to increase methane production from two agricultural substrates. *Appl. Energy* **2013**, *104*, 62–70. [CrossRef]
52. Gustafson, V. Combined Production of Bioethanol and Biogas from Wheat Straw. Master's Thesis, Lund University, Lund, Sweden, 2015.
53. Liu, X.; Zicari, S.M.; Liu, G.; Li, Y.; Zhang, R. Pretreatment of wheat straw with potassium hydroxide for increasing enzymatic and microbial degradability. *Bioresour. Technol.* **2015**, *185*, 150–157. [CrossRef]
54. Pavlostathis, S.G.; Gossett, J.M. Alkaline treatment of wheat straw for increasing anaerobic biodegradability. *Biotechnol. Bioeng.* **1985**, *27*, 334–344. [CrossRef] [PubMed]
55. Bolado-Rodríguez, S.; Toquero, C.; Martín-Juárez, J.; Travaini, R.; García-Encina, P.A. Effect of thermal, acid, alkaline and alkaline-peroxide pretreatments on the biochemical methane potential and kinetics of the anaerobic digestion of wheat straw and sugarcane bagasse. *Bioresour. Technol.* **2016**, *201*, 182–190. [CrossRef] [PubMed]

56. Pode, R. Potential applications of rice husk ash waste from rice husk biomass power plant. *Renew. Sustain. Energy Rev.* **2016**, *53*, 1468–1485. [CrossRef]
57. An, D.; Guo, Y.; Zhu, Y.; Wang, Z. A green route to preparation of silica powders with rice husk ash and waste gas. *Chem. Eng. J.* **2010**, *162*, 509–514. [CrossRef]
58. Mor, S.; Chhoden, K.; Ravindra, K. Application of agro-waste rice husk ash for the removal of phosphate from the wastewater. *J. Clean. Prod.* **2016**, *129*, 673–680. [CrossRef]
59. Gonçalves, M.R.F.; Bergmann, C.P. Thermal insulators made with rice husk ashes: Production and correlation between properties and microstructure. *Constr. Build. Mater.* **2007**, *21*, 2059–2065. [CrossRef]
60. De Sensale, G.R. Effect of rice-husk ash on durability of cementitious materials. *Cem. Concr. Compos.* **2010**, *32*, 718–725. [CrossRef]
61. Zain MF, M.; Islam, M.N.; Mahmud, F.; Jamil, M. Production of rice husk ash for use in concrete as a supplementary cementitious material. *Constr. Build. Mater.* **2011**, *25*, 798–805. [CrossRef]
62. Munshi, S.; Sharma, R.P. Utilization of Rice Straw Ash as A Mineral Admixture in Construction Work. *Mater. Today Proc.* **2019**, *11*, 637–644. [CrossRef]
63. Mor, S.; Manchanda, C.K.; Kansal, S.K.; Ravindra, K. Nanosilica extraction from processed agricultural residue using green technology. *J. Clean. Prod.* **2017**, *143*, 1284–1290. [CrossRef]
64. Kazmi SM, S.; Abbas, S.; Munir, M.J.; Khitab, A. Exploratory study on the effect of waste rice husk and sugarcane bagasse ashes in burnt clay bricks. *J. Build. Eng.* **2016**, *7*, 372–378. [CrossRef]
65. Hubadillah, S.K.; Othman MH, D.; Ismail, A.F.; Rahman, M.A.; Jaafar, J.; Iwamoto, Y.; Yusop MZ, M. Fabrication of low cost, green silica based ceramic hollow fibermembrane prepared from waste rice husk for water filtration application. *Ceram. Int.* **2018**, *44*, 10498–10509. [CrossRef]
66. Hossain, S.S.; Roy, P.K. Waste rice husk ash derived sol: A potential binder in high alumina refractory castables as a replacement of hydraulic binder. *J. Alloys Compd.* **2019**, *817*, 152806. [CrossRef]
67. Hossain, S.S.; Mathur, L.; Roy, P.K. Rice husk/rice husk ash as an alternative source of silica in ceramics: A review. *J. Asian Ceram. Soc.* **2018**, *6*, 299–313. [CrossRef]
68. Song, S.; Cho, H.B.; Kim, H.T. Surfactant-free synthesis of high surface area silica nanoparticles derived from rice husks by employing the Taguchi approach. *J. Ind. Eng. Chem.* **2018**, *61*, 281–287. [CrossRef]
69. Costa JA, S.; Paranhos, C.M. Systematic evaluation of amorphous silica production from rice husk ashes. *J. Clean. Prod.* **2018**, *192*, 688–697. [CrossRef]
70. Ramezanianpour, A.A.; Mahdikhani, M.; Ahmadibeni, G. The effect of rice husk ash on mechanical properties and durability of sustainable concretes. *Int. J. Civ. Eng.* **2009**, *7*, 83–91.
71. Chopra, D.; Siddique, R. Strength, permeability and microstructure of self-compacting concrete containing rice husk ash. *Biosyst. Eng.* **2015**, *130*, 72–80. [CrossRef]
72. Ameri, F.; Shoaei, P.; Bahrami, N.; Vaezi, M.; Ozbakkaloglu, T. Optimum rice husk ash content and bacterial concentration in self-compacting concrete. *Constr. Build. Mater.* **2019**, *222*, 796–813. [CrossRef]
73. Zareei, S.A.; Ameri, F.; Dorostkar, F.; Ahmadi, M. Rice husk ash as a partial replacement of cement in high strength concrete containing micro silica: Evaluating durability and mechanical properties. *Case Stud. Constr. Mater.* **2017**, *7*, 73–81. [CrossRef]
74. Toufigh, V.; Abyaneh, M.J.; Jafari, K. Study of Behavior of Concrete under Axial and Triaxial Compression. *ACI Mater. J.* **2017**, *114*, 56–163. [CrossRef]
75. Jarrah, M.; Khezrzadeh, H.; Mofid, M.; Jafari, K. Experimental and numerical evaluation of piston metallic damper (PMD). *J. Constr. Steel Res.* **2019**, *154*, 99–109. [CrossRef]
76. Zareei, S.A.; Ameri, F.; Bahrami, N.; Shoaei, P.; Moosaei, H.R.; Salemi, N. Performance of sustainable high strength concrete with basic oxygen steel-making (BOS) slag and nano-silica. *J. Build. Eng.* **2019**, *25*, 100791. [CrossRef]
77. Zhu, H.; Liang, G.; Zhang, Z.; Wu, Q.; Du, J. Partial replacement of metakaolin with thermally treated rice husk ash in metakaolin-based geopolymer. *Constr. Build. Mater.* **2019**, *221*, 527–538. [CrossRef]
78. Korotkova, T.G.; Ksandopulo, S.J.; Donenko, A.P.; Bushumov, S.A.; Danilchenko, A.S. Physical properties and chemical composition of the rice husk and dust. *Orient. J. Chem.* **2016**, *32*, 3213–3219. [CrossRef]
79. Sathawane, S.H.; Vairagade, V.S.; Kene, K.S. Combine effect of rice husk ash and fly ash on concrete by 30% cement replacement. *Procedia Eng.* **2013**, *51*, 35–44. [CrossRef]
80. Torres-Carrasco, M.; Reinosa, J.J.; de la Rubia, M.A.; Reyes, E.; Peralta, F.A.; Fernández, J.F. Critical aspects in the handling of reactive silica in cementitious materials: Effectiveness of rice husk ash vs nano-silica in mortar dosage. *Constr. Build. Mater.* **2019**, *223*, 360–367. [CrossRef]
81. Chen, X.; Yu, J.; Zhang, Z.; Lu, C. Study on structure and thermal stability properties of cellulose fibers from rice straw. *Carbohydr. Polym.* **2011**, *85*, 245–250. [CrossRef]
82. He, Y.; Pang, Y.; Liu, Y.; Li, X.; Wang, K. Physicochemical characterization of rice straw pretreated with sodium hydroxide in the solid state for enhancing biogas production. *Energy Fuels* **2008**, *22*, 2775–2781. [CrossRef]
83. Song, Z.L.; Yag, G.H.; Feng, Y.Z.; Ren, G.X.; Han, X.H. Pretreatment of rice straw by hydrogen peroxide for enhanced methane yield. *J. Integr. Agric.* **2013**, *12*, 1258–1266. [CrossRef]

84. Amnuaycheewa, P.; Hengaroonprasan, R.; Rattanaporn, K.; Kirdponpattara, S.; Cheenkachorn, K.; Sriariyanun, M. Enhancing enzymatic hydrolysis and biogas production from rice straw by pretreatment with organic acids. *Ind. Crops Prod.* **2016**, *87*, 247–254. [CrossRef]
85. Mirmohamadsadeghi, S.; Karimi, K.; Zamani, A.; Amiri, H.; Horváth, I.S. Enhanced solid-state biogas production from lignocellulosic biomass by organosolv pretreatment. *BioMed Res. Int.* **2014**, *2014*, 350414. [CrossRef] [PubMed]
86. Jarabo, R.; Monte, M.C.; Fuente, E.; Santos, S.F.; Negro, C. Corn stalk from agricultural residue used as reinforcement fiber in fiber-cement production. *Ind. Crops Prod.* **2013**, *43*, 832–839. [CrossRef]
87. Adesanya, D.A.; Raheem, A.A. A study of the workability and compressive strength characteristics of corn cob ash blended cement concrete. *Constr. Build. Mater.* **2009**, *23*, 311–317. [CrossRef]
88. Adesanya, D.A.; Raheem, A.A. Development of corn cob ash blended cement. *Constr. Build. Mater.* **2009**, *23*, 347–352. [CrossRef]
89. Pinto, J.; Vieira, B.; Pereira, H.; Jacinto, C.; Vilela, P.; Paiva, A.; Varum, H. Corn cob lightweight concrete for non-structural applications. *Constr. Build. Mater.* **2012**, *34*, 346–351. [CrossRef]
90. Tong, X.; Smith, L.H.; McCarty, P.L. Methane fermentation of selected lignocellulosic materials. *Biomass* **1990**, *21*, 239–255. [CrossRef]
91. Li, X.; Li, L.; Zheng, M.; Fu, G.; Lar, J.S. Anaerobic co-digestion of cattle manure with corn stover pretreated by sodium hydroxide for efficient biogas production. *Energy Fuels* **2009**, *23*, 4635–4639. [CrossRef]
92. Zheng, M.; Li, X.; Li, L.; Yang, X.; He, Y. Enhancing anaerobic biogasification of corn stover through wet state NaOH pretreatment. *Bioresour. Technol.* **2009**, *100*, 5140–5145. [CrossRef]
93. Song, Z.; Liu, X.; Yan, Z.; Yuan, Y.; Liao, Y. Comparison of seven chemical pretreatments of corn straw for improving methane yield by anaerobic digestion. *PLoS ONE* **2014**, *9*, 93801. [CrossRef] [PubMed]
94. Khatri, S.; Wu, S.; Kizito, S.; Zhang, W.; Li, J.; Dong, R. Synergistic effect of alkaline pretreatment and Fe dosing on batch anaerobic digestion of maize straw. *Appl. Energy* **2015**, *158*, 55–64. [CrossRef]
95. Hassan, M.; Ding, W.; Bi, J.; Mehryar, E.; Talha, Z.A.A.; Huang, H. Methane enhancement through oxidative cleavage and alkali solubilization pre-treatments for corn stover with anaerobic activated sludge. *Bioresour. Technol.* **2016**, *200*, 405–412. [CrossRef] [PubMed]
96. Cobreros, C.; Reyes-Araiza, J.L.; Manzano-Ramírez, A.; Nava, R.; Rodríguez, M.; Mondragón-Figueroa, M.; Rivera-Muñoz, E.M. Barley straw ash: Pozzolanic activity and comparison with other natural and artificial pozzolans from Mexico. *BioResources* **2015**, *10*, 3757–3774. [CrossRef]
97. Belhadj, B.; Bederina, M.; Montrelay, N.; Houessou, J.; Quéneudec, M. Effect of substitution of wood shavings by barley straws on the physico-mechanical properties of lightweight sand concrete. *Constr. Build. Mater.* **2014**, *66*, 247–258. [CrossRef]
98. Schneider, L.; Haverinen, J.; Jaakkola, M.; Lassi, U. Pretreatment and fractionation of lignocellulosic barley straw by mechanocatalysis. *Chem. Eng. J.* **2017**, *327*, 898–905. [CrossRef]
99. Lam, P.S.; Sokhansanj, S.; Bi, X.; Lim, C.J.; Naimi, L.J.; Hoque, M.; Ye, X.P. Bulk density of wet and dry wheat straw and switchgrass particles. *Appl. Eng. Agric.* **2008**, *24*, 351–358. [CrossRef]

agriculture MDPI

Review

Biostimulant Effects of Waste Derived Biobased Products in the Cultivation of Ornamental and Food Plants

Enzo Montoneri [1,*], Andrea Baglieri [1] and Giancarlo Fascella [2]

1 Dipartimento di Agricoltura, Alimentazione e Ambiente, Università di Catania, 95123 Catania, Italy; abaglie@unict.it
2 CREA Research Centre for Plant Protection and Certification, 90011 Firenze, Italy; giancarlo.fascella@crea.gov.it
* Correspondence: enzo.montoneri@gmail.com

Abstract: Soluble bio-based substances (SBS) may be isolated from the anaerobic digestate of the organic humid fraction of urban waste; from the whole vegetable compost made from gardening residues and from the compost obtained after aerobic digestion of a mixture of urban waste digestate, gardening residues and sewage sludge. These SBS can be used as sustainable and efficient plant biostimulants in alternatives to the commercial products based on fossil sources such as the Leonardite. The present review summarizes the main findings obtained from recent studies accomplished with the SBS applied on several ornamental (Euphorbia; Lantana; Murraya; Hibiscus) and vegetable species (tomato; red pepper; spinach; maize; bean; wheat; tobacco; oilseed rape) with the aim to evaluate their effect on plant growth; fruit and ornamental quality. The main results from these studies show that the non-commercial SBS are more efficient than commercial fossil-based products; at equal applied doses; in enhancing plant growth; leaf chlorophylls; photosynthetic activity; fruit ripening and yield and aesthetic effect; improving flower and fruit quality and optimizing water use efficiency. Depending upon the plant species, increases of the plant performance indicators ranging from zero to 1750% are reported for the plants cultivated in the presence of SBS, relatively to the control plants cultivated in absence of SBS added to the cultivation substrate. The review suggests that biowaste recycling is a sustainable and environmentally friendly source of plant biostimulants, as an alternative to existing fossil sourced agrochemicals.

Keywords: bio-based substances; biostimulants; antifungal agents; municipal biowastes; ornamental plants; food plants

Citation: Montoneri, E.; Baglieri, A.; Fascella, G. Biostimulant Effects of Waste Derived Biobased Products in the Cultivation of Ornamental and Food Plants. *Agriculture* **2022**, *12*, 994. https://doi.org/10.3390/agriculture12070994

Academic Editor: Maria Roulia

Received: 15 June 2022
Accepted: 8 July 2022
Published: 10 July 2022

1. Introduction

Cultivating plants is a human activity involving several sectors. Agriculture deals with cultivation of crops for human consumption as well as animal production. Horticulture strictly involves the cultivation of plants for food consumption, as well as plants not for human consumption. Horticulture differs from floriculture. The former involves different types of garden crops, while the latter involves flowering and foliage plants. Ornamental horticulture is the cultivation of decorative plants of all kinds, including not only plants with attractive flowers, but also plants with decorative leaves, stems, bark, or fruit. Basically, floriculture and ornamental horticulture have decorative and aesthetic purposes. Aside from categories' definitions and differences in the cultivated species, all these categories' activities share similar problems.

Common farming practice is to boost plant production with a fertilizer dose higher than that adsorbed by soil and plant. Thus, noxious fertilizers' components accumulate in soil, reach the food chain, leach through soil into ground water, and ultimately affect human and animal health. Mineral and organic fertilizers are used. The global fertilizer market is 156 billion USD/year. [1]. Major ones are urea and mineral phosphates (80% of the EU fertilizers' market value), with 0.11–0.46 €/kg production cost. They are based on energy-intensive production processes or manufactured from non-renewable feedstock imported from third countries [2]. Organic fertilizers belong to a niche market (0.15% of the total fertilizer market) [3]. The world consumption of mineral fertilizers containing N, P and K is ca. 200 Mt/year [4]. EU consumption of mineral fertilizers is 16 Mt/year [5]. From 70 to 250 kg/ha nitrates leaching may occur depending on fertilizer dose, soil, and plant type [6]. Based on average 51 kg/ha applied surplus and total 175 Mha cultivated area, 9 kt/year nitrate leach through soil and water. To improve the balance between fertilizers dose and crop requirement, the max EU ruled dose is 150–350 kg/ha. Major organic fertilizers are composts of biowastes from urban, animal, or agriculture sources, manure, peat and leonardite hydrolysates. Composts are commonly applied to soil at 10–30 t/ha.year [7]. High doses that obtain the desired effects are due to compost insolubility causing slow nutrients' uptake by plants. This causes leaching of excess major and trace metal components through soil and water. Similarly, manure is applied at 70 t/ha dose. In addition to leaching, manure causes greenhouse-gases emission due to fermentation in soil. For example, typical aerial NH_3 concentration in a pig farm is 5–35 ppm against a 25 ppm threshold level [8]. A higher NH_3 level harms both animal and human health. Emission of 420,000 t/year NH_3 is estimated from a total 1400 Mt/year EU manure production [9]. Peat and leonardite hydrolysates contain soluble organic and mineral matter. EU consumption is 240 kt/year. These hydrolysates are obtained from fossil source. Based on average 40% C content, their use causes 355 kt/year CO_2 emission from fossil C and depletion of fossil sources. Except for municipal biowastes (MBW), a common problem of all fertilizers is that their sources are found in restricted sites, not available worldwide. This poses the problem of product supply and cost. The problem is highly relevant in Europe, which imports most of its mineral consumption from third countries.

One other important restraint on plant productivity is pests and diseases. These are highly relevant for food plants. Food production loss due to plant diseases is estimated to be 10–50%/year [10]. Plant protection relies on pesticides use, which increases food cost and may cause hormonal disruption in human. A common problem of all fertilizers is the need to use them together with pesticides. Together with lowering cost, there is much concern for decreasing the exploitation and depletion of natural resources to produce fertilizers.

In the last ten years, relevant research has been focused on bio-stimulants. These belong to a new functional product category (FPC6) contemplated in the New European Fertiliser Regulation [11,12]. Bio-stimulants are supposed to stimulate the plant metabolism, regardless of their nutrient content, and so improve plant growth, even under abiotic stress, and resistance to diseases [13]. Applied at much lower doses than common mineral and organo-mineral fertilisers, bio-stimulants are expected to induce plant resistance to pathogens and provide high crop productivity at the same time. In this fashion, they reduce/minimize the negative environmental impact of the excessive application of commercial fertilisers and pesticides.

Work by the authors in the past fifteen years proves that soluble bioorganic substances (SBS) obtained from urban and agriculture biowastes have both biostimulant and antifungal properties [14]. No other known products have both properties. The research hypothesis was that, by virtue of the solubility properties and organo-mineral composition, the SBS could increase plant growth and crop production compared to current commercial mineral and organo-mineral agrochemicals. The present paper reports the critical review of work performed by the authors with SBS for the cultivation of food and non-food plants. The review proves the research hypothesis. It demonstrates the SBS biostimulant properties for all tested plants and discusses the economic and environmental benefits for agriculture and horticulture.

2. SBS Composition and Properties

The SBS are obtained by hydrolysis at 60–90 °C and pH 13 of several different mixes of urban food, green and sewage sludge wastes fermented under anaerobic and aerobic conditions [14]. Under these conditions, the SBS were obtained together with the secondary insoluble (IR) product. The fermented wastes yielding the SBS described in the present review were sampled from different streams of the Italian ACEA Pinerolese MBW treatment plant. The SBS contain organic and mineral matter. The organic matter is a mix of molecules with molecular weight from 5 to over 750 kDa. These molecules are constituted by several different organic moieties made by aliphatic and aromatic C substituted by acid and basic functional groups of different strengths. Mineral elements of groups 1 to 4 are bonded to or complexed by the organic moieties. These chemical features are inherited from the pristine biowastes. The molecules contained in SBS are water soluble memories of the native recalcitrant lignocellulosic polysaccharides, proteins, fats, and lignin proximates still present in the biowastes after anaerobic and aerobic fermentation. It is no wonder that, due to their origin, richness of mineral elements, organic functional groups and acquired water solubility, the SBS molecules exhibit a wide range of properties as plant biostimulants, plant resistance inducers, bio-photosensitizers, oxidation catalysts, polymers for manufacturing mulch films, composite pellets, composite plastic articles, and high performance surfactants. Tables 1–4 report the compositional details of the SBS, IR and the pristine fermented biowastes. Table 5 list the plants cultivated with the SBS and summarizes the main SBS effects on the cultivated plants. All data in Tables 1–5 are extrapolated from the references cited in Table 5. The data reported in Tables 1–4 were obtained through a specifically designed analytical protocol [14]. This included calculation of moisture, ash and volatile solids (VS) contents from the sample weight losses determined after heating to 105 and 650 °C, inorganic elements analysis by AAS and/or ICP, microanalyses for C, H, N determination performed with a C. Erba (Rodano, Milan, Italy) NA-2100 elemental analyser. The C types and functional groups reported in Table 4 were determined by solid-state 13C NMR spectroscopy. Solid-state 13C NMR spectra were acquired at 67.9 MHz on a JEOL GSE 270 spectrometer equipped with a Doty probe. The cross-polarization magic angle spinning (CPMAS) technique was employed, and for each spectrum, about 104 free induction decays were accumulated. The pulse repetition rate was set at 0.5 s, the contact time at 1 ms, the sweep width was 35 KHz, and MAS was performed at 5 kHz. Signals assignment as a function of the resonance range were: 0–53 ppm aliphatic C, 53–63 ppm O-Me or N-alkyl C, 63–95 ppm O-alkyl C, 95–110 ppm di-O-alkyl C, 110–140 ppm aromatic C, 140–160 ppm phenol or phenyl ether C, 160–185 ppm carboxyl C, and 185–215 ppm ketone C.

Table 1. Waste ingredients in pristine biowastes (PFB).

PFB	Ingredients
D	Digestate from anaerobic fermentation of unsorted food wastes
CV	Compost of private gardening and public park trimming residues (V)
CVD	Compost of D and V mix in 2/1 weight respective ratio
CVDF	Compost of D, V and sewage sludge (F) mix in 5.5/3.5/1 respective ratio
ETP	Exhausted tomato plants at the end of the crop harvesting season

Table 2. Mineral elements and ash content (*w/w*%) in the pristine biowastes (PFB), in the soluble (SBS) product and insoluble (IR) hydrolysates obtained.

	Si	Fe	Al	Mg	Ca	K	Na	Ash
CVDF PFB	6.27 ± 0.04	1.02 ± 0.01	1.06 ± 0.02	0.83 ± 0.01	3.23 ± 0.05	1.32 ± 0.03	0.07 ± 0.01	59.4
CVDF SBS	0.92 ± 0.03	0.53 ± 0.02	0.44 ± 0.02	0.49 ± 0.01	2.59 ± 0.03	5.49 ± 0.04	0.15 ± 0.01	27.3
CVDF IR	7.68 ± 0.06	1.23 ± 0.03	1.05 ± 0.01	1.15 ± 0.02	3.20 ± 0.03	1.32 ± 0.02	0.04 ± 0.01	77.6
D PFB	3.46 ± 0.05	0.77 ± 0.03	0.40 ± 0.02	0.88 ± 0.02	7.16 ± 0.08	0.53 ± 0.03	0.22 ± 0.02	34.5
D SBS	0.36 ± 0.03	0.16 ± 0.00	0.78 ± 0.04	0.18 ± 0.01	1.32 ± 0.05	9.15 ± 0.06	0.39 ± 0.01	15.4
D IR	4.73 ± 0.03	0.48 ± 0.01	0.47 ± 0.06	1.07 ± 0.02	9.54 ± 0.05	3.44 ± 0.05	0.16 ± 0.01	49.0
CVD PFB	10.70 ± 0.03	1.07 ± 0.02	0.71 ± 0.03	1.12 ± 0.01	4.27 ± 0.14	1.09 ± 0.03	0.08 ± 0.01	56.1
CVD SBS	2.49 ± 0.04	0.88 ± 0.02	0.60 ± 0.06	0.93 ± 0.02	4.70 ± 0.08	3.76 ± 0.07	0.17 ± 0.01	28.3
CVD IR	12.60 ± 0.05	0.95 ± 0.01	0.75 ± 0.03	1.13 ± 0.02	4.96 ± 0.05	2.13 ± 0.06	0.07 ± 0.01	56.8
CV PFB	12.14 ± 0.07	1.03 ± 0.02	0.59 ± 0.01	1.67 ± 0.25	4.86 ± 0.61	1.18 ± 0.07	0.06 ± 0.01	57.1
CV SBS	2.55 ± 0.01	0.77 ± 0.04	0.49 ± 0.04	1.13 ± 0.06	6.07 ± 0.38	3.59 ± 0.21	0.16 ± 0.01	27.9
CV IR	15.04 ± 0.33	1.10 ± 0.05	0.67 ± 0.01	1.45 ± 0.01	4.19 ± 0.09	1.49 ± 0.02	0.06 ± 0.01	71.3
ETP PFB	0.98 ± 0.03	0.30 ± 0.02	0.27 ± 0.02	0.42 ± 0.02	4.65 ± 0.03	3.30 ± 0.02	0.22 ± 0.01	20.2
ETP SBS	0.22 ± 0.03	0.33 ± 0.02	0.34 ± 0.03	0.80 ± 0.04	2.10 ± 0.02	9.15 ± 0.06	0.24 ± 0.01	23.3
ETP IR	0.85 ± 0.03	0.25 ± 0.01	0.17 ± 0.01	0.27 ± 0.01	4.41 ± 0.02	4.49 ± 0.06	0.15 ± 0.01	36.9

Table 3. Total C, N and P content (*w/w*%) in pristine biowastes (PFB), and in soluble (SBS) and insoluble (IR) hydrolysates obtained.

	C	N	C/N	P_2O_5
CVDF PFB	24.36 ± 0.16	2.25 ± 0.11	10.83	1.30 ± 0.22
CVDF SBS	35.47 ± 0.09	4.34 ± 0.17	8.17	1.44 ± 0.03
CVDF IR	11.72 ± 0.22	1.02 ± 0.05	11.49	0.53 ± 0.05
D PFB	29.99 ± 0.20	3.81 ± 0.12	7.87	3.27 ± 0.15
D SBS	45.07 ± 0.12	7.87 ± 0.12	5.73	1.14 ± 0.10
D IR	27.68 ± 0.08	1.80 ± 0.05	15.38	2.75 ± 0.03
CVD PFB	27.07 ± 0.78	2.45 ± 0.07	11.05	0.75 ± 0.05
CVD SBS	37.51 ± 0.04	4.89 ± 0.03	7.67	0.84 ± 0.04
CVD IR	22.11 ± 0.24	1.64 ± 0.01	13.48	1.14 ± 0.18
CV PFB	22.43 ± 0.42	1.91 ± 0.03	11.74	0.39 ± 0.02
CV SBS	38.25 ± 0.09	4.01 ± 0.03	9.54	0.53 ± 0.05
CV IR	18.44 ± 0.67	1.15 ± 0.09	16.03	0.37 ± 0.02
ETP PFB	36.44 ± 0.24	3.51 ± 0.18	10.38	
ETP SBS	47.30 ± 0.09	6.52 ± 0.13	7.25	
ETP IR	28.83 ± 0.08	2.52 ± 0.10	11.44	

Table 4. Carbon types and functional groups content (*w/w*% of total C) [a].

	Cal	OMe + NR	OR	OCO	Ph	PhOY	COX	CO
CVDF PFB	31.81	8.59	27.67	6.18	10.72	5.90	8.17	1.96
CVDF SBS	31.17	7.88	19.13	6.73	16.58	7.69	10.49	0.34
CVDF IR	28.90	8.32	27.14	7.46	13.23	7.01	6.79	1.16
D PFB	33.60	9.10	26.61	5.99	8.94	4.27	10.53	0.97
D SBS	43.38	9.86	14.01	3.37	9.60	3.23	15.89	0.66
D IR	50.80	5.52	18.95	4.00	8.54	3.28	7.23	1.68
CVD PFB	37.25	9.75	28.14	4.35	8.03	5.20	6.67	0.62
CVD SBS	40.90	7.34	14.18	3.85	12.27	5.97	12.92	2.56
CVD IR	31.73	9.39	29.32	6.39	9.78	6.21	5.87	1.31
CV PFB	32.86	8.33	23.85	6.34	12.30	6.73	8.21	1.37
CV SBS	36.90	7.24	13.22	4.18	13.39	6.84	13.53	4.69
CV IR	31.70	8.43	24.58	6.14	11.49	7.23	7.74	2.68
ETP PFB	14.34	7.22	49.60	11.62	6.89	3.44	6.28	0.61
ETP SBS	47.38	9.39	10.39	2.19	11.50	3.81	14.37	0.97
ETP IR	5.00	7.97	58.98	13.19	7.00	3.66	2.97	1.22

[a] Aliphatic (Cal), aromatic (Ph), anomeric (OCO), carboxylic and/or amide (COX), ketone (CO) carbon, and carbon bonded to amine (NR), methoxy (OMe), alkoxy (OR), and phenolic and/or phenoxy (PhOY) groups.

Table 5. Plants cultivated with SBS and SBS effects: increase (*w/w*% relative to control) of total biomass or crop production, unless otherwise indicated. Data for specific indicators in Section 3.

	CVDF	CVD	CV	D	ETP
Euphorbia [15]	331			117	
Lantana [16]	143			85	
Hibiscus [13,17] [a]			15	23	
Murraya [18]	67			35	
Tomato Micro-Tom [19] [b]					
Tomato Lycopersicon [7,20] [c]	16	4–13	21	5	
Tomato Micro-Tom [21]	46		1	16	
Red pepper [22]		66			
Maize [23]	89				
Bean [24]					109–1750 [d,e]
Grain [21]	10		9	9	
Tobacco [21]	6		0	0	
Spinach [25]				24–40 [f]	
Oilseed Rape [10]	56 [g]			42 [g]	

[a] Reference 13 for CV and 17 for D. [b] Used only as model plant (see below for results). [c] Reference 20 for CVD and 21 for CVDF, CV and D. [d] Increase of enzyme activities and soluble proteins concentration in leaves and roots. [e] Increase of root diameter (66%) by ETP PFB, SBS and IR, and of chlorophyll b (135%) by ETP SBS and IR. [f] Reduction of nitric to total N ratio in leaves. [g] Reduction of plant lesions due to Leptosphaeria maculans.

3. Demonstration of SBS Effects as Biostimulant/Photosensitizers in the Cultivation of Tomato, Bean, Euphorbia, Lantana and Hibiscus Plants

3.1. Tomato Solanum Lycopersicum

The first trials to investigate SBS performance in the cultivation of food plants were reported in 2012 [7]. The tomato *Solanum Lycopersicum* was used as a probe plant. The plant was cultivated in a farm greenhouse using a commercial product obtained from animal residues (RCP) as organic fertilizer. The experimental plan was designed to measure the plant growth, and the crop production and quality in the cultivation soil treated with RCP (the control treatment), in comparison with the soil treated alternatively with CVD SBS, IR and PFB. Due to the different composition of the control RCP fertilizer and the SBS test materials, the four materials were applied in different amounts to the soil in order to contribute to each soil plot the same 1.1–1.2 organic matter t/ha dose.

The results showed that the plants grown on the CVD SBS treated soil performed better than all others. The former ones exhibited 5–19 day earlier tomato ripening, 4–13% higher production of per plant fruit and per cluster number of fruits, and 7–8% higher leaves chlorophyll content. This result was achieved in spite of the fact that RCP contribute more organic N to the cultivation soil (120 N kg/ha) than CVD SBS (80 N kg/ha). The superior effect of CVD SBS was ascribed to two specific product properties. First, the CVD SBS highest water solubility allowed faster nutrient uptake by the plant, compared to RCP and the other two CVD PCB and IR products. Secondly, CVD SBS had a peculiar photosensitizing properties. These properties had been reported in previous work where the SBS was used to promote the photo oxidation in industrial organic waste waters [26]. In the case of the above tomato cultivation trials, the highest leaf chlorophyll content for the plant cultivated in the CVD SBS indicated enhanced photosynthesis, compared to that taking place in the plants cultivated in the soil treated with RCP, and with CVD PCB and IR. The results of the tomato cultivation trials, coupled to those obtained for the photo oxidation in industrial organic waste water, suggested the fascinating belief that SBS might promote either C fixation or mineralization, according to operating conditions.

Further work was performed by cultivating tomato plants with the D SBS, CV SBS and CVDF SBS [20] described in Tables 1–4. The cultivation trials were carried out in the same farm and conditions as the CVD trials [7], except for the fact that the applied D SBS, CV SBS and CVDF SBS doses were much lower and at three levels: 30 kg/ha, 145 kg/ha and 500 kg/ha [20]. The organic matter doses applied with the D, CV and CVDF SBS ranged from 22–25 kg/ha to 360–420 kg/ha, from 46 to 3 times lower than the organic matter applied with the CVD SBS in the previous work [7]. The control soil in the D, CV, CVDF SBS trials was the same as for the CVD SBS trials. Compared to control, the following most significant effects were measured for the plants grown on the soil treated with SBS: 9.4% increase of plant diameter by 145 kg/ha CV SBS; 9.7% increase of chlorophyll content by 500 kg/ha CV SBS, not significantly different from the increase by the 145 kg/ha treatment; 21% increase of fruit yield by145 kg/ha CV SBS. Generally, for all other SBS treatments and for the control soil, plant performance resulted not significantly different from or lower than that measured for the 145 kg/ha CV SBS treated soil. Compared to the CVD SBS trials carried out at 1.1 organic matter applied dose [7], the 104 kg/ha CV SBS organic matter dose applied in the 145 kg/ha CV SBS treatment [20] gave almost double crop production increase, relative to the control plants. At this regard, the data in Table 6 show that the applied doses of the plant nutrient N, P, K element with the different CVDF, D and CV SBS treatments are far lower than those applied with the CVD SBS [7] treatment, and that the doses applied with the CV SBS are the lowest ones. Yet, the 145 kg/ha CV SBS treatment produced the best performing plants. This fact added further argument to support the belief that the SBS activity was not merely fertilizing the soil with mineral nutrient, but most likely stimulating the plant metabolism. The analysis of the results [7] indicated the CV SBS as the most efficient biostimulant, due to its specific organo-mineral composition and solubility properties.

Table 6. Supplied mineral elements (ME) and soil nutrients (kg/ha) in cultivation trials with CVD SBS 1.1 t/ha organic matter [7] and CV, CVDF, D SBS 145 kg/ha dose [20].

	Total ME	Total NPK	N	P	K
CVDF SBS	15.5	15.2	6.29	0.91	7.96
D SBS	17.8	25.4	11.4	0.72	13.3
CVD SBS	221	162	80	7.4	75
CV SBS	21.5	12.7	5.8	0.33	5.2

3.2. *Red Pepper Capsicum Annuum, F1 Barocco*

The CVD SBS was tested also in the cultivation of red pepper [22]. Compared to the first previous tomato study [7], in the cultivation of red pepper [22] the same CVD SBS and soil were used, but CVD SBS was applied to the soil at much lower doses. These were 7, 35, 70, 140, 350, and 700 kg/ha, corresponding to organic matter ranging from 5 to 500 kg/ha. The SBS organic matter application range in the red pepper study [22] was slightly wider than that tested for the CVDF, D and CV SBS in the second tomato study [20], but still from 46 to 3 times lower than that applied with the CVD SBS in the first tomato study [7].

In the red pepper study [22], plant size, leaves' chlorophyll content and crop production over the growing cycle were measured. Compared to the control cultivation, the most significant effect in the presence of CVD SBS were shown at 140 kg/ha applied dose from the leaves' chlorophyll content and fruit production. The leaves chlorophyll content reached a peak upon increasing the SBS dose up to 140 kg/ha and then decreased upon increasing further the SBS dose to 700 kg/ha. Relatively to the control plant, for the plants grown in the 140 kg/ha treated soil the following increases were observed: 12% for the chlorophyll content, 90% for the 1st harvesting week crop, 66% for the total crop production, 17% for the per fruit weight. These effects were far higher than those obtained for the tomato plant cultivations [7,20].

Particularly interesting in the red pepper study [22] was the trend of the leaf chlorophyll content and crop production, both showing a peak at 140 kg/ha CVD SBS applied to the soil. The same trend was observed for the photodegradation of organic pollutants in the presence of different concentrations of SBS [27,28]. In this case, the provided explanation was that the added SBS to the polluted waste water catalysed the photo degradation of the organic pollutant up to a peak added content. At higher SBS concentration, the self-photo degradation of SBS occurred. This lowered the concentration of the pristine SBS, and therefore its availability to catalyse the photo degradation of the organic pollutant in the waste water. In the case of the red pepper study [22], the data confirm the correlation of chlorophyll formation, photosynthesis and crop production. However, understanding the mode of action of the CVD SBS in the biochemical system under investigation is far harder than in the sterile homogenous aqueous system of the industrial waste water [27], where no biochemical reaction took place. By comparison, in the red pepper cultivation system [22], both chemical and biochemical reactions took place. There, the complex processes of photosynthesis and chlorophyll formation was regulated by the availability of enzymes and light. In this respect, the red pepper study [22] did not provide data that could support the effect of the CVD SBS on the natural enzymatic pool present in the system.

3.3. *Bean Phaseolus Vulgaris*

The biochemical response to the SBS was addressed in the study of the effects of ETP SBS (Tables 1–6) on bean plants [24]. Specifically, the plant nitrogen metabolism was studied by determining the nitrate reductase, glutamine synthetase, and glutamate synthase activities and their relation with the content of soluble proteins in the plant leaves and roots. In the bean study, the ETP SBS, ETP PFB and ETP IR were applied separately to a substrate consisting of peat and sand in 14 cm × 14 cm × 15 cm pots. Four grams ETP SBS per pot were applied. Eight grams of ETP PFB and IR per pot were applied.

In this fashion, the three ETP materials contributed nearly equal N and C content to the substrate. The results showed no statistically significant effect on the plant shoot and root weights and leaves' chlorophyll content by the substrate treatments, compared to the control substrate. However, the leaves and roots of the plants grown on the ETP SBS treated substrate exhibited the highest enzyme activities, compared to the control and the other treatments. The increases of enzymes content by ETP SBS were remarkably high. Relatively to the control plant, they ranged from 109 to1750%. The content of the soluble proteins in the plant leaves and roots was also measured. Consistently with the data for the enzyme activities, the plant grown on the substrate treated with ETP SBS had the highest protein content. Relatively to the control plant, the protein increases were 77% in the leaves and 226% in the roots. The data demonstrated that the ETP SBS promoted the highest N assimilation by the plant. This fact was proposed as a possible indication of auxin-like effect by ETP SBS.

3.4. *Euphorbia x lomi Rauh*

Further studies were carried out on the SBS in comparison with commercial products claimed by the vendors as natural organic amendments or biostimulants. The D SBS and CVDF SBS were tested for the cultivation of *Euphorbia x lomi* Rauh [15] in comparison with a Leonardite derived commercial product (LND). The plants cultivation was carried out in pots containing a substrate of sphagnum peat and perlite. Compared to the SBS, the chemical composition of LND was qualitatively similar, but quantitatively very different. The latter had much lower N and P, and much higher K than the SBS. Under these circumstance, the authors chose to apply the LND at the dose recommended by the vendor, and the SBS at the nearly equal weight dose of the as-purchased LND. Coherently with this criterion, the three products were applied in aqueous solution at the following doses (g/pant): 4.6 for CVDF SBS applied as substrate drench, 3.1 and 1.5 for D SBS applied as substrate drench and foliar spray, respectively, and 1.9 and 0.94 for LND applied as substrate drench and foliar spray, respectively. This dose application scheme allowed comparing the CVDF SBS versus LND at close crude product doses, and the two SBS, one with the other, at the same applied N doses.

The following statistically significant effects were measured in the Euphorbia study [15]. The CVDF SBS treatment yielded the highest number of leaves per plant, leaf area, number of flowers per plant, total stem, leaves, roots biomass production, water use efficiency, leaf chlorophyll content and gas exchange than the control plant and the D SBS and LND treatments. Increases relative to the control plants were: 95% for the number of leaves, 78% for the leaf area, 233% for the number of flowers, 331% for total biomass production and water use efficiency, 33% for leaves' chlorophyll content, and 258% for leaf gas exchange. The other treatments also gave significantly higher increases relative to the control plants, but lower increases compared to CVDF SBS. The different effects arose from the different applied products, regardless of the application modes being by foliar spray or substrate application. For these reasons, the highest effects of CVDF SBS could lie on the highest supplied N per plant and the highest content Fe, Ca, P, carboxylic, phenolic and amino groups. Fe ions could have an important role for the plant photosynthesis. By virtue of its organic acid and basic groups, the CVDF kept Fe ions in solution at circumneutral pH. These were inferred responsible for the photosensitizing properties of SBS [21,27,28].

3.5. *Lantana Camara and L. sellowiana*

The results obtained in the Euphorbia study were replicated for the cultivation of Lantana [16] in the presence of the same CVDF and D SBS, and LND products. Two different plant species, *Lantana camara* (CAM) and *L. sellowiana* (SEL), were cultivated. The same ranking order of performance by the tested products was reported for the two plant species. Relative to the control plants, the increases of the measured indicators by the CVDF SBS were: 45% for CAM and SEL plant height; 92% for CAM and 105% for SEL number of leaves; 234% for CAM and 171% for leaf area; 176% for CAM and 326%

for SEL number of flowers; 35% for CAM and 25% for SEL root length; 184% for CAM and 176% for SEL root dry weight; 101% for CAM and 114% for SEL stem and leaves total dry weight; 140% for CAM and SEL water use efficiency; 31% for CAM and 26% for SEL leaf chlorophyll content; 190% for CAM and 181% for SEL leaf gas exchange. The other treatments gave also significantly higher increases relatively to the control plants, but lower increases compared to CVDF SBS. The greater effect of CVDF SBS on the plant photosynthesis and leaf chlorophyll content of the Lantana plants, compared to the plants treated with D SBS and with LND, could be related to the higher content of Si, Mg, Fe and N in the CVDF SBS. Indeed, it is known that Fe and Mg play significant direct roles in photosynthesis, whereas N is present in chlorophyll molecules. By virtue of the organic acid and basic functional groups bonding and/or complexing the mineral elements, the CVDF SBS improved the plant take up and availability of the mineral elements necessary for chlorophyll biosynthesis and photosynthetic activity.

3.6. *Hibiscus Moscheutos L. Subsp. Hibiscus Palustris*

Other authors [13,17] applied the D and CV SBS in the cultivation of Hibiscus. In the first study [17], the D and CV SBS were compared with the D and CV PFB and IR materials, and with a commercial biostimulant (CB). According to the description of the vendor, CB was a plant extract containing fulvic and humic substances, amino acids and glycine betaine claimed to perform as biostimulants. As a consequence of the different products' sources, CB contained 74% organic matter, 24% C and 5.3% N, against 41–66% organic matter, 23–39% C and 1.4–6.6% N for the CV and D materials. Due to the large difference in the chemical composition of the investigated products, different crude product doses were applied in order to guarantee that the amount of added organic carbon contributed by the D and CV products in the cultivation substrate was the possible closest to the amount of C (0.42 kg/m^3) contributed by the commercial product as suggested by its vendor. The control (no added SBS or CB) and treated substrates received the same basic standard mineral fertilization. Relevant for this study is the fact that the applied doses of the tested products contributed to the substrate 5–15% of the minimum dose for common organic fertilizers normally applied in agriculture.

The results of the first Hibiscus study [17] showed that all treated substrates gave significant increases of the plant biomass production and biometric parameters, compared to the control substrate. The D SBS treatment gave the highest effects. It yielded 22–33% increases for the dry weights of leaf, stem and flower, total shoot, and leaf, leaf area index. The increase of the leaf chlorophyll was 8% and not statistically significant. The other treatments gave equal or lower increases, compared to the D SBS treatment. The treatments ranking order for the gas exchange activity order was different. In this case, the three CV products gave the highest statistically significant increases, compared to the control substrate: 24% for the photosynthetic activity rate, 46% for stomatal conductance, 31% for the evapotranspiration rate.

Based on the results of the studies on tomato [7,20], Euphorbia [15] and Lantana [16], in the above Hibiscus study [17] two important issues were discussed. First, the Hibiscus data further support the properties of the municipal biowaste derived products as photosensitizers, promoters of photosynthetic activity. Secondly, when products are applied to the cultivation soil or substrate previously treated with the same basic standard mineral fertilization, the mineral composition differences among the added products are likely to be levelled out by the higher relatively amount of nutrients supplied by the conventional chemical fertilizers. Thus, the relationship of the observed effects with the applied products chemical nature and/or composition is likely to be dimmed.

To challenge more the performance of the CV and D SBS as biostimulants, the second Hibiscus study [13] was carried out under nutrient stress conditions. Pot trials were performed a substrate containing peat and pumice at pH 6 by calcium carbonate. Osmocote was used as controlled release fertilizer (CRF). The experimental plan comprised four substrate treatments: the standard fertilization (SF) treatment at 6 kg/m^3 CRF dose, and the low fertilization (LF), the LF with added D SBS (LFD) and the LF with added CV SBS (LFCV) treatments, all last three at 3 kg/m^3 CRF dose. The nutrients' supply (kg/m^3) was 1.2 N, 0.8 P_2O_5, 0.7 K_2O for SF, and 0.6 N, 0.3 P_2O_5, 0.3 K_2O for LF, LFD and LFCV. Plant performance indicators were biomass accumulation, biometric parameters, leaf gaseous exchanges and elemental composition, and nitrogen (N)-use efficiency.

As expected, relatively to SF treatment, the performance indicators of the plant grown in the LF substrate were measured significantly lower by 47% for plant dry weight, 19% for plant height, 58% for plant volume, 46% for leaf area, 17% for relative growth rate, 22% for N assimilation rate, 23% for the photosynthetic rate, 27% for the chlorophyll content, 60% for the stomatal conductance, and 50% for the evapotranspiration rate. However, the values of the plant indicators measured for the plants cultivated in the LFCV treated substrate were significantly different from those of the plant grown in the SF. The LFD substrate performed significantly better than the LF substrate, but not as well as LFCV. Only in the case of the evapotranspiration rate, the measured values for the plants grown in both the LFD and LFCV treated substrates resulted significantly higher by 35%, relative to the values measured for the SF plants. The treatments significantly also affected the N use efficiency indexes. The LFD treatment gave 17% higher N physiological use efficiency than the other treatments. Compared to the SF and LF treatments, the LFD and LFCV treatments, respectively, enhanced the agronomic N use efficiency by 62% and 117%, and the N recovery use efficiency by 50 and 134%.

The second Hibiscus study [13] established that the Hibiscus plant performance is negatively affected by the low nutrient doses applied in the LF treatment, compared to the SF treatment. It also confirmed that the negative effect of the LF treatment could be well compensated by the SBS addition to the substrate, particularly in the case of the LFCV treatment. Since, the added SBS in the LF substrate did not alter the N, P, and K nutrient content, the positive effects of the added SBS could only arise from their property to stimulate the plant metabolism. This result supported the auxin-like effect in the bean cultivation study [24] proposed for the ETP SBS.

4. Replicability of SBS Effects in the Cultivation of Other Food and Ornamental Plants

4.1. Murraya Paniculata L. Jacq

Murraya paniculata L. Jacq was cultivated [18] under similar experimental conditions as reported for the Euphorbia [15] and Lantana studies [16]. In the Murraya study, the D SBS, CVDF SBS, and LND commercial product were applied at 3.1, 4.5 and 2 g per plant, respectively. The CVDF SBS treatments resulted the most efficient. Relative to the control plant, the plant grown in the CVDF SBS substrate exhibited increases of 61% in plant height, 72% number of stems, 116% number of flowers, 242% number of fruits, 63% number of leaves, 95% leaf area, 67% total stem, leaf, root dry weight, 54% root length, 147% water use efficiency, 88% leaf chlorophyll content, 196% net photosynthesis, and 933% stomatal conductance. The D SBS and LND treatments ranked second and third, respectively, in the order of decreasing efficiency. The products ranking order and the observed increases replicated the results obtained in the previous Euphorbia [15] and Lantana studies [16]. The strong correlation between the plant biometric parameters and the leaf chlorophyll content, net photosynthesis and stomatal conductance, particularly evidenced for the CVDF SBS treatment, strongly supports the hypothesis of the product "auxin-like effect", which was demonstrated in the previous hibiscus [17] and bean [24] studies, respectively, for the D and CV, and for the ETP SBS.

4.2. Tomato cv. Microtom, Grain cv. Abate and Tobacco cv. Burley

The CV, CVDF and D SBS were also tested [21] in the cultivation tomato *Micro-Tom*, a model cultivar for plant research [19]. The neat SBS were applied in pots of 15 cm diameter at dose of 240 mg/pot corresponding to 140 kg/ha as the dose tested in the cultivation of tomato *Lycopersicon* in green house farm soil [20]. The experimental plant also included pots containing mixtures of SBS and NPK 20-20-20 mineral commercial fertilizer applied at 7 NPK/SBS ratio. The control was a sterile substrate. The D SBS + NPK and CV SBS + NPK treatments gave, respectively, 53% and 79% fruit production increment relative to the control soil, followed by 46% increment by the plain CVDF SBS, 40% increment by the plain NPK, 16% by the plain D SBS and CVDF SBS + NPK, and 1% plain CV SBS treatments. The same experiments were performed for the cultivation of grain and tobacco with the plain SBS only. Production increments were much lower: 10% for grain by the three SBS, 6% for tobacco only by CVDF SBS and no increment by the other two treatments.

Comparing the plain SBS treatments in the tomato *Microtom* study [21] to those with the same doses of SBS in the tomato *Lycopersicon* study [20], the former ones gave much higher fruit production. Additionally, in the *Micro-Tom* study [21], the CVDF SBS was the most efficient (46% increment), whereas in the *Lycopersicon* study the CV SBS was the most efficient (21%). These results and those obtained in grain and tobacco cultivation [21] evidence how SBS effects strongly depend on the type of cultivar; i.e., different SBS produce different effects in different cultivars. The high *Micro-Tom* fruit production increments by the D SBS + NPK and CV SBS + NPK treatments reveals a strong synergy between SBS and NPK mineral nutrients. This arises most likely from the biostimulant properties of SBS coupled to their capacity to transfer faster and more efficiently the mineral nutrients in soluble readily available form from the cultivation substrate to the plant. Particularly relevant in the *Micro-Tom* study is the higher performance of plain CVDF SBS compared to plain NPK, although the latter is applied at a dose seven times higher than the CVDF SBS. This is in line with the SBS biostimulant properties demonstrated in the studies on the food and ornamental plants reported in Section 3.

4.3. *Zea Mays Maize*

The CVDF SBS performance was also tested in the cultivation of maize [23], in comparison with urea, CVDF PFB and IR. The study was carried out in a farm in the province of Torino (Italy). The plants were grown in a non-irrigated silty-loamy sol in the summer season. Before seeding, the soil was fertilized with N-P-K (15-15-15) fertilizer at 260 kg/ha dose to each parcel. The three SBS materials were applied to the farm soil at 7–9078 dry matter kg/ha. Urea was applied at 200 kg/ha dose.

The results of the maize trials [23] performed in the farm field showed that all treatments gave significant large increases of kernel production, compared to the control untreated soil. The highest 89% increase were recorded for the CVDF SBS treatment at 50 kg/ha dose. The urea 200 kg/ha treatment gave 38% increase. The other treatments gave increases, which were lower than, although not significantly different from the CVDF SBS 50 kg/ha treatment. Remarkably, the plants grown in the soil treated with 7 kg/ha CVDF SBS dose exhibited the highest photosynthetic activity. Compared to the results obtained in the tomato [7,20] and red pepper [22] cultivation studies, where the highest crop production increases were obtained at 140 kg/ha SBS doses, in the maize study [23] the most efficient SBS dose was lower by almost 3× factor. For all practical purposes, the most remarkable result was the demonstration that 50 kg/ha CVDF SBS yielded higher crop production than 200 kg/ha urea. This prospected high environmental and economic benefits for farmers deriving from the substitution of the commercial urea fertiliser with CVDF SBS. The same perspectives were offered by the tomato *Micro-Tom* study [21], which pointed out the higher performance of CVDF SBS applied at dose seven times lower than the commercial NPK fertilisers.

5. Other Effects of SBS in Agriculture: Healthy Plants and Food Crop Production

Two other studies carried out for the cultivation of spinach [25] and oil seed rape [10] have disclosed other important effects of SBS, which are associated to the bio-stimulant properties.

5.1. *Spinacia Oleracea L. "Gigante d'Inverno"*

For the spinach studies, composite materials pellets containing D-SBS, sun flower protein concentrate (SPC) and urea were fabricated and tested [25] for their performances as controlled release fertilizers (CRFs). The reason was that urea is a largely used fertilizers worldwide. However, it has negative environmental effects due to release of excess nitrogen over the plant uptake rate. In soil urea is hydrolysed to ammonia and then transformed into nitrates, which accumulate in soil and the plant leaves. Excess nitrates leach from soil into ground water and cause eutrophication. Excess nitrates in food plants may have carcinogenic effects for humans. To overcome these drawbacks, CRFs are commercialized. A typical largely used CRF is Osmocote in form of granules containing urea coated with synthetic polymers. The composite D SBS-SPC-urea were fabricated and tested as biobased CRFs materials, which could potentially substitute synthetic organic materials derived from fossil sources.

The first study [25] disclosed the D-SBS property to retard the formation of ammonia from urea hydrolysis and enhance the release of organic nitrogen from SPC. This effect was explained to derive from a plausible chemical interaction of D-SBS functional groups with urea and SPC. As these findings prospected the D-SBS-SPC-urea composites as potential new biobased CRFs, the second study [25] was undertaken to test the above composites in the cultivation of spinach. In this case, a commercial material (Evergreen TS) was used as cultivation substrate. Several formulations containing different amounts of D-SBS, SPC and urea were tested, in comparison with neat D-SBS, SPC, urea, and Osmocote. The cultivation substrate containing no added products was used as control. The trials were carried out in 2 L pots. The test pots contained the same 280–285 mg amount of total N, against 28 mg in the control pot. The plant weight, leaf chlorophyll content, total N and nitrate uptake in leaves and roots were determined at the end of the cultivation trials.

The results of the spinach study [25] showed no statistically significant differences by the substrate treatments compared to the control substrate in leaves and roots weight. The neat SPC and SPC-urea pellets gave significantly the highest leaf chlorophyll content, compared to the control substrate. The other treatments gave lower or not significantly different values, compared to the neat SPC and SPC-urea treatment. The most relevant results were for the nitrate content and nitrate/total N ratio in leaves. The leaves of the plants grown in the substrate treated with the pellets containing D-SBS together SPC and urea had high total N uptake with significantly lower nitric to total N ratio (9.6–12.0), compared to that (15.3–16.5) for the plant grown in the substrates treated with the pellets containing SPC and/or urea, but no D-SBS. The best plants containing high total N content and low nitrates accumulation were those grown in the substrates treated with the SPC-BP, SPC-BP-U, urea-BP and Osmocote® formulation. The nitrate concentration in the spinach leaves of all these plant was below the limit of 2 g/kg recommended for preserved frozen spinach by the European Commission. The results confirmed that all composites containing D-SBS yield the safest crop coupled with high biomass production. These findings proved that, although not supplying the plant as much nitrogen as SPC and urea, D-SBS strongly affects the process of organic nitrogen mineralization in soil. Based on the results of the two studies [25] carried out on the D-SBS-SPC-urea composites, a reaction scheme was proposed encompassing the biochemical and chemical interaction properties of D-SBS.

5.2. Oilseed Rape Brassica napus L. cv. Columbus

The oil seed rape study [10] revealed the properties of D and CVDF SBS as plant disease suppressant. Oilseed rape (*Brassica napus* L.) cv. Columbus plants were infected with the fungal pathogen *Leptosphaeria maculans*. Plant cotyledons and roots were sprayed with 2 and 0.02% SBS aqueous solutions, respectively. For comparison with the SBS, Benzothiadiazole (BTH) commercialized by Syngenta under the trade name Bion®, a widely used plant disease suppressant, was also applied at the dose suggested be the vendor. Compared to the control plan (no applied SBS or BTH), the plants treated with SBS showed the following effects. The 2% D and CVDF SBS solutions caused 42 e 56% lower leaf necrosis, respectively. The 0.02% D and CVDF SBS solutions caused 31 and 37% lower leaf necrosis, respectively. By comparison, the BTH treatment caused 80–90% lower leaf necrosis, compared to the control. The study assessed that the SBS induced plant defence by ethylene dependent signalling pathway. The results showed that the SBS effects were lower than BTH's. On the other hand, the SBS are bio-based products. On this ground, the study pointed out that, in spite their lower effects compared to BTH's, the SBS were environmentally suitable for utilization in organic farming, whereas synthetic chemicals as BTH are not.

The findings of the oilseed rape study [10] spurred further R&D to produce SBS with empowered antifungal properties. The D SBS was oxidized [28] to yield the Dox SBS. Antimicrobial assays are being carried out to assess the Dox SBS power to reduce the mycelial growth of nine targeted fungal phyto-pathogens, which represent serious threats for food and ornamental plants. The efficiency of SBS as potential plant disease suppressant, coupled to their properties as plant growth bio-stimulants and regulators of mineral nutrients release in CRFs, prospect new farming practices with high environmental and economic benefits.

6. SBS Economic and Environmental Benefits, and Perspectives for Agriculture and Horticulture

Mineral and organic products are marketed as fertilizers, plant biostimulants, and plant disease suppressing agents. Prices of these products cover a wide range. Benzothiadiazole, used as plant disease suppressants [29], is the most expensive product. Its price is at 800 USD/kg level [30]. By comparison, the production cost of mineral fertilizers is in the 0.11–0.46 €/range. The increasing demand of mineral fertilizers depletes fossil sources. The excessive applied doses to boost crop production causes accumulation in and leaching through soil into natural waters, and consequent eutrophication. In the last few decades, biostimulants have emerged as a new product category for agriculture [31]. This category includes substance or microorganism that, regardless of their mineral nutrients content, are supposed to enhance plant nutrition efficiency, abiotic stress tolerance and/or crop quality traits. They are supposed to modify the plant physiology, and so to enhance the plant growth and stress response. Compared with biofertilizers, biostimulants act at much lower applied doses. Humic substances (HS), extracted from soil and fossil deposits, belong to the biostimulants' category.

The SBS, described in Sections 3–5 of the present review, bear similar origin and chemical features as humic substances [32]. The advantages of SBS compared to HS and other commercial products claimed or reported in the literature as biostimulants is that the SBS are obtained from municipal biowastes available worldwide [33,34]. They do not cause depletion of soil organic matter or fossil deposits, and their production cost is very low. Thus, new eco-friendly and low cost perspectives are opening for novel SBS-based farm practices to replace and/or decrease mineral fertilizers consumption in agriculture.

At the present time, the market turnover of organic fertilizers is small, compared to the mineral fertilizers'. The US total fertilizer market is around 40 billion USD, with only 60 million USD contributed by organic fertilizers. Prices for various organic fertilizers range [3,35–37] range from 140 USD/t for solid products containing 10% soluble organics to 3000 USD/t for products sold in solution containing 35% organics and other mineral elements. Based on information collected by the authors of the present review, through interviews with major Italian distributors of peat derived organic fertilizers, the European market turnover is 20–25 million EUR/year., the minimum sale price is 1000 EUR/t, equivalent to 20–25 kt/year. sale. By comparison, the Euphorbia [15], Lantana [16] and Murraya [18] studies demonstrate that SBS are more efficient biostimulants than commercial products derived from Leonardite. The latter products containing 30% dry matter are sold for 7 EUR/kg [15], which corresponds to over 23 EUR/kg dry matter. The SBS production cost has been estimated about 0.1–0.5 EUR/kg [32]. The figures prospect attracting economic benefits deriving from the allocation of SBS in the organic fertilizer market. Further commercial opportunity for SBS may derive from the growth of the bio-stimulants market [38,39], estimated to reach 5 billion euros in the current decade.

To fully appreciate the economic perspectives of marketing SBS in biostimulants' product category, it should be considered that SBS contain all mineral nutrients needed by plants (see Section 2 above). These are bonded to the soluble lignocellulosic matter. The research results (see Sections 3–5) point out that the reason of the observed effects on plant growth and productivity is that the SBS supply the plants with the mineral nutrients in a readily available soluble form, thus facilitating the nutrients uptake by the plant. Thus, the SBS fall into the high price organic fertilizers' category. It is also important to be aware of the following fact exemplified for the Italian market. The SBS are obtained from composted urban bio-wastes. Italy produces 4.2 million t/year. organic humid bio-waste [40]. This can potentially yield 300–400 kt/year. SBS. This potential production exceeds the above estimated organic fertilizers market size. It is evident that, at the present time, this market cannot absorb all organic fertilizers that can be obtained from the produced compost.

It should also be considered that the SBS have been proven efficient plant disease suppressants (see Section 5). The capacity to induce plant protection against pathogens adds significant higher value to the potential SBS market [30], in comparison with fertilizers that only enhance plant growth [36,37], but do not have at the same time antifungal properties.

The above literature survey however points out that the organic fertilizers market is in the early stage. In this context, the SBS might be favoured for their capability to provide an integrated complete plant nourishment, which contains both mineral and organic matter of renewable sources. In principle, these products could replace current commercial mineral and organic fertilizers, and also antifungal agents. To appreciate the full potential of SBS uses in agriculture, it should be taken also in consideration the work [41–44] reporting SBS as potential components of new composite mulch films. Used in agriculture, these films might have multiple function, i.e., protecting plants against negative external influences, creating an ideal microclimate, and slowly releasing the SBS into the soil to stimulate plant and crop growth.

Environmental benefits from using of SBS derive mainly from the substitution of mineral fertilizers. The tomato Micro-Tom [21] and maize [23] cultivation studies showed that performance-wise 1 kg SBS is equivalent to 5–7 kg NPK fertilizers. The Euphorbia [15], Lantana [16] and Murraya [18] studies showed that 1 kg SBS yields equal or better plant productivity of at least 1 kg of organic fertilizers derived from fossil source. On this basis, using 1 kg SBS in place of 5–7 kg mineral fertilizers or 1 kg of organic fertilizers from fossil source would allow large reductions of nitrate leaching into natural waters and 100% CO_2 emission in air, respectively.

7. Results' Summary and Discussion

Five SBS have been tested as organo-mineral fertilizers in the cultivation of thirteen food and ornamental plants (Table 5). The studies have been carried out in comparison with their IR co-products and the pristine sourcing PFB materials (Tables 2–4), with conventional NPK fertilizers, commercial organo-mineral fertilizers claimed by the vendor for their biostimulant properties, and with synthetic controlled release fertilizers and antifungal agents. The collected data demonstrate the performance of SBS as biostimulants (Section 3) and antifungal agents (Section 5.2), and the replicability of their effects over the different tested plants (Section 4).

The results of the trials described in Sections 3–5 evidenced that the performance ranking order of the applied SBS products depended on the cultivated plant species. Overall, ten not commercial SBS and IR research products, five different PFB pristine sourcing materials, and several commercial fertilizers and biostimulants were used for the cultivation of thirteen plants. For each plant, several performance indicators were measured. The ranking order of the applied products also depended on the plant performance indicators. Under these circumstances, it is not possible to summarize the results of the present review results in form of comprehensive figures. Specific data plots are given in each of the references cited above for each trial. The authors feel that the results of the present review are more easily and clearly summarized in form of the following text.

The first trial carried out for the cultivation of tomato [7] demonstrated that CVD SBS performed better than CVD IR, CVD PFB and the commercial RCP product. The second tomato trial [20] demonstrated that CV SBS performed better than D and CVDF SBS, and CVD SBS in the first trial, even though the applied N, P and K doses with CV SBS were the lowest ones. Thus, the CV SBS seems the best choice for farmers adopting the SBS-based practice for the cultivation of tomato *Lycopersicon* as part of their business activity. On the other hand, potential stakeholders of the SBS-based practice should take in consideration the results of the trials carried out for the cultivation of tomato Microtom and red pepper. In the former case [21], the CVDF SBS performed better than the CV and CVD SBS. In the case of red pepper [22], the results demonstrated that the plants cultivated in presence of CVD SBS performed much better than tomato Lycopersicon cultivated in the presence of CVD SBS [7], and tomato Lycopersicon cultivated in the presence of CV, D and CVDF SBS [20], even though the CVD SBS dose applied in the red pepper study was much lower than that applied in the first tomato study [7], and the same as the CV, D and CVDF SBS doses applied in the second tomato study [20]. The four case studies [7,20–22] definitely prove that the all three CV, CVD and CVDF SBS obtained from composted biowastes represent the best choices depending on the type of food plant to cultivate.

The trials performed for the cultivation of maize, spinach and of the ornamental plants disclosed a somewhat different product hierarchy, particularly in relation to D SBS that never ranked first in the tomato and red pepper studies. The Euphorbia [15], Lantana [16] and Murray [18] studies showed that the CVDF SBS performed better than the D SBS and the commercial LND product. However, the first Hibiscus study [17] showed that D SBS performed better than D IR and PFB, CV SBS, IR and PFB, and the commercial CB product. The second Hibiscus study [10] showed that under nutrient deficiency conditions, both the CV and D SBS treatments compensated the negative effects of the nutrient deficiency on the plant performance indicators well, and that the CV SBS was more effective than the D SBS treatment. The maize cultivation study [23] showed that CVDF SBS performed better than CVDF IR and PFB. The spinach [26] and oilseed rape [10] studies demonstrated D and/or CVDF SBS as regulators of the N release and oxidation and/or antifungal agents. As for tomato and red pepper [7,20–22], the Euphorbia [15], Lantana [16], Murray [18], maize [23] studies confirmed the optimum performance of the CVDF and CV SBS obtained from the composted biowastes, whereas the hibiscus [10,17], spinach [26] and oilseed rape [10] studies disclosed useful important effects by the D SBS.

Table 5 summarizes the SBS ranking order, based on the indicator that for each plant was mostly affected by the applied SBS. It may be observed that, generally, the CVDF SBS produced the highest increase of total biomass and crop production in all plants, which were cultivated in its presence. The biomass and crop production increase ranged from 6% in the case of tobacco to 331% in the case of Euphorbia. Generally, the ornamental plants were more sensitive to the SBS application than the food plants. The other CV, CVD and D SBS were also effective, although at lower level than the CVDF SBS. The former ones produced biomass and crop production increases, which ranged from zero in the case of tobacco to 117% in the case of Euphorbia cultivated in the presence of D SBS.

Although not as effective as CVDF SBS on biomass and crop production, the D SBS exhibited other relevant effects, such as reduction of nitric to total N ratio in spinach leaves and of lesions caused by Leptosphaeria maculans in oilseed rape. In the spinach trial [25], the D SBS was used as a component of a composite pellet also containing urea and sunflower protein concentrate. Although the SBS was a minor component in the pellet and was not a significant source of N, relatively to N supplied by the other two components, the D SBS was found to cause a number of effects. It slowed down the formation of ammonia from urea hydrolysis and enhanced the release of organic nitrogen from the sunflower protein concentrate in the cultivation substrate, and strongly affected the mineralization of the total N supplied by the pellet. These effects were ascribed to the interaction of the D SBS functional groups with urea and the protein concentrate. The hypothesis is quite plausible considering the relative high content of carboxylate functional group coupled to the other NR, OR and PhOY in the D SBS (Table 4). From the practical point of view, the final effect of D SBS was 24–40% reduction of nitric to total N ratio in leaves and the production of the safest crop coupled with high biomass production.

The oilseed rape study [10] investigated the mechanism of the disease suppression effect reported by both CVDF and D SBS. The two SBS did not show any antimicrobial effect against *L. maculans* in vitro. This suggested different mechanisms underlying the observed reduction of lesions caused by *Leptosphaeria maculans* reported in Table 5, among which the induced resistance characterized by an increased resistance to infection occurring after a previous pathogen attack.

Quantitatively, the most remarkable effects were exhibited by the ETP SBS. This product produced 109–1750% increase of the of the enzyme activities and soluble proteins concentration in the leaves and roots of bean plants. This effect demonstrated undoubtedly the biostimulant properties of the applied product.

Reasons for the SBS performances have been proposed. These are based on chemical and biochemical interactions/reactions catalysed by SBS thanks to their chemical composition. The SBS contain 15–30% minerals together with organic matter. They can, therefore, add soluble plant nutrients to soil. They can also perform as bio-effectors, stimulate the uptake from roots of soil nutrients with a hormone-like effect and/or plant growth by promoting rhizobacteria, and catalyse the plant photosynthetic activity (see all references in Section 3). Applied in mixtures [22,26] with urea, other organic fertilizers and NPK conventional mineral fertilizers, they can regulate the release of N, P and K to the soil and taken up by the plant, and so reduce the amount of nitrates leaching through the soil into natural waters and taken up by the plant crop.

Whereas from the basic science point of view, the collected data do not allow demonstrating the action mechanism for the observed performances of SBS in agriculture, from the practical point of view, the most relevant result is that the highest SBS effect on the plant performance indicators occurs at about 140 kg/ha applied dose to the cultivation soil or substrate. Depending on the plant and the type of applied SBS, increases up to three order of magnitude for the plant performance indicators are measured relatively to the control plants. At higher dose levels, no further increases are observed. The remarkable high effects occurring at relatively low treatment dose prospect using the SBS to augment plant growth and productivity, and at the same time reduce the consumption and negative environmental impact of conventional fertilizers applied at high dose.

The collected data offer ground to attempt establishing case-by-case correlations between the composition and the effects of the different SBS used for the cultivation of the same plants. However, the SBS are complex mixtures of organic molecules differing for molecular weight and chemical features. These molecules are in turn bonded to different mineral elements. Under these circumstances, correlations between SBS chemical composition and effects as given in Sections 2–5 do not help much to identify the active molecules, which are responsible for the observed effects. These difficulties are inherent to products obtained from materials of biological origin.

8. Conclusions

Undoubtedly, the present review offers ground for scientists to pose many questions needing explanations, particularly in relation to the action mechanism underlying the SBS effects. Based on the different compositions of the 15 materials used to cultivated the 13 plant species (Tables 2–5), any scientist is aware that the answer to the many questions required identification and isolation of the active principles in each of the 15 materials, and testing them individually for each plant species. Such task would involve experimental work requiring cost and time, which make rather doubtful achieving success. Under these circumstances, the authors think that further tentative explanations of the many questions posed by the present review would be only speculative. Yet, the data obtained from the authors over the past 11 years are very useful to guide farmers in their new SBS-based farm practices. Whereas the present review leaves many scientific questions answered, it demonstrates the replicability of the SBS effects under different environmental conditions.

In absence of the present review, potential stakeholders should search, read, study and evaluate each of the single papers cited in the reference section of the present review, and try to draw the conclusions that allowed them adopting successfully the SBS-based farm practice. On the other hand, in spite of its mainly empirical nature, the present review offers interesting and useful scope for scientific research.

At the present time, the implementation of SBS to operation level in real farm practices must rely on the empirical findings reviewed in the present paper. To this end, the most important factor is the replicability of the composition and performance of each SBS dedicated to the cultivation of specific plant species. Potential stakeholders may rely on the fact that a waste treatment plant may produce a wide variety of SBS tailored for the cultivation of specific plants, depending on the variety of bio-waste sources and treatments types. Further studies should assess how the variability of the bio-waste source, as a function of the environmental and operational conditions where it is produced, may affect the composition and performance of the intended SBS.

Author Contributions: Conceptualization, E.M; data curation, E.M., G.F. and A.B.; writing-original draft preparation, E.M.; writing-review and editing, E.M., G.F and A.B.; funding acquisition, E.M. All authors have read and agreed to the published version of the manuscript.

Funding: This review was carried out to establish the baseline for the LIFEEBP LIFE19 ENV/IT/000004 project funded by the European Commission under the LIFE 2019 programme.

Institutional Review Board Statement: Not applicable.

Informed Consent Statement: Not applicable.

Data Availability Statement: Not applicable.

Conflicts of Interest: The authors declare no conflict of interest.

References

1. Mordor Intelligence. Fertilizers Market–Growth, Trends, COVID-19 Impact, and Forecasts (2022–2027). Available online: https://www.mordorintelligence.com/industry-reports/fertilizers-market (accessed on 1 May 2022).
2. Bio-Based Industries. Amended Annual Work Plan and Budget 2017 for the Bio-Based Industries Joint Undertaking. Available online: https://www.bbieurope.eu/sites/default/files/awp_2017.pdf (accessed on 1 May 2022).
3. Diffen. Chemical Fertilizer vs. Organic Fertilizer. Available online: https://www.diffen.com/difference/Chemical_Fertilizer_vs_Organic_Fertilizer#Cost (accessed on 1 May 2022).
4. FAO. World Fertilizer Trends and Outlook to 2018. Available online: https://www.fao.org/3/a-i4324e.pdf (accessed on 1 May 2022).
5. Eurostat. Agri-Environmental Indicator-Mineral Fertiliser Consumption. Available online: https://ec.europa.eu/eurostat/statistics-explained/index.php/Agri-environmental_indicator_mineral_fertiliser_consumption (accessed on 1 May 2022).
6. IFA. About Fertilizers. Available online: https://www.fertilizer.org/ (accessed on 1 May 2022).
7. Sortino, O.; Dipasquale, M.; Montoneri, E.; Tomasso, L.; Perrone, D.G.; Vindrola, D.; Negre, M.; Piccone, G. Refuse derived soluble bio-organics enhancing tomato plant growth and productivity. *Waste Manag.* **2012**, *32*, 1792–1801. [CrossRef] [PubMed]
8. Foged, H.L.; Flotats, X.; Blasi, A.B.; Palatsi, J.; Magri, A.; Schelde, K.M. Inventory of Manure Processing Activities in Europe. Technical Report No. I Concerning "Manure Processing Activities in Europe" to the European Commission, Directorate-General Environment. 2018. 138p. Available online: https://agro-technology-atlas.eu/docs/21010_technical_report_I_inventory.pdf (accessed on 1 May 2022).
9. Ji, F.; McGlone, J.J.; Kim, S.W. Effects of dietary humic substances on pig growth performance, carcass characteristics, and ammonia emission. *J. Anim. Sci.* **2006**, *84*, 2482–2490. [CrossRef] [PubMed]
10. Jindrichova, B.; Burketova, L.; Montoneri, E.; Francavilla, M. Biowaste-derived hydrolysates as plant disease suppressants for oilseed rape. *J. Clean. Prod.* **2018**, *183*, 335–342. [CrossRef]
11. Jisa. Agricultural Biostimulants in the New European Fertiliser Regulation. Available online: https://www.fertilizantesyabonos.com/english/agricultural-biostimulants-in-the-new-european-fertiliser-regulation/on (accessed on 1 May 2022).
12. EUR-LEX. Regulation (EU) 2019/1009 of the European Parliament and of the Council of 5 June 2019 Laying Down Rules on the Making Available on the Market of EU Fertilising Products and Amending Regulations (EC) No 1069/2009 and (EC) No 1107/2009 and repealing Regulation (EC) No 2003/2003 (Text with EEA relevance). Available online: https://eur-lex.europa.eu/legal-content/EN/ALL/?uri=CELEX%3A32019R1009 (accessed on 1 May 2022).
13. Massa, D.; Lenzi, A.; Montoneri, E.; Ginepro, M.; Prisa, D.; Burchi, G. Plant response to biowaste soluble hydrolysates in hibiscus grown under limiting nutrient availability. *J. Plant Nutr.* **2017**, *41*, 396–409. [CrossRef]

14. Montoneri, E. Municipal Waste Treatment, Technological Scale Up and Commercial Exploitation: The Case of Bio-Waste Lignin to Soluble Lignin-like Polymers. In *Food Waste Reduction and Valorisation*; Chapter 6; Morone, P., Papendiek, F., Tartiu, V.E., Eds.; Springer: Berlin/Heidelberg, Germany, 2017. [CrossRef]
15. Fascella, G.; Montoneri, E.; Ginepro, M.; Francavilla, M. Effect of urban biowaste derived soluble substances on growth, photosynthesis and ornamental value of Euphorbia x lomi. *Sci. Hortic.* **2015**, *197*, 90–98. [CrossRef]
16. Fascella, G.; Montoneri, E.; Francavilla, M. Biowaste versus fossil sourced auxiliaries for plant cultivation: The Lantana case study. *J. Clean. Prod.* **2018**, *185*, 322–330. [CrossRef]
17. Massa, D.; Prisa, D.; Montoneri, E.; Battaglini, D.; Ginepro, M.; Negre, M.; Burchi, G. Application of municipal biowaste derived products in Hibiscus cultivation: Effect on leaf gaseous exchange activity, and plant biomass accumulation and quality. *Sci. Hortic.* **2016**, *205*, 59–69. [CrossRef]
18. Fascella, G.; Montoneri, E.; Rouphael, Y. Biowaste-derived Humic-like Substances Improve Growth and Quality of Orange Jasmine (*Murraya paniculata* L. Jacq.) Plants in Soilless Potted Culture. *Resources* **2021**, *10*, 80. [CrossRef]
19. Shikata, M.; Ezura, H. Micro-Tom Tomato as an Alternative Plant Model System: Mutant Collection and Efficient Transformation. *Methods Mol. Biol.* **2016**, *1363*, 47–55.
20. Sortino, O.; Montoneri, E.; Patanè, C.; Rosato, R.; Tabasso, S.; Ginepro, M. Benefits for agriculture and the environment from urban waste. *Sci. Total Environ.* **2014**, *487C*, 443–451. [CrossRef]
21. Isagro SpA. Efficacia Biologica di Compost Forniti Dall'università di Torino. 2014, unpublished confidential report delivered to the author. Available online: www.isagro.com (accessed on 1 May 2022).
22. Sortino, O.; Dipasquale, M.; Montoneri, E.; Tomasso, L.; Avetta, P.; Bianco Prevot, A. 90% yield increase of red pepper with unexpectedly low doses of compost soluble substances. *Agron. Sustain. Dev.* **2013**, *33*, 433–441. [CrossRef]
23. Rovero, A.; Vitali, M.; Rosso, D.; Montoneri, E.; Chitarra, W.; Tabasso, S.; Ginepro, M.; Lovisolo, C. Sustainable maize production by urban biowaste products. *Int. J. Agron. Agric. Res.* **2015**, *6*, 75–91.
24. Baglieria, A.; Cadilia, V.; Mozzetti Monterumici, C.; Gennari, M.; Tabasso, S.; Montoneri, E.; Nardi, S.; Negre, M. Fertilization of bean plants with tomato plants hydrolysates. Effect on biomass production, chlorophyll content and N assimilation. *Sci. Hortic.* **2014**, *176*, 194–199. [CrossRef]
25. Padoan, E.; Montoneri, E.; Bordiglia, G.; Boero, V.; Ginepro, M.; Evon, P.; Vaca-Garcia, C.; Fascella, G.; Negre, M. Waste biopolymers for eco-friendly agriculture and safe food production. *Coatings* **2022**, *12*, 239. [CrossRef]
26. Bianco Prevot, A.; Avetta, P.; Fabbri, D.; Laurenti, E.; Marchis, T.; Perrone, D.G.; Montoneri, E.; Boffa, V. Waste derived bio,-organic substances for light induced generation of reactive oxygenated species. *ChemSuschem* **2011**, *4*, 85–90. [CrossRef]
27. Gomis, J.; Bianco Prevot, A.; Montoneri, E.; Gonzalez, M.C.; Amat, A.M.; Martire, D.O.; Arques, A.; Carlos, L. Waste sourced bio-based substances for solar-driven wastewater remediation: Photodegradation of emerging pollutants. *Chem. Eng. J.* **2014**, *235*, 236–243. [CrossRef]
28. Montoneri, E.; Fabbri, G.; Quagliotto, P.L.; Baglieri, A.; Padoan, E.; Boero, V.; Negre, M. High molecular weight biosurfactants from mild chemical reactions of fermented municipal biowastes. *ChemistrySelect* **2020**, *5*, 2564–2576. [CrossRef]
29. Burketova, L.; Trda, L.; Ott, P.G.; Valentova, O. Bio-based resistance inducers for sustainable plant protection against pathogens. *Biotechnol. Adv.* **2015**, *33*, 994–1004. [CrossRef]
30. eBiochem. Wholesale 2,1,3-Benzothiadiazole. 2022. Available online: http://www.ebiochem.com/product/2-1-3-benzothiadiazole-71535 (accessed on 11 June 2022).
31. du Jardin, P. Plant biostimulants: Definition, concept, main categories and regulation. *Sci. Hortic.* **2015**, *196*, 3–14. [CrossRef]
32. Montoneri, E.; Mainero, D.; Boffa, V.; Perrone, D.G.; Montoneri, C. Biochemenergy: A project to turn an urban wastes treatment plant into biorefinery for the production of energy, chemicals and consumer's products with friendly environmental impact. *Int. J. Global Environ. Issues* **2011**, *11*, 170–196. [CrossRef]
33. Sheldon-Coulson, G.A. Production of Levulinic Acid in Urban Biorefineries. Master's Thesis, Massachusetts Institute of Technology, Engineering Systems Division, Cambridge, MA, USA, 2011. Available online: https://dspace.mit.edu/bitstream/handle/1721.1/68450/769021899-MIT.pdf?sequence\protect$\relax\protect{\begingroup1\endgroup\@@over4}$2 (accessed on 11 June 2022).
34. Ellen Macarthur Foundation. Cities in the Circular Economy: An Initial Exploration. 2017. Available online: https://ellenmacarthurfoundation.org/cities-in-the-circular-economy-an-initial-exploration (accessed on 20 June 2019).
35. Made in China. Organic Fertilizer Price. Organic Fertilizer Price. Available online: https://www.made-in-china.com (accessed on 11 June 2022).
36. Alibaba. Leonardite Humic Acid Fulvic Acid Fertilizer. 2022. Available online: https://www.alibaba.com/trade/search?fsb=y&IndexArea=product_en&CatId=&SearchText=leonardite+humic+acid+fulvic+acid+fertilizer (accessed on 11 June 2022).
37. Ebay. Leonardite. 2022. Available online: https://www.ebay.it/sch/i.html?_from=R40&_trksid=p2380057.m570.l1313&_nkw=leonardite&_sacat=0 (accessed on 11 June 2022).
38. Hazra, D.K.; Purkait, A. Role of biostimulant formulations in crop production: An overview. *Int. J. Agric. Sci. Vet. Med.* **2020**, *8*, 38–46.

39. Marketsandmarkets. Biostimulants Market by Active Ingredients (Humic Substances, Seaweed Extracts, Microbial Amendments, Amino Acids), Mode of Application (Folier, Soil Treatment, Seed Treatment), Form (Liquid, and Dry), Crop Type, & by Region-Global Forecast to 2026. Available online: https://www.marketsandmarkets.com/pdfdownloadNew.asp?id=1081 (accessed on 11 June 2022).
40. Bastioli, C. Bioplastics: An Italian Case Study of Bioeconomy in Italy. 2013. Available online: https://www.kyotoclub.org/docs/Bruxelles_060313_Bioplastics_Bastioli.pdf (accessed on 11 June 2022).
41. Franzoso, F.; Tabasso, S.; Antonioli, D.; Montoneri, E.; Persico, P.; Laus, M.; Mendichi, R.; Negre, M. Films made from poly (vinyl alcohol-co-ethylene) and soluble biopolymers isolated from municipal biowaste. *J. Appl. Polym. Sci.* **2015**, *132*, 1301. [CrossRef]
42. Franzoso, F.; Causone, D.; Tabasso, S.; Antonioli, D.; Montoneri, E.; Persico, P.; Laus, M.; Mendichi, R.; Negre, M. Films made from polyethylene-co-acrylic acid and soluble biopolymers sourced from agricultural and municipal biowaste. *J. Appl. Polym. Sci.* **2015**, *132*, 5803. [CrossRef]
43. Franzoso, F.; Antonioli, D.; Montoneri, E.; Persico, P.; Tabasso, S.; Laus, M.; Mendichi, R.; Negre, M.; Vaca-Garcia, C. Films made from poly (vinyl alcohol-co-ethylene) and soluble biopolymers isolated from post-harvest tomato plant. *J. Appl. Polym. Sci.* **2015**, *132*, 6006.
44. Franzoso, F.; Vaca-Garcia, C.; Rouilly, A.; Evon, P.; Montoneri, E.; Persico, P.; Mendichi, R.; Nisticò, R.; Francavilla, M. Extruded versus solvent cast blends of poly(vinyl alcohol-co-ethylene) and biopolymers isolated from municipal biowaste. *J. Appl. Polym. Sci.* **2016**, *133*, 43009. [CrossRef]

MDPI AG
Grosspeteranlage 5
4052 Basel
Switzerland
Tel.: +41 61 683 77 34

Agriculture Editorial Office
E-mail: agriculture@mdpi.com
www.mdpi.com/journal/agriculture

www.ingramcontent.com/pod-product-compliance
Lightning Source LLC
Chambersburg PA
CBHW061434210926
51850CB00008B/1513
* 9 7 8 3 7 2 5 8 1 9 8 9 8 *